Heat Shock Proteins

Volume 15

Series editors

Alexzander A. A. Asea
Professor, Department of Medicine and
Director, Precision Therapeutics Proteogenomics Diagnostic Center
Eleanor N. Dana Cancer Center
University of Toledo College of Medicine and Life Sciences
Toledo, United States of America

Stuart K. Calderwood
Professor and Director, Division of Molecular and Cellular Radiation Oncology
Department of Radiation Oncology
Beth Israel Deaconess Medical Center and Harvard Medical School
Boston, United States of America

Heat Shock Proteins: key mediators of Health and Disease. Heat shock proteins (HSP) are essential molecules conserved through cellular evolution required for cells to survive the stresses encountered in the environment and in the tissues of the developing and aging organism. These proteins play the essential roles in stress of preventing the initiation of programmed cell death and repairing damage to the proteome permitting resumption of normal metabolism. Loss of the HSP is lethal either in the short-term in cases of acute stress or in the long-term when exposure to stress is chronic. Cells appear to walk a fine line in terms of HSP expression. If expression falls below a certain level, cells become sensitive to oxidative damage that influences aging and protein aggregation disease. If HSP levels rise above the normal range, inflammatory and oncogenic changes occur. It is becoming clear that HSP are emerging as remarkably versatile mediators of health and disease. The aim of this series of volumes is to examine how HSP regulation and expression become altered in pathological states and how this may be remedied by pharmacological and other interventions.

More information about this series at http://www.springer.com/series/7515

Alexzander A. A. Asea • Punit Kaur
Editors

Heat Shock Proteins and Stress

Springer

Editors
Alexzander A. A. Asea
College of Medicine and Life Sciences
University of Toledo
Toledo, OH, USA

Punit Kaur
Department of Experimental
Radiation Oncology
The University of Texas MD
Anderson Cancer
Houston, TX, USA

ISSN 1877-1246 ISSN 1877-1254 (electronic)
Heat Shock Proteins
ISBN 978-3-030-08084-6 ISBN 978-3-319-90725-3 (eBook)
https://doi.org/10.1007/978-3-319-90725-3

This Springer imprint is published by the registered company Springer Nature Switzerland AG
The registered company address is: Gewerbestrasse 11, 6330 Cham, Switzerland

Preface

The clinical definition of stress is a physical, mental, or emotional factor that causes bodily or mental tension. On a whole-body level, stresses can be induced by environmental, psychological, or social situations or by illness, or from a medical procedure. At the cellular level, stress can be induced by a wide variety of conditions including heat shock, cold shock, pH shift, hypoxia, UV light, and during wound healing or tissue remodeling. Increased expression of heat shock proteins (HSP) protects the cell by stabilizing unfolded proteins, giving the cell time to repair or resynthesize damaged proteins.

The book *Heat Shock Proteins and Stress* provides the most comprehensive review on contemporary knowledge on the role of HSP in stress. Using an integrative approach to understanding the regulation of HSP responses, the contributors provide a synopsis of novel mechanisms by which HSP responses are regulated under normal physiological and pathophysiological conditions.

To enhance the ease of reading and comprehension this book has been subdivided into various sections: Section I reviews current progress on our understanding of HSP in cellular stress; Section II evaluates the role of HSP in oxidative stress; Section III focuses the reader on the role of HSP in stress response pathway in invertebrates, vertebrate, plants, and aquatic organisms.

Key basic and clinical research laboratories from major universities and academic medical hospitals around the world contribute chapters that review present research activity and importantly project the field into the future. The book is a must read for researchers, postdoctoral scholars, and graduate students in the fields of Translational Medicine, Clinical Psychology, Human Physiology, Zoology, Botany, Biotechnology, Molecular Medicine, Infectious Diseases, and Pathology.

Toledo, OH, USA — Alexzander A. A. Asea
Houston, TX, USA — Punit Kaur

Contents

About the Editors

Alexzander A. A. Asea is a highly innovative and accomplished world-renowned clinical and basic research scientist and visionary executive leader who has exceptional experience spearheading clinical and basic science research, training, education, and commercialization initiatives within top-ranked academic biomedical institutes. Prof. Asea's initial findings studying the effects of Hsp72 on human monocytes led to the proposal of a novel paradigm that Hsp72, previously known to be as intracellular molecular chaperones, can be found in the extracellular milieu where it has regulatory effects on immunocompetent cells – a term now called chaperokine. Prof. Asea has authored over 255 scientific publications including peer-reviewed articles, reviews, books, book chapters, editorials, and news headliners in a wide range of biomedical-related disciplines. Prof. Asea is the series editor of the widely successful book series *Heat Shock Proteins* (Springer Nature Publications) and is an editorial board member of 13 other scientific peer-reviewed journals. Currently, Prof. Asea is at the University of Toledo College of Medicine and Life Sciences in Toledo, USA.

Punit Kaur is an expert in onco-proteogenomics, with extensive training and experience in quantitative mass spectrometry imaging, protein chemistry, and biomarker discovery. Dr. Kaur's main research focus is on the use of heat-induced nanotechnology in combination with radiotherapy and chemotherapy in the cancer stem cell therapy. Dr. Kaur has published more than 40 scientific articles, book chapters, and reviews, and currently serves as editorial board member for the *European Journal of Cancer Prevention* and the *Journal of Proteomics and Bioinformatics*. Dr. Kaur is an editor of five books in the highly successful *Heat Shock Proteins* book series by Springer Nature Publishers. Currently, Dr. Kaur is a Visiting Scientist Professor at the University of Texas MD Anderson Cancer Center in Houston, USA.

Part I
Cellular Stress

Chapter 1
Molecular Chaperones and the Nuclear Response to Stress

Lynn Boyd and Katherine M. Sampuda

Abstract Chaperones are a well conserved class of proteins that reside in many different cellular compartments. The nucleus is a compartment of special interest because it houses the genetic material and allows for the expression and maintenance of genes. Many chaperones localize to the nucleus under stress conditions. The current body of evidence indicates that the nuclear function of chaperones is similar to chaperone function in the cytoplasm. Emerging evidence on the nuclear import pathway for chaperones suggests that novel pathways exist that allow chaperones to enter the nucleus under conditions of environmental stress. One such pathway, the Hikeshi pathway, is responsible for the transport of HSP70 and possibly other molecular chaperones.

Keywords Chaperone · HSP · HSP70 · Nuclear import · Protein aggregation · Stress

Abbreviations

CHIP	c-terminus of Hsc70-interacting protein
HOP	HSP70/HSP90 organizing protein
HSF1	Heat shock factor 1
HSP	Heat shock protein
PQC	Protein quality control
sHSP	Small heat shock protein
SING	Stress induced nuclear granule
TPR	Tetratricopeptide repeat
UPS	Ubiquitin protein system
UV	Ultraviolet

L. Boyd (✉) · K. M. Sampuda
Department of Biology, Middle Tennessee State University, Murfreesboro, TN, USA
e-mail: Lynn.Boyd@mtsu.edu

A. A. A. Asea, P. Kaur (eds.), *Heat Shock Proteins and Stress*,
Heat Shock Proteins 15, https://doi.org/10.1007/978-3-319-90725-3_1

1.1 Introduction

The different compartments of the eukaryotic cell each have their own characteristic proteome. Each compartment possesses a protein quality control (PQC) system that functions to monitor and repair damage to the proteome. The nucleus is a compartment of special interest since it houses the genetic material which must be maintained, replicated, and protected from damage. Several human diseases, such as Huntington's disease, are associated with protein misfolding in the nucleus. It has been reported that the nucleus is the most susceptible compartment in regards to protein damage following a heat insult and this damage can be mitigated by overexpression of a nuclear targeted molecular chaperone, HSP70 (Hageman et al. 2007). This suggests that chaperones can work similarly in the nucleus and cytosol. The nucleus differs from other membrane-bound compartments in that trafficking between the nucleus and cytosol is more dynamic than that of the other organelles such as the ER or mitochondria. Thus, it has been more difficult to discern which PQC pathways might be specific to the nucleus versus cytosol. The nuclear pore provides a channel through which small proteins can travel freely and in both directions between the nucleus and cytosol. Larger proteins must use a transport pathway to travel through the nuclear pore. Many molecular chaperones are localized to both the nucleus and cytosol. Molecular chaperones are known mediators of protein folding and thus may be required during times of proteotoxic stress. Chaperones recognize exposed hydrophobic patches on a protein's surface and help the protein to refold into its proper conformation. Several chaperones are known to relocate to the nucleus under stress conditions (discussed below). Although, most studies have looked at the stress of elevated temperature, some recent studies have shown that other types of stress can also induce this nuclear response.

Several neurodegenerative diseases are associated with nuclear protein aggregates. A link between these aggregation events and chaperones is well established. First, chaperones have been shown to localize to these aggregates (see Table 1.1). Second, overexpression of chaperones is known to reduce the level of nuclear aggregates (reviewed in Nath and Lieberman 2017). Third, reduced expression or mutations in chaperones makes cells more susceptible to nuclear aggregation events (Nath and Lieberman 2017). In addition to nuclear protein aggregates, several other nuclear bodies have been described. In several cases, chaperones have been shown to localize to these nuclear bodies. Table 1.1 shows known nuclear bodies and nuclear aggregates that are associated with localization of chaperones.

There are a large number of individual chaperones and co-chaperones that have been identified. Generally, these can be divided up into several different classes: the HSP60 family, the HSP70 family, the HSP90 family, the HSP100 family, the HSP110 family, and the small heat shock proteins (sHSP). Additionally, there are several co-chaperone families including HSP40 (DNAJ) proteins and the TPR domain containing proteins (such as HOP/Sti1). First, we discuss the entry of these proteins into the nucleus. Next, we look at the evidence surrounding the actual function of these chaperones inside the nucleus.

Table 1.1 Nuclear bodies and aggregates with chaperone localization

Nuclear bodies with chaperone localization		
Nuclear body	**Chaperone protein**	**Reference**
Clastome	Hsc70, Hsp70	(Lafarga et al. 2002)
Cajal body	NOPP140, Telomere Cajal body protein 1 (TCAB1), Survival Motor Neuron (SMN)	(Isaac et al. 1998; Raimer et al. 2016)
Nuclear speckles	HSPB1, alpha B-crystallin (HSPB5), HSPB7, Hsp27 (Drosophila), HSPA6 (HSP70B'), HSPA1A (HSP70–1)	(Bao et al. 2002; Van den IJssel et al. 2003; Michaud et al. 2008; Vos et al. 2009)
Nuclear stress body	HSF1, HSF2	(Morimoto and Boerkoel 2013)
Nuclear stress granules	HSF1, HSF2, and Hsp70	(Alastalo et al. 2003; Sarge et al. 1993)
PML body	DEK proto-oncogene, death domain-associated protein 6 (DAXX), ATP-dependent helicase ATRX (ARTX), and histone cell cycle regulator (HIRA)	(Ivanauskiene et al. 2014)
Nuclear aggregates with chaperone localization		
Disease	**Chaperone protein**	**Reference**
Huntington's disease (Poly (Q) aggregate)	Hsp70, Hsc70, Hsp26 and Hsp104	(Jana et al. 2000; Walter et al. 2011)
Spinobulbar muscular atrophy (Poly(Q) aggregate)	Hsp70, Hsp90, HDJ-2/HSDJ	(Cummings et al. 1998; Stenoien et al. 1999)
Dentatorubral Pallidoluysian atrophy (Poly(Q) aggregate)	HDJ-2/HSDJ, HSP70	(Cummings et al. 1998)
Spinocerebellar Ataxia (Poly(Q) aggregates)	HDJ-2/HSDJ, Hsc70 (Hsp73), Hsp70 (Hsp72)	(Chai et al. 1999; Cummings et al. 1998)

1.1.1 Nuclear Protein Quality Control

The presence of chaperones and ubiquitin proteasome pathway components in the nucleus suggests that PQC in the nucleus may work similarly to the cytosolic system. Although it has been suggested that misfolded nuclear proteins might be exported from the nucleus for degradation (Chen and Madura 2014), other reports have suggested just the opposite, that some proteins are imported into the nucleus specifically for degradation. The most detailed description of nuclear protein quality control comes from studies in the budding yeast, *Saccharomyces cerevisiae*. The ubiquitin pathway enzymes Cdc34p, Ubc1p, and San1p target four temperature sensitive nuclear proteins for degradation by the 26S proteasome (Gardner et al. 2005). Further studies in *S. cerevisiae* identified a chaperone-assisted degradation pathway where Hul5, an E3 ligase associated with the 26S proteasome, HSP70, and Bag102, a co-chaperone, mediate the tagging and the removal of the nuclear kinetochore component Spc7–23 (Kriegenburg et al. 2014). In mammalian COS cell culture, modified β-galactosidase from *Escherichia coli* was shown to be degraded rapidly in the nucleus whereas its unmodified form remained stable (Tsuneoka and Mekada 1992). However, the molecular pathways of nuclear PQC in mammalian cell culture have yet to be discovered. In the nematode, *C. elegans*, expression of chaperones can repress the formation of stress induced nuclear granules (SINGs) (Sampuda et al. 2017). These nuclear granules contain high concentrations of ubiquitin and proteasome and occur after exposure to proteotoxic stress such as high salt or oxidative stress, but not in response to heat shock. The SINGs may be sites of localized protein degradation in the nucleus. That the expression of chaperones can repress SING formation suggests that SINGs are a nuclear response to protein misfolding.

Many chaperones like heat shock protein 70 (HSP70), heat shock protein 90 (HSP90) and small heat shock proteins (sHSP) are induced by heat shock and other stress conditions. Heat shock induced expression of chaperones is mediated by the heat shock factor 1 (HSF1) transcription factor (Brunquell et al. 2016; Shibata and Morimoto 2014). Since many key chaperones are above the size of free diffusion through the nuclear pore transport into the nucleus would require some sort of transport mechanism such as that involving the Ran and importin proteins (Weis 2003). The topic of yeast nuclear chaperones and their role in nuclear PQC has recently been reviewed (Jones and Gardner 2016). In this chapter, we will first discuss the circumstances and mechanisms for delivery of chaperones into the nucleus. In the following section, we discuss the known functions of those chaperones in the nucleus.

1.1.2 Nuclear Import of Chaperones

In addition to responding to heat stress, chaperones are constitutively expressed under normal physiological conditions. They help to fold newly synthesized proteins as they leave the ribosome and aid in protein translocation across the membrane. However, during stress conditions such as heat shock, chaperones shift to refolding thermally damaged proteins to prevent them from aggregating. Chaperone expression is upregulated by HSF1. In unstressed states, cytosolic HSF1 monomers are suppressed by a chaperone complex consisting of molecular chaperones HSP70 and HSP90. During stressed conditions like heat shock, cadmium sulfate, or azetidine, HSP70 and HSP90 release HSF1 and interact with misfolded proteins. Unbound HSF1 is then translocated into the nucleus to form a transcriptionally active HSF1 trimer that binds to heat shock elements found upstream of HSP genes. Upon recovery from heat shock, HSP70 binds to HSF1 and translocates to the cytosol where HSP90 binds to form the HSP70/HSP90 complex to inhibit HSF1 (Trinklein et al. 2004).

After transcription, heat shock proteins are translated in the cytosol where many of them function. As previously mentioned, a portion of these chaperones translocate to the nucleus. Nuclear import of many of these chaperones requires active transport. In some cases, chaperones contain a nuclear localization sequence that promotes transport through a classical nuclear import pathway. In other cases, chaperones have been known to piggyback into the nucleus as they transport proteins from the cytosol into the nucleus (Melchior and Gerace 1995).

The classical nuclear import pathway revolves around the Importin α/β family. In this pathway, importins bind to cargo proteins and help transport them across the membrane and into the nucleus. Once the cargo protein is in the nucleus, Ran GTPase binds to the importin prompting the release of the cargo protein. Importin is then shuttled back into the cytosol by Ran (Melchior and Gerace 1995). This pathway is reliant on the Ran gradient, which is higher in the nucleus and lower in the cytosol. Heat shock, oxidative stress, and UV irradiation downregulate the classical importin α/β-mediated pathway by altering the distribution of Ran (Czubryt et al. 2000; Kodiha et al. 2004; Kose et al. 2012; Yasuda et al. 2006).

The yeast chaperone HSP104 has a nuclear localization signal and is imported into the nucleus after heat shock (Tkach and Glover 2008). Interestingly, nuclear import under heat stress does not depend upon the nuclear localization signal and must depend upon some non-canonical import pathway. In the yeast, *Saccharomyces cerevisiae,* both starvation and ethanol stress induced nuclear accumulation of the HSP70 family member Ssa4p (Chughtai et al. 2001; Quan et al. 2004). Nuclear localization following starvation is dependent upon a short hydrophobic region near the N-terminus on the protein and requires the activity of β importin. Nuclear localization following ethanol stress also depends upon β importin and the N-terminal domain.

CHIP (C-terminus of Hsc70-interacting protein) is a TPR family co-chaperone and E3 ligase that ubiquitinates proteins that HSP70 and HSP90 are unable to fold. After heat stress, CHIP localizes predominantly to the nucleus (Dai et al. 2003). This localization coincides with the translocation of HSF1 into the nucleus and it has been proposed that CHIP functions as a regulator of HSF1 activity in the nucleus (Dai et al. 2003).

The co-chaperone HOP (HSP70/HSP90 organizing protein) serves as a linker for the HSP70 and HSP90 chaperones. Thus, it plays a crucial role in linking these two major chaperone activities. HOP contains a bipartite nuclear localization signal required for its nuclear localization under non-stress conditions. However, its translocation into the nucleus following stress does not depend upon this NLS (Daniel et al. 2008).

The classical import pathway is reduced during stress conditions (Kodiha et al. 2004). However, chaperones still localize to the nucleus. This implied that a non-canonical import pathway was utilized by molecular chaperones. Kose et al. (2012) showed that a carrier protein, Hikeshi, does not bind to Ran and functions under heat shock to transport HSP70 into the nucleus. This pathway was termed the Hikeshi-mediated nuclear import pathway. It is currently unclear if chaperones other than HSP70 are transported by the Hikeshi pathway.

1.1.3 Role of Nuclear Chaperones in Stress

Since many human diseases are associated with protein aggregates, there has been much interest in investigating the potential of nuclear chaperones to alleviate protein aggregation in the nucleus. The HSP100 family member, HSP104, is a nuclear localized chaperone in *Saccharomyces cerevisiae*. HSP104 accumulates in the nucleus and interacts with HSP70 to help solubilize and rescue aggregated proteins (Parsell et al. 1994; Yokom et al. 2016). Metazoans do not have HSP100 but other chaperones may fulfill this role in those organisms.

HSP110 chaperones are homologous to HSP70 but they are larger and tend to lack chaperone activity on their own. HSP110 proteins serve as nucleotide exchange factors for HSP70. In mammalian cells, HSP110 chaperones localize to nuclear protein aggregates and partner with HSP70 to function as disaggregases (Rampelt et al. 2012). Thus, this chaperone class may be important for protecting against diseases which involve nuclear protein aggregates. In yeast, HSP110 homologs assist the HSP104 disaggregases with their localization to aggregates and their disaggregase activity (Kaimal et al. 2017).

The co-chaperones also play a role in nuclear PQC. CHIP is a TPR family co-chaperone and E3 ligase that ubiquitinates proteins that HSP70 and HSP90 are unable to refold. CHIP forms a complex with HSF1 in the nucleus and is required for HSF1 trimerization and transcriptional activation (Dai et al. 2003) In addition, CHIP, via its E3 ubiquitin ligase activity, participates in the ubiquitination of substrates within the nucleus. Nuclear substrates of CHIP include dioxin receptor, mutant androgen receptor, p65, RUNX1, RUNX2, and a sirtuin family member (SirT6) (Adachi et al. 2007; Lees et al. 2003; Li et al. 2008; Ronnebaum et al. 2013; Shang et al. 2009).

HSP40 is a co-chaperone family also known as the DNAJ family. HSP40 proteins localize to the cytosol and nucleus where they regulate the substrate binding and ATPase activity of HSP70 chaperones and thus play a pivotal role in protein quality control in the cell. One HSP40 homolog, Mrj, regulates transcriptional activity in the nucleus by regulating the recruitment of histone deacetylase to chromatin (Dai et al. 2005). Also, it has been shown that aggregation of the TDP-43 protein is prevented by its association with HSP40/HSP70 in the nucleus (Udan-Johns et al. 2014). TDP-43 is found in nuclear aggregates in patients with the neurodegenerative disease, ALS. It is normally complexed with chaperones, but when this association is disrupted, the protein forms nuclear aggregates. This result clearly demonstrates the importance of a properly functioning nuclear chaperone system.

1.2 Conclusions

Much remains to be learned about the role chaperones in the nucleus and how they help the cell cope with stress. Studies on nuclear bodies and nuclear protein aggregates provided some of the first evidence showing that chaperones localize to the nucleus. Several chaperones are known to accumulate in the nucleus under stress conditions. Import of chaperones into the nucleus occurs via a variety of mechanisms. Once in the nucleus, some chaperones can work to break apart nuclear aggregates, while other presumably perform other functions. What is the specific role of nuclear chaperones in the response to stress? Is their main function to refold proteins or to target them for proteasomal degradation? It seems likely that they move into the nucleus to deal with the protein damage that is associated with stress. However, it is not clear how the nuclear chaperone activity compares with what is happening in the cytoplasm. Although some evidence suggests that protein turnover may occur in the nucleus (Gardner et al. 2005; Nielsen et al. 2014), the relationship between the chaperones and the proteasome in the nucleus is not clear. Does Chip function in the nucleus to pull chronically misfolded proteins into the UPS for degradation? This is one of the many fundamental questions about nuclear PQC that remain to be answered.

Acknowledgements The authors would like to acknowledge the community of researchers whose excellent work has brought us to our current understanding of the stress response. We have tried to site the current research regarding the nuclear stress response and we apologize to any researchers whose work might not have been included in this chapter.

References

Adachi H, Waza M, Tokui K, Katsuno M, Minamiyama M, Tanaka F, Doyu M, Sobue G (2007) CHIP overexpression reduces mutant androgen receptor protein and ameliorates phenotypes of the spinal and bulbar muscular atrophy transgenic mouse model. J Neurosci 27:5115–5126

Alastalo T-P, Hellesuo M, Sandqvist A, Hietakangas V, Kallio M, Sistonen L (2003) Formation of nuclear stress granules involves HSF2 and coincides with the nucleolar localization of Hsp70. J Cell Sci 116:3557–3570

Bao YP, Cook LJ, O'Donovan D, Uyama E, Rubinsztein DC (2002) Mammalian, yeast, bacterial, and chemical chaperones reduce aggregate formation and death in a cell model of Oculopharyngeal muscular dystrophy. J Biol Chem 277:12263–12269

Brunquell J, Morris S, Lu Y, Cheng F, Westerheide SD (2016) The genome-wide role of HSF-1 in the regulation of gene expression in Caenorhabditis elegans. BMC Genomics 17:559

Chai Y, Koppenhafer SL, Bonini NM, Paulson HL (1999) Analysis of the role of heat shock protein (Hsp) molecular chaperones in Polyglutamine disease. J Neurosci 19:10338–10347

Chen L, Madura K (2014) Degradation of specific nuclear proteins occurs in the cytoplasm in *Saccharomyces cerevisiae*. Genetics 197:193–197

Chughtai ZS, Rassadi R, Matusiewicz N, Stochaj U (2001) Starvation promotes nuclear accumulation of the hsp70 Ssa4p in yeast cells. J Biol Chem 276:20261–20266

Cummings CJ, Mancini MA, Antalffy B, DeFranco DB, Orr HT, Zoghbi HY (1998) Chaperone suppression of aggregation and altered subcellular proteasome localization imply protein misfolding in SCA1. Nat Genet 19:148–154
Czubryt MP, Austria JA, Pierce GN (2000) Hydrogen peroxide inhibition of nuclear protein import is mediated by the mitogen-activated protein kinase, Erk2. J Cell Biol 148:7–16
Dai Q, Zhang C, Wu Y, McDonough H, Whaley RA, Godfrey V, Li H-H, Madamanchi N, Xu W, Neckers L et al (2003) CHIP activates HSF1 and confers protection against apoptosis and cellular stress. EMBO J 22:5446–5458
Dai Q, Qian S-B, Li H-H, McDonough H, Borchers C, Huang D, Takayama S, Younger JM, Ren HY, Cyr DM et al (2005) Regulation of the cytoplasmic quality control protein degradation pathway by BAG2. J Biol Chem 280:38673–38681
Daniel S, Bradley G, Longshaw VM, Söti C, Csermely P, Blatch GL (2008) Nuclear translocation of the phosphoprotein Hop (Hsp70/Hsp90 organizing protein) occurs under heat shock, and its proposed nuclear localization signal is involved in Hsp90 binding. Biochim Biophys Acta 1783:1003–1014
Gardner RG, Nelson ZW, Gottschling DE (2005) Degradation-mediated protein quality control in the nucleus. Cell 120:803–815
Hageman J, Vos MJ, van Waarde MAWH, Kampinga HH (2007) Comparison of intra-organellar chaperone capacity for dealing with stress-induced protein unfolding. J Biol Chem 282:34334–34345
Isaac C, Yang Y, Meier UT (1998) Nopp140 functions as a molecular link between the nucleolus and the coiled bodies. J Cell Biol 142:319–329
Ivanauskiene K, Delbarre E, McGhie JD, Küntziger T, Wong LH, Collas P (2014) The PML-associated protein DEK regulates the balance of H3.3 loading on chromatin and is important for telomere integrity. Genome Res 24:1584–1594
Jana NR, Tanaka M, Wang G, Nukina N (2000) Polyglutamine length-dependent interaction of Hsp40 and Hsp70 family chaperones with truncated N-terminal huntingtin: their role in suppression of aggregation and cellular toxicity. Hum Mol Genet 9:2009–2018
Jones RD, Gardner RG (2016) Protein quality control in the nucleus. Curr Opin Cell Biol 40:81–89
Kaimal JM, Kandasamy G, Gasser F, Andréasson C (2017) Coordinated Hsp110 and Hsp104 activities power protein disaggregation in Saccharomyces cerevisiae. Mol Cell Biol 37:e00027
Kodiha M, Chu A, Matusiewicz N, Stochaj U (2004) Multiple mechanisms promote the inhibition of classical nuclear import upon exposure to severe oxidative stress. Cell Death Differ 11:862–874
Kose S, Furuta M, Imamoto N (2012) Hikeshi, a nuclear import carrier for Hsp70s, protects cells from heat shock-induced nuclear damage. Cell 149:578–589
Kriegenburg F, Jakopec V, Poulsen EG, Nielsen SV, Roguev A, Krogan N, Gordon C, Fleig U, Hartmann-Petersen R (2014) A chaperone-assisted degradation pathway targets kinetochore proteins to ensure genome stability. PLoS Genet 10:e1004140
Lafarga M, Berciano MT, Pena E, Mayo I, Castaño JG, Bohmann D, Rodrigues JP, Tavanez JP, Carmo-Fonseca M (2002) Clastosome: a subtype of nuclear body enriched in 19S and 20S proteasomes, ubiquitin, and protein substrates of proteasome. Mol Biol Cell 13:2771–2782
Lees MJ, Peet DJ, Whitelaw ML (2003) Defining the role for XAP2 in stabilization of the dioxin receptor. J Biol Chem 278:35878–35888
Li X, Huang M, Zheng H, Wang Y, Ren F, Shang Y, Zhai Y, Irwin DM, Shi Y, Chen D et al (2008) CHIP promotes Runx2 degradation and negatively regulates osteoblast differentiation. J Cell Biol 181:959–972
Melchior F, Gerace L (1995) Mechanisms of nuclear protein import. Curr Opin Cell Biol 7:310–318
Michaud S, Lavoie S, Guimond M-O, Tanguay RM (2008) The nuclear localization of Drosophila Hsp27 is dependent on a monopartite arginine-rich NLS and is uncoupled from its association to nuclear speckles. Biochim Biophys Acta BBA - Mol Cell Res 1783:1200–1210
Morimoto M, Boerkoel CF (2013) The role of nuclear bodies in gene expression and disease. Biology 2:976–1033

Nath SR, Lieberman AP (2017) The Ubiquitination, disaggregation and proteasomal degradation machineries in Polyglutamine diseases. Front Mol Neurosci 10:78

Nielsen SV, Poulsen EG, Rebula CA, Hartmann-Petersen R (2014) Protein quality control in the nucleus. Biomolecules 9:646–661

Parsell DA, Kowal AS, Singer MA, Lindquist S (1994) Protein disaggregation mediated by heat-shock protein Hsp104. Nature 372:475–478

Quan X, Rassadi R, Rabie B, Matusiewicz N, Stochaj U (2004) Regulated nuclear accumulation of the yeast hsp70 Ssa4p in ethanol-stressed cells is mediated by the N-terminal domain, requires the nuclear carrier Nmd5p and protein kinase C. FASEB J 18:899–901

Raimer AC, Gray KM, Matera AG (2016) SMN - a chaperone for nuclear RNP social occasions? RNA Biol 14:701–711

Rampelt H, Kirstein-Miles J, Nillegoda NB, Chi K, Scholz SR, Morimoto RI, Bukau B (2012) Metazoan Hsp70 machines use Hsp110 to power protein disaggregation. EMBO J 31:4221–4235

Ronnebaum SM, Wu Y, McDonough H, Patterson C (2013) The ubiquitin ligase CHIP prevents SirT6 degradation through noncanonical Ubiquitination. Mol Cell Biol 33:4461–4472

Sampuda KM, Riley M, Boyd L (2017) Stress induced nuclear granules form in response to accumulation of misfolded proteins in Caenorhabditis elegans. BMC Cell Biol 18:18

Sarge KD, Murphy SP, Morimoto RI (1993) Activation of heat shock gene transcription by heat shock factor 1 involves oligomerization, acquisition of DNA-binding activity, and nuclear localization and can occur in the absence of stress. Mol Cell Biol 13:1392–1407

Shang Y, Zhao X, Xu X, Xin H, Li X, Zhai Y, He D, Jia B, Chen W, Chang Z (2009) CHIP functions an E3 ubiquitin ligase of Runx1. Biochem Biophys Res Commun 386:242–246

Shibata Y, Morimoto RI (2014) How the nucleus copes with proteotoxic stress. Curr Biol 24:R463–R474

Stenoien DL, Cummings CJ, Adams HP, Mancini MG, Patel K, DeMartino GN, Marcelli M, Weigel NL, Mancini MA (1999) Polyglutamine-expanded androgen receptors form aggregates that sequester heat shock proteins, proteasome components and SRC-1, and are suppressed by the HDJ-2 chaperone. Hum Mol Genet 8:731–741

Tkach J, Glover J (2008) Nucleocytoplasmic trafficking of the molecular chaperone Hsp104 in unstressed and heat-shocked cells. Traffic 9:39–56

Trinklein ND, Murray JI, Hartman SJ, Botstein D, Myers RM (2004) The role of heat shock transcription factor 1 in the genome-wide regulation of the mammalian heat shock response. Mol Biol Cell 15:1254–1261

Tsuneoka M, Mekada E (1992) Degradation of a nuclear-localized protein in mammalian COS cells, using Escherichia coli beta-galactosidase as a model protein. J Biol Chem 267:9107–9111

Udan-Johns M, Bengoechea R, Bell S, Shao J, Diamond MI, True HL, Weihl CC, Baloh RH (2014) Prion-like nuclear aggregation of TDP-43 during heat shock is regulated by HSP40/70 chaperones. Hum Mol Genet 23:157–170

Van den IJssel P, Wheelock R, Prescott A, Russell P, Quinlan RA (2003) Nuclear speckle localisation of the small heat shock protein alpha B-crystallin and its inhibition by the R120G cardiomyopathy-linked mutation. Exp Cell Res 287:249–261

Vos MJ, Kanon B, Kampinga HH (2009) HSPB7 is a SC35 speckle resident small heat shock protein. Biochim Biophys Acta 1793:1343–1353

Walter GM, Smith MC, Wisén S, Basrur V, Elenitoba-Johnson KSJ, Duennwald ML, Kumar A, Gestwicki JE (2011) Ordered assembly of heat shock proteins, Hsp26, Hsp70, Hsp90, and Hsp104, on expanded Polyglutamine fragments revealed by chemical probes. J Biol Chem 286:40486–40493

Weis K (2003) Regulating access to the genome: Nucleocytoplasmic transport throughout the cell cycle. Cell 112:441–451

Yasuda Y, Miyamoto Y, Saiwaki T, Yoneda Y (2006) Mechanism of the stress-induced collapse of the ran distribution. Exp Cell Res 312:512–520

Yokom AL, Gates S, Jackrel ME, Mack KL, Su M, Shorter J, Southworth DR (2016) Spiral architecture of the Hsp104 disaggregase reveals the structural basis for polypeptide translocation. Nat Struct Mol Biol 23:830–837

Chapter 2
Extracellular Hsp70 and Low-Grade Inflammation- and Stress-Related Pathologies

Eduardo Ortega, Isabel Gálvez, and Leticia Martín-Cordero

Abstract Inflammation is part of, but not limited to, the innate immune response against pathogens. Availability of large amounts of nutrients, low level of physical activity, or increased longevity are conditions that can lead to tissue malfunction; they are associated with many chronic inflammatory diseases which are generally characterized by the presence of parainflammation and low-grade systemic inflammation. Unlike the mechanisms of inflammation induced by infection, the mechanisms of systemic inflammatory states are poorly understood. Disruption of the cytokine-HPA axis and of sympathetic nervous system regulation predisposes to chronic inflammatory pathologies. Thus, anomalous immune/inflammatory responses accompanied by an altered neuroendocrine response underlie a variety of immune- and stress-related diseases. The capacity of eHsp72 to affect and to be affected by immune and neuroendocrine mechanisms makes eHsp72 an intrinsic component of the immune-neuro-endocrine network. Since inflammation is one of the stress stimuli that result in the release of eHsp72, circulating levels of eHsp72 may also be considered as a biomarker for the diagnosis of chronic low-level inflammatory states. In this context, and based on recent investigations of our research group, this chapter reviews and discusses the potential role of eHsp72 in inflammation- and stress-related pathologies such as obesity-induced metabolic syndrome, fibromyalgia, and osteoarthritis. Finally, the role of eHsp72 in the physiological adaptation to inflammation and stress during aging is also discussed.

Keywords Aging · Fibromyalgia · Hsp72 · Metabolic syndrome · Osteoarthritis · Systemic inflammation

E. Ortega (✉) · I. Gálvez
Department of Physiology (Immunophysiology Research Group), Faculty of Sciences, University of Extremadura, Badajoz, Spain
e-mail: orincon@unex.es

L. Martín-Cordero
Department of Nursing (Immunophysiology Research Group), University Center of Plasencia, University of Extremadura, Plasencia, Spain

A. A. A. Asea, P. Kaur (eds.), *Heat Shock Proteins and Stress*,
Heat Shock Proteins 15, https://doi.org/10.1007/978-3-319-90725-3_2

Abbreviations

CRP	C-reactive protein
CVD	Cardiovascular disease
DAMP	Damage-associated molecular pattern
ECM	Extracellular cartilage matrix
eHsp	Extracellular heat shock protein
FLS	Fibroblast-like synoviocyte
FM	Fibromyalgia
HPA	Hypothalamus-pituitary-adrenal axis
Hsc	Constitutive heat shock protein
iHsp	Intracellular heat shock protein
IL	Interleukin
MetS	Metabolic syndrome
NA	Noradrenaline
NF-κB	Nuclear factor-kappa B
OA	Osteoarthritis
RA	Rheumatoid arthritis
SNS	Sympathetic nervous system
TLR	Toll-like receptor
TNF	Tumor necrosis factor

2.1 Introduction

Inflammation is part of, but not limited to, the innate immune response mechanisms against pathogens. Different aspects and concepts of physiological and pathophysiological inflammatory responses have been very adequately reviewed, explained and discussed by Medzhitov (2008). It is widely known that when the acute inflammatory response is triggered, in response to an infection (particularly a bacterial infection) or tissue damage, a coordinated release of plasma and leukocytes (mainly neutrophils) into the infected or damaged area occurs. An appropriate and satisfactory inflammatory response results in the elimination of the infectious agents followed by a resolution and repair phase which is mediated primarily by resident and recruited macrophages. If the acute inflammatory response fails to eliminate the pathogen, the inflammatory process persists and acquires new characteristics. Infiltrated neutrophils are replaced by macrophages and, in the case of infection, also by T lymphocytes. If the effect of these cells is not sufficient, a chronic inflammatory state including the formation of granulomas and tertiary lymphoid tissues take place. In addition, chronic inflammation may arise from other causes such as tissue damage resulting from an autoimmune response (Kumar et al. 2003; Medzhitov 2008). Additionally, under adverse conditions tissues suffer from stress and can malfunction. Depending on the magnitude of this adversity, adaptation to the new condition may require the help of resident and recruited macrophages in the tissue together with the release, on a small scale, of leukocytes and plasma proteins. This adaptive response is an intermediate state between the basal state and inflammation, and is known as parainflammation. The parainflammatory response is graded: at one end it is very close to the basal state, while at the opposite end it is close to inflammation. Induction of parainflammation does not require infection or tissue damage; rather, this state is generated by tissue malfunction and aims to restore tissue functionality and homeostasis. If tissue disruption persists for a long time, parainflammation may become chronic. This occurs as a result of mutations, environmental factors, or as a consequence of some "modern human diseases". In fact, many chronic inflammatory diseases seem to be associated with conditions such as availability of large amounts of nutrients, low level of physical activity, or increased longevity (Medzhitov 2008). Diseases related to these conditions are generally characterized by the presence of low grade inflammation (or parainflammation).

Unlike the mechanisms of inflammation induced by infection, the mechanisms of systemic inflammatory states are poorly understood, but it is clear that they do not seem to follow the classical patterns of transition from acute inflammation to chronic inflammation (Medzhitov 2008). Chronic low-grade systemic inflammation is a term used to reflect increments in the systemic concentration of TNF-α, IL-1β, IL-6, and C-reactive proteins (CRP) among other inflammatory and anti-inflammatory mediators. Despite the fact that changes in acute-phase reactants are much smaller than those in acute infections, chronicity of low-grade inflammation is strongly associated with advancing age and lifestyle factors such as obesity, and consequently with increased risk of cardiovascular disease (CVD) and type 2 diabetes (Bruusgaard 2005; Mathur and Pedersen 2008). Thus, chronic low-grade systemic inflammation has generally been circumscribed to obesity and metabolic syndrome, where adipose tissue shows a high infiltration of macrophages that are the main source of inflammatory cytokines such as TNF-α (Wellen and Hotamisligil 2005). Nevertheless, low level systemic elevations (and/or elevated production by cultured immune cells) of inflammatory cytokines such as IL-1β, TNF-α, IL-8 or IL-6 are also present in other pathologies, especially those in which an inflammatory/stress feedback dysregulation can contribute to the aetiology, such as inflammation- and stress-related diseases. Inflammatory and stress responses are regulated by neuroendocrine-immune mechanisms. It is well known that the immune system interacts with the neuroendocrine system, so this interaction can be regarded as a dynamic network that tends to reach adaptive equilibrium in both health and disease (Del Rey et al. 2012). During the immune response, pro-inflammatory cytokines activate the stress system through stimulation of the hypothalamus-pituitary-adrenal (HPA) axis, resulting in the release of glucocorticoids and catecholamines in order to prevent an excessive immune response and to protect the organism from an "overshoot" of pro-inflammatory cytokines (Elenkov 2008; Besedovsky and Del Rey 2000). Since the HPA axis is stimulated during the immune response to non-pathogenic antigens and also during certain pathological conditions in which the immune system is involved, an immunoregulatory circuit involving immune-derived cytokines and the HPA axis may well operate during certain pathological conditions. Indeed, disruption of the cytokine-HPA axis circuit can predispose to chronic inflammatory diseases (Del Rey and Besedovsky 2000), including chronic low-grade inflammatory pathologies (Martín-Cordero et al. 2011; Bote et al. 2012; Gálvez et al. 2017). Thus, anomalous immune/inflammatory responses accompanied by an altered neuroendocrine response constitute a mechanism that underlies a variety of immune- and stress-related diseases.

The Hsp70 family consists of a class of heat shock proteins that includes the constitutively expressed Hsp70 (Hsc70, 73 kDa) and the stress-inducible Hsp70 (Hsp72, 72 KDa). Under normal physiological conditions, Hsp72 is expressed at low levels. However, following stress stimuli, intracellular synthesis of Hsp72 and release of extracellular Hsp72 (eHsp72) release increase markedly, giving rise to a stress response that plays a role in cytoprotection as an intracellular molecular chaperone, and also as an extracellular chaperokine that induces the release of pro-inflammatory cytokines (Asea et al. 2000; Asea 2006). Thus, Hsp70 can act both as a pro- and anti-inflammatory molecule in the production and release of inflammatory mediators. As an anti-inflammatory effector, intracellular Hsp70 prevents the expression of inflammatory mediators by inhibiting the translocation of the transcription factor nuclear factor-kappa B (NF-κB) to the nucleus, avoiding the pro-inflammatory cytokine signal transduction and gene expression and, consequently, increasing endotoxin- and cytokine-mediated cytotoxicity tolerance. As a pro-inflammatory mediator, extracellular Hsp70 (eHsp72) produces a multifaceted immune/inflammatory response involving the activation of a variety of immune effector cells and cytokine release (Amorim and Moseley 2010), particularly inflammatory cells and pro-inflammatory cytokines with the participation of NF-κB (Ortega et al. 2009a; Giraldo et al. 2010a). Therefore, the ability of Hsp70 to regulate inflammation is an important aspect of the progression of a variety of pathophysiologic states characterized by a dysregulated inflammatory response. In fact, it has been proposed that "the highly inducible Hsp70 provides a useful platform to discuss the role of these primitive and ubiquitous proteins in modulating human inflammatory disease" (Amorim and Moseley 2010). In addition, the capacity of eHsp72 to affect and to be affected directly or indirectly by immune and neuroendocrine mechanisms indicates that this heat shock protein is part of the signaling system that can modulate the activity of the neuro-endocrine-immune network. Thus, due to its capacity to affect the production of cytokines that in turn induce neuroendocrine responses, eHsp72 is an intrinsic component of the immune-neuro-endocrine network, which can also mediate neuroendocrine metabolic adjustments and influence adaptive brain functions (Ortega et al. 2012a).

Since inflammation is one of the stress stimuli that result in the release of eHsp72 (Campisi et al. 2003), high circulating levels of eHsp72 may also be considered as a biomarker for reinforcing the diagnosis of a chronic low-level inflammatory status. As discussed above, the ability of Hsp70 to modulate the inflammatory response in the context of stress responses, with a protective anti-inflammatory activity of intracellular Hsp70 and, paradoxically, a pro-inflammatory action of extracellular Hsp70, is widely accepted. However, the magnitude of the pro-inflammatory response induced by eHsp72 (through the induction of IL-1β, TNF-α, and IL-6 release), as well as the induced compensatory anti-inflammatory response through the release of the anti-inflammatory cytokine IL-10 must be taken into account, since they may be different in healthy individuals and in individuals with low-level inflammatory conditions, particularly those associated with aging and obesity.

The main objective when treating patients with inflammatory diseases is to diminish inflammation and manage secondary consequences in order to improve their quality of life. In the therapeutic management of low-grade inflammation- and stress-related pathologies, different strategies focus on reducing the circulating levels of inflammatory and stress biomarkers, as well as on rebalancing the interactions between the cytokine-mediated inflammatory response and the stress responses mediated by the sympathetic nervous system (SNS) and the HPA axis. Determination of circulating levels of eHsp72 may then constitute a good biomarker of inflammation and stress for the evaluation of effectiveness of these strategies. Nevertheless, studies of the behavior of inflammatory cells in response to eHsp72 in each situation and condition can also be crucial, with particular attention to the different responses depending on the baseline inflammatory status of each patient.

This chapter reviews and discusses the potential role of eHsp72 in a classical low-grade inflammatory disease, obesity-induced metabolic syndrome, and also in fibromyalgia and osteoarthritis, two pathologies in which recent investigations have reported that low-level systemic inflammation together with a dysregulated stress response also seem to contribute to their etiology. Finally, although aging cannot be considered a pathology, it can also be drawn into the spectrum of physiological conditions in which there is an underlying low-grade systemic inflammation and dysregulated stress response; hence the potential role of extracellular Hsp70 in the physiological adaptation to inflammation and stress during aging is briefly discussed.

2.1.1 Extracellular Hsp70 and Metabolic Syndrome

Metabolic syndrome (MetS), or insulin resistance syndrome, is a metabolic disorder associated with obesity that involves risk factors for type 2 diabetes mellitus and arteriosclerosis, with concomitant increased risk of cardiovascular events (Hansen 1999; Eckel et al. 2005; Zimmet et al. 2005). It is well known that obesity is associated with low-grade inflammation (Das 2002, 2004), and inflammation appears to be a common link between atherosclerosis, obesity, and insulin resistance (Bastard et al. 2006). In fact, a model of interaction between the immune and metabolic systems has been proposed, in which inflammation induces obesity and insulin resistance, and in turn obesity induces low-grade inflammation and also promotes insulin resistance and all of the other characteristics associated with MetS (Wellen and Hotamisligil 2005). As has been mentioned, low-grade inflammation is a condition that involves increased systemic concentrations of TNF-α, IL-1β, IL-6, IL-1ra, and CRP (Petersen and Pedersen 2005). Inflammatory cytokines can also stimulate the

HPA axis, increasing levels of glucocorticoids and catecholamines and affecting inflammatory and immune processes. In this way, disruption of the cytokine-HPA axis feedback loop can aggravate inflammatory pathologies (Besedovsky and Del Rey 2007). It is also well established that noradrenaline (NA) is involved in regulating both metabolism and most of the mechanisms of the immune response, including the innate response and the systemic and local release of inflammatory cytokines (Ortega et al. 2000; Elenkov and Chrousos 2002; García et al. 2003; Elenkov et al. 2005; Sanders 2006; Besedovsky and Del Rey 2007). Indeed, noradrenergic dysfunction, including over-activity of the SNS, is today also recognized as a characteristic of obesity-related MetS, contributing to the pathophysiology and clinical prognosis of this disorder (Straznicky et al. 2010a, 2010b). In this context, studies in our laboratory found that Zucker rats (an experimental model of MetS) present high circulation levels of NA, CRP and IL-6, leading us to hypothesize that animals with MetS present dysregulation in the negative feedback mechanism between IL-6 and NA that can contribute to the systemic low-grade inflammation and/or hyperglycaemia of MetS (Martín-Cordero et al. 2011).

Catecholamines play a key role in the extracellular release of Hsp72. Adrenergic receptors are involved in the stress-induced increase of eHsp72, and NA could stimulate a receptor-mediated exocytotic pathway of Hsp72 release during stressor exposure (Fleshner et al. 2004). Noradrenaline thus participates as a stress signal for the release of eHsp72 during stress (Ortega et al. 2010), increasing the expression and release of Hsp72 by human immune cells such as neutrophils (Ortega et al. 2010). Therefore, a wide variety of pathological (including inflammatory) and physiological stressful stimuli can induce a marked increase in intracellular synthesis (Lindquist and Craig 1988) and extracellular release (Campisi et al. 2003; Fleshner et al. 2004) of Hsp72; pathologies with noradrenergic dysregulation together with low-grade inflammation, as is the case of MetS, can also affect the extracellular concentration of Hsp72. A high basal concentration of eHsp72 is associated with hypertension, arteriosclerosis (Johnson and Fleshner 2006), and MetS (García et al. 2013). This agrees with the notion that eHsp72 could be a good biomarker of inflammation and stress, and conceptually it could also be considered as a potentially good biomarker of the regulation and progression of MetS. Since eHsp72 can stimulate the release of IL-1β, IL-6, and TNF-α, theoretically it can also contribute to the low-grade systemic inflammation underlying MetS and therefore to insulin resistance and type 2 diabetes mellitus.

As has already been widely and very well reviewed, intracellular Hsp72 is involved in the regulation of type 2 diabetes in human and animal models, protecting against obesity-induced insulin resistance (Chung et al. 2008; Whitham and Febbraio 2010). In fact, patients with type 2 diabetes mellitus can present a reduced gene expression of Hsp72 (Kaur et al. 2010). In the cascade of molecules involved in inflammation and obesity, the pro-inflammatory cytokine TNF-α plays a prominent role in mediating downstream transduction cascades that affect insulin signaling. It has been observed that TNF-α is overexpressed and overproduced in the adipose tissue of rodent models of obesity and obese humans (Hotamisligil et al. 1993, 1995; Wellen and Hotamisligil 2005). Intracelullar Hsp72 confers protection against disturbed metabolic homeostasis via multiple modes of action including, but not limited to, reducing inflammation (Chung et al. 2008; Morino et al. 2008; Gupte et al. 2009), protecting cells against TNF-α-induced insulin resistance (Chung et al. 2008), and improving skeletal muscle oxidation (Chung et al. 2008; Gupte et al. 2009; Liu and Brooks 2012). Nevertheless, in contrast with intracellular Hsp70 (iHsp70), extracellular Hsp70 (eHsp70) has potent pro-inflammatory effects; thus recently it has been suggested that imbalances in Hsp70 status, described by the [eHsp70]/[iHsp70] ratio, may be determinant in triggering a chronic pro-inflammatory state that leads to insulin resistance and type 2 diabetes mellitus development, suggesting that changes in this ratio value could be used as a biomarker for the management of inflammatory response in insulin resistance and diabetes (Krause et al. 2015).

In addition, eHsp72 can also elicit an autoimmune response with the production of anti-Hsp70 antibodies (Pockley et al. 2003). A relationship between circulating anti-Hsp70 antibody levels and deregulated parameters associated with MetS, such as hypertension, dyslipidemia, atherosclerosis, and obesity has been reported (Pockley et al. 2002, 2003; Ghayour-Mobarhan et al. 2007). In fact, obese subjects present significantly higher plasma anti-Hsp70 concentrations compared with overweight and normal weight subjects (Ghayour-Mobarhan et al. 2007). Altered levels of circulating antibodies against Hsp70 have been also associated with type 2 diabetes and diabetic microvascular/macrovascular complications, and MetS (Wu et al. 2001; Ghayour-Mobarhan et al. 2005, 2007; Gruden et al. 2009, 2013; Kaur et al. 2010). However, the underlying mechanisms of this association are still unclear (Gruden et al. 2013). Participation of Hsp70 in the response and physiological regulation of type 2 diabetes and MetS is not simply a protective effect of intracellular Hsp70 and a potential inflammatory-induced response mediated by extracellular Hsp70. Not only the [eHsp70]/[iHsp70] ratio, but also the levels of anti-Hsp70 antibodies in response to the increased concentration of eHsp72, and probably different pro- or anti-inflammatory effects of eHsp72 in healthy or MetS-suffering individuals, play an important role in physiological regulation by Hsp70 in obesity-associated MetS.

Release of TNF-α by adipose tissue is primarily produced by non-fat cells present in the tissue (Fain et al. 2004); inflammatory factors released by macrophages could induce inflammation in adipocytes by stimulating the transcription of inflammatory genes (Odrowaz-Sypniewska 2007). In this context, eHsp72 stimulates the release of inflammatory cytokines from healthy lean Zucker rats (García et al. 2013), thus confirming the pro-inflammatory activity of eHsp72. The question now is whether macrophages from obese animals respond to eHsp72 in the same way as those from healthy ones. An altered eHsp72-induced release of inflammatory cytokines by macrophages from genetically obese rats (fa/fa) is shown by a decrease in eHsp72-induced release of IL-1β and TNF-α, and an increase in release of IL-6 with respect to the macrophage response from the lean rats (Fa/fa). The lower TNF-α release in response to eHsp72 by macrophages from genetically obese rats could be explained by the high eHsp72-induced release of IL-6, since IL-6 can inhibit TNF-α release by macrophages in a paracrine context. Also physiologically relevant is the fact that eHsp72 inhibited the release of IL-1β and IL-6 in macrophages from obese rats with respect to their constitutive release, thus confirming a different or altered response to eHsp72 in macrophages from obese animals as compared to macrophages from healthy ones. These results suggest that eHsp72 participates in the adaptation/regulation processes in MetS; the elevated circulating levels of eHsp72 in obese animals (which seems to be a consequence of the previously mentioned inflammatory and stress dysregulation underlying MetS) could also be viewed as a physiological adaptation for counterbalancing the increased release of inflammatory cytokines in MetS (García et al. 2013). Nevertheless, this adapted response could also have a cost in terms of susceptibility to infections in obese individuals with type 2 diabetes, which has also been reported (Bray 1985; Knight et al. 2000).

Increasing Hsp expression could be considered a therapeutic strategy to protect from insulin resistance and diabetes, and to reinforce vascular defense and delay or avoid clinical complications associated with cardiovascular diseases (Kaur et al. 2010), in both genetic- and diet-induced obesity (Whitham and Febbraio 2010). Interventions may include heat therapy (Chung et al. 2008; Gupte et al. 2009), electrical therapy (Morino et al. 2008), physical exercise (Morino et al. 2008), or pharmacological molecules that stimulate heat shock response (McCarty 2006; Gupte et al. 2009; Literati-Nagy et al. 2009). Heat response elevates intracellular Hsp72 in white and brown adipose tissue, gut, heart, and brain; it attenuates inflammatory markers (JNK, IKK-b and NF-kB) in multiple organs and tissues, it improves insulin stimulated signaling in muscle and liver, and it restores impaired vasodilatation in vessels (revised by Henstridge et al. 2014). Thus, induction of intracellular Hsp72 is a novel approach to prevent insulin resistance-induced vascular complications and insulin sensitivity in obesity (Karpe and Tikoo 2014).

Although the benefits of increasing the concentration of iHsp70 in obesity and type 2 diabetes are well documented, little is known about the strategies which focus on modifying concentration and function of eHsp2. It is plausible to speculate that strategies aimed to reduce elevated systemic levels of eHsp72 in MetS could contribute to reduce potential eHsp72-induced inflammatory response. Habitual aerobic exercise is an accepted non-pharmacological therapeutic strategy in the management of MetS to improve diabetic status and insulin sensitivity, thus reducing the risk of CVD (Bassuk and Manson 2005). However, in the context of danger theory, increased circulating concentration of eHsp72 during exercise is known to facilitate the innate/inflammatory immune response (Fleshner et al. 2003; Ortega et al. 2009a; Giraldo et al. 2010a), for example by stimulating the release of inflammatory cytokines by macrophages. Consequently, exercise which increases systemic eHsp72 could elevate inflammation and insulin resistance, thus exacerbating disorders associated to MetS. In fact, studies in our laboratory found that habitual exercise-induced stress in obese Zucker rats impaired the inflammatory/stress feedback response contributing to the metabolic, inflammatory, and stress disorders associated with MetS. Nevertheless, although exercise drastically increased the circulating concentration of eHsp72 in the obese animals, it improved the deregulated macrophage release of inflammatory cytokines (IL-1β, IL-6, and TNF-α) in response to eHsp72, thus approaching the behavior determined in lean rats as reference (García et al. 2013). Hence, strategies aimed to modify eHsp72 concentration in

MetS must take into account not only the concentration of this protein but also the different potential adaptive immunophysiological functions in the obese individuals as compared to the functions in healthy individuals (Ortega 2016). Thus, Hsp70 is considered a master regulator for controlling the immune system, inflammation and insulin resistance (Molina et al. 2016), and exercise is able to modulate both extracellular and intracellular Hsp70 levels; it is therefore an efficient and powerful tool for normalizing and/or maintaining the [eHsp70]/[iHsp70] ratio at appropriate levels (Krause et al. 2015).

2.1.2 Extracellular Hsp70 and Fibromyalgia

Fibromyalgia (FM) is a form of non-articular rheumatism defined by the presence of chronic widespread pain (on both the left and the right sides of the body, both above and below the waist, and obligatorily with accompanying axial skeletal pain) and allodynia to pressure in more than 11 of 18 specified sites or tender points (Wolfe et al. 1990). FM usually appears at an age of 30–40 years (only rarely after the age of 55 years), mainly in women (more than 90% of patients). FM is often associated with symptoms such as reduced physical activity strength, fatigue, sleep disorders, and psychiatric disorders (e.g. depression). In particular, FM patients have great difficulty in performing everyday activities such as walking, working, or remaining seated or standing for long periods of time. Despite this symptom-based definition, the lack of objective criteria has led to doubts about the diagnosis of the pathology and even about its existence. Nevertheless, although the biophysiology of FM remains elusive and the treatment of the syndrome is generally empirical, recent hypotheses of the aetiology of this syndrome have included inflammatory disorders accompanied by changes in the neuro-immune-endocrine system (Van West and Maes 2001); it has been proposed that it belongs not only to the spectrum of inflammatory pathologies but also to the spectrum of “stress-related illnesses”. In this context, studies in our laboratory found that FM patients show an inflammatory state accompanied by an altered stress response. It was therefore proposed that an inflammatory/stress feedback dysregulation also underlies FM. This is manifested by high circulating levels of IL-8 (the main pro-inflammatory cytokine involved in FM) and CRP, high circulating levels of cortisol, and increased systemic levels of NA and eHsp72. This was the first time (and to the best of our knowledge the only time) that an elevated circulating concentration of eHsp72 has been reported in FM patients. Since inflammation also results in the release of eHsp72, high circulating levels of eHsp72 may also contribute to reflect a “chronic low-level inflammatory status” in these patients (Bote et al. 2012).

In healthy women, circulating eHsp72 is particularly involved in the activation of neutrophils (Ortega et al. 2009a; Giraldo et al. 2010a). Human neutrophils can express and release Hsp72, both constitutively and in response to NA. This suggests that NA contributes to the expression and release of Hsp72 under stress situations, with a role for neutrophils as a source for the release of eHsp72 to the blood, thus increasing the inflammatory state of the organism (Giraldo et al. 2010b). Elevated circulating concentrations of NA could therefore act as a "stress signal" for the expression and release of Hsp72 in inflammatory cells in FM patients (Bote et al. 2012). In fact, we have found that neutrophils from FM patients release and express constitutively more Hsp72 than neutrophils from healthy women (Hinchado et al. 2011), and monocytes from FM patients release more Hsp72 than those from healthy women both constitutively and in response to NA (unpublished data). Increased release of inflammatory cytokines (such as IL-1β, TNF-α, and IL-6) by monocytes and enhanced functional capacity of neutrophils (chemotactic, phagocytic, and fungicidal activities) in FM patients can also be justified, at least partially, by their increased systemic concentration of eHsp72 (Bote et al. 2012): This is because stress-induced physiological concentrations of eHsp72 stimulate chemotaxis, phagocytosis, and microbicidal capacities of neutrophils in women (Ortega et al. 2009a; Giraldo et al. 2010a), and they induce the release of IL-1β, TNF-α, and IL-6 by monocytes (Asea et al. 2000), which has also been confirmed in our laboratory in isolated monocytes from FM patients. IL-8 is the pro-inflammatory cytokine regarded as the main systemic inflammatory marker for this pathology (Ortega et al. 2009b; Bote et al. 2012; Wallace et al. 2001; Bazzichi et al. 2007); it is mainly produced by monocytes and endothelial cells (Iizasa and Matsushima 2001). Although no relevant differences have yet been found between women with FM and age-matched healthy women in the constitutive release of IL-8 by monocytes *in vitro*, recent unpublished data from our laboratory reveal that IL-8 release induced by eHsp72 is of a greater magnitude in monocytes from patients with FM, a circumstance which could reinforce a role for eHsp72 in the syndrome. A proposed model of immune-neuroendocrine dysregulation in fibromyalgia is shown in Fig. (2.1).

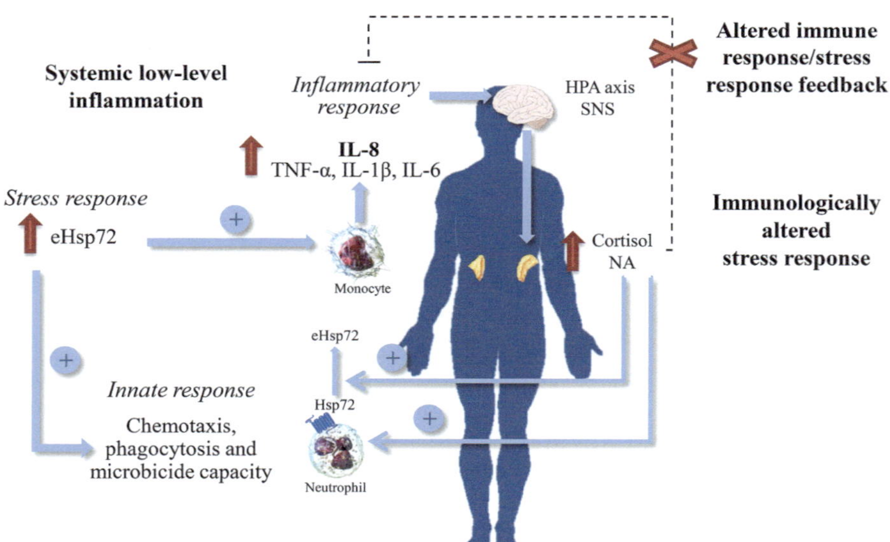

Fig. 2.1 A proposed model of immune-neuroendocrine dysregulation in fibromyalgia

Non pharmacological anti-inflammatory and anti-stress strategies are frequently used as therapeutic aids in the management of FM syndrome. Acceptance of the hypothesis of the anti-inflammatory effects of exercise makes it especially interesting as a therapeutic aid in treating chronic low-grade inflammatory pathologies, such as FM. In this context, habitual exercise (particularly aquatic exercise in warm water) is a good strategy for improving pain and other symptoms associated to FM by inducing an improvement in the altered inflammatory and stress biomarkers associated with this syndrome (Ortega et al. 2009b; Ortega et al. 2012b; Bote et al. 2014). However, paradoxically, it is known that exercise activates the innate and inflammatory response in healthy people (Ortega et al. 1993; Northoff et al. 1994; Suzuki et al. 2002; Giraldo et al. 2009), an effect that is mediated by stress hormones (Ortega 2003; Ortega et al. 2007) and stress proteins, such as eHsp72, even during acute moderate exercise (Fleshner et al. 2003; Ortega et al. 2006, 2009a, 2010). Nevertheless, the effects of exercise may be different in patients with chronic inflammatory diseases and in healthy people. A role for eHsp72 in FM syndrome was confirmed when comparing the effect, between FM patients and age-matched healthy women, of a single bout of moderate cycling (45 min at 55% VO_2 max) on the pro-inflammatory cytokine IL-8, neutrophil activity (chemotaxis and superoxide anion production), and production of inflammatory cytokines by monocytes, together with the systemic concentration of eHsp72 (with NA and cortisol) (Bote et al. 2013). In FM patients, the exercise induced a decrease in the systemic concentration of IL-8, NA, and eHsp72, as well as in neutrophil activation and in inflammatory cytokine release by monocytes (IL-1β, TNF-α, IL-6, IL-10, and IL-18). This was contrary to the fully expected exercise-induced increase in all the above-mentioned biomarkers in healthy women. This opposite behavior in FM patients with respect to healthy women after acute exercise in the inflammatory response seems to be mediated, at least partially, by opposite behavior in the stress response mediated by NA and eHsp72. Thus, the decrease in NA might contribute to the decrease in circulating levels of eHsp72 in FM patients after exercise. In addition, the decreased concentration of eHsp72 could contribute to the decrease in activation of neutrophils and in release of inflammatory cytokines by monocytes. Inflammatory and stress responses to exercise may then differ according to each individual's particular set-point, and these opposite responses may even be working each in a positive regulatory direction (Bote et al. 2013). While moderate exercise-induced activation of the innate and/or inflammatory responses can be considered to be an "alert state" in healthy people for avoiding and defending against pathogenic attacks, given that it is a positive physiological adaptation in a situation of vulnerability for the organism (Ortega 1994), the anti-inflammatory effects induced by the same acute moderate exercise in FM patients is also a particular positive effect, with the focus now being on homeostatic adjustment, and involving inflammation-stress feedback mechanisms. In this homeostatic adjustment, eHsp72 seems to play an important role in FM syndrome. To the best of our knowledge, this paradoxical opposite response to acute moderate exercise in FM patients with respect to healthy people has not been found in any other chronic inflammatory pathologies (e.g., those revised by Ploeger et al. 2009). Thus, circulating levels of eHsp72 in response to moderate sessions of exercise could also contribute as an inflammatory and stress biomarker for the diagnosis of FM syndrome.

2.1.3 Extracellular Hsp70 and Osteoarthritis

Osteoarthritis (OA) is the most prevalent arthritic disease and one of the most common chronic health conditions, affecting mainly elderly people. However, not only age but also genetic, epigenetic, environmental, biomechanical, metabolic, and biochemical factors are involved in the onset and progression of OA (Musumeci et al. 2015). OA affects all the tissues that form synovial joints, the most common being the knees and hips, and it is mainly characterized by cartilage destruction, subchondral bone remodelling, osteophyte formation, and synovial inflammation, leading to joint stiffness, swelling, pain, and loss of mobility (Glyn-Jones et al. 2015; Kapoor 2015). While previously characterized as a non-inflammatory degenerative arthropathy of purely mechanical cartilage degradation, several inflammatory and immune processes within the joint are now recognized to be strongly implicated in the pathogenesis, progression, and burden of OA (Scanzello and Goldring 2012; Rahmati et al. 2016). There is general agreement that it is a complex condition in which inflammatory mediators are released by cartilage, bone, and synovium (Kapoor et al. 2011; Loeser et al. 2012). The multifactorial pathophysiological mechanisms of OA, however, have not yet been clearly elucidated. In this context, it is plausible to speculate that Hsp70, as a stress and inflammatory mediator, could play a role in the pathophysiology of OA.

OA seems to be initiated not only by tissue damage from mechanical injury, but also by infiltration of inflammatory mediators or defects in joint metabolism and homeostasis (Kapoor 2015). Thus, during cellular stress and cartilage damage, extracellular cartilage matrix (ECM) fragments and heat shock proteins (Foell et al. 2007) may act as endogenous damage-associated molecular patterns (DAMP) and trigger a local innate immune response via pattern recognition receptors (PRR), including Toll-like receptors (TLR) (Scanzello et al. 2008; Liu-Bryan and Terkeltaub 2015). This suggests a potential role for stress proteins in OA. The signaling cascade following the activation of TLR-2 and TLR-4 in response to "danger signals" would ultimately result in activation of transcription factors such as NF-κB, which stimulates the inflammatory and innate immune responses in the OA joint (Scanzello et al. 2008) with the release of pro-inflammatory and catabolic mediators from synovial macrophages, fibroblast-like synoviocytes (FLS) and chondrocytes (Kapoor et al. 2011; Sokolove and Lepus 2013; Orlowsky and Kraus 2015). These events contribute to further tissue damage and cartilage breakdown either directly or indirectly, and thus perpetuate and amplify a vicious circle of inflammatory and catabolic activity, resulting in increased synovial inflammation and cartilage degradation (Liu-Bryan 2013; Sokolove and Lepus 2013; Orlowsky and Kraus 2015).

Of the few studies on Hsp70 in OA that exist, the majority were carried out on articular cartilage tissue or chondrocytes *in vitro*. Madreperla and co-workers (1985) first reported the induction of synthesis of Hsp70 in calf chondrocytes that had been incubated at temperatures similar to those reached in cartilage during walking. Soon after, Kubo and co-workers (1985) demonstrated for the first time the synthesis of stress proteins in human chondrocytes incubated at elevated temperatures. In healthy human chondrocytes Hsp70 synthesis was evident after exposure to elevated temperatures (44–46 °C), while in OA chondrocytes Hsp70 synthesis occurred at physiologic temperatures (37 °C), although there was an apparent increase in the synthesis of this protein at higher temperatures. This demonstrated active synthesis of Hsp70 at physiologic temperatures in OA cartilage, suggesting that the cells of this tissue are under stress *in vivo*. Takahashi and co-workers (1997a) also reported that Hsp70 is overexpressed in cartilage from OA subjects. These authors speculated that a possible cause for this may be that the damage of cartilage surface that occurs in OA may create significant friction and heat stress that could enhance Hsp70 synthesis in order to protect the cells against excessive stress (Kubo et al. 1985; Takahashi et al. 1997b). Indeed, chondrocytes from OA joints are subjected to several stressful stimuli that may lead them to initiate a stress response by overexpression of Hsp70 and cytokine genes such as non-physiological high levels of mechanical stress –high weight loading– and heat stress –high temperature in the joint– (Kubo et al. 2001; Takahashi et al. 1997b), and elevated concentrations of catabolic and inflammatory mediators like cytokines in cartilage tissue and synovial fluid (Cruz et al. 1991; Sohn et al. 2012). Moreover, OA cartilage expression levels of Hsp70 have even been reported to correlate positively with the clinical severity of the disease (Kubo et al. 1985; Takahashi et al. 1997b), and thus Hsp70 has been proposed as a parameter showing OA severity and progression (Takahashi et al. 1997b). However, in spite of the fact that Hsp70 is highly expressed in OA cartilage and exerts cytoprotective functions by acting as a chaperone within chondrocytes, a considerable proportion of these cells still die in OA (Terauchi et al. 2003), and thus cartilage degradation and OA progression are not completely prevented by the chondroprotective effects of Hsp70. This protection may not be sufficient due to the release of Hsp72 to the extracellular milieu as well as the presence of pro-inflammatory, catabolic, and apoptotic factors that may counteract the beneficial effects of intracellular Hsp70 in OA.

Surprisingly, despite the important role of synovitis and low-grade systemic inflammation in OA, there are hardly any studies assessing eHsp72 concentrations in synovial fluid or serum in this pathology. In addition to local response, inflammation- and damage-induced vascular leakage result in a subsequent influx of inflammatory mediators and plasma proteins also capable of functioning as DAMPs (Loeser et al. 2012; Sohn et al. 2012; Liu-Bryan 2013), further propagating the local inflammatory response and cartilage degradation (Sokolove and Lepus 2013). In this way, synovitis is a key factor in OA pathophysiology because of the increased presence of several soluble inflammatory mediators in the synovial fluid, such as IL-1β, TNF-α, IL-6, IL-8, IL-15, and IL-17 (Sellam and Berenbaum 2010; Loeser et al. 2012); and infiltration of macrophages, T cells, B cells, mast cells, and natural killer cells in the synovial tissue (De Lange-Brokaar et al. 2012).

In fact, Sohn and co-workers (2012) have reported high serum levels of inflammatory cytokines in OA, in agreement with recent results found in our laboratory showing higher systemic levels of the inflammatory cytokines IL-1β, TNF-α, IL-8, and IL-6 in OA patients with respect to age-matched healthy individuals (Gálvez et al. 2017). In addition, OA patients also presented elevated circulating concentrations of eHsp72 together with low systemic levels of cortisol, thus reflecting an inflammatory state accompanied by an altered stress response and an immune-neuroendocrine dysregulation in OA (Gálvez et al. 2017). These results contribute to confirm the hypothesis that a low-level chronic systemic inflammation underlies OA and that eHsp72 could be a new potentially good systemic biomarker of stress and inflammation in this pathology. Thus, not only could local inflammatory events within joint tissues be reflected systemically, but synovial tissue would also be exposed to the influence of low-grade systemic inflammation (Hügle and Geurts 2016). The presence of low-grade systemic inflammation could also contribute to initiate and/or aggravate OA and, in turn, locally produced inflammatory mediators released into the bloodstream may have an impact on the perpetuation of this chronic systemic inflammatory status (Berenbaum 2013; Gálvez et al. 2017). However, the exact sequence of inflammatory events and how they are initiated currently remains unclear, and whether systemic inflammation induces OA or is a consequence of this pathology has not yet been completely elucidated (Berenbaum 2013; Goldring and Otero 2011; Gálvez et al. 2017). In this sense, the high serum levels of eHsp72 may be at least partially a systemic reflection of its overproduction in the OA joint. Under the effect of diverse stress stimuli, such as pro-inflammatory cytokines, OA chondrocytes could initiate a cellular protection response by increasing the production and probable release of Hsp70. Furthermore, owing to its chaperokine activity, it can be speculated that eHsp72 could induce inflammatory cells to release more pro-inflammatory cytokines, thus creating a positive feedback that would contribute to perpetuate the increased inflammatory response. Hence, since inflammation is one of the stress stimuli that results in the release of eHsp72 (Campisi et al. 2003), high circulating levels of eHsp72 in OA *per se* may reflect a "chronic low-grade inflammatory status". Although locally overexpressed Hsp70 may have a protective role in the intracellular milieu in OA, an increased extracellular release of this protein may contribute to the perpetuation and progression not only of this pathology, but also of other pathologies and conditions influenced by a high systemic inflammatory status in OA patients.

Currently, OA treatments mainly target symptoms, but there are no available therapies that modify OA progression in the long term. Therefore, the goals of OA treatment are to reduce pain and stiffness, and to preserve and improve joint mobility and quality of life (Hochberg et al. 2012). Regarding Hsp, some of the above-mentioned experiments have suggested that Hsp70 expressed in OA chondrocytes could protect the cells against excessive stress (Kubo et al. 1985; Takahashi et al. 1997a, 1997b). Thus, when high levels of Hsp70 are expressed in chondrocytes, cells would be more protected against stress, cytotoxicity, cell death, ECM destruction and thus prevent OA progression (Kubo et al. 2001); for this reason strategies focused on increasing intracellular expression levels of Hsp70 can be an effective therapeutic aid in the management of this pathology. Nevertheless, direct administration of Hsp70 into chondrocytes from outside the body is infeasible. Consequently, it is necessary to stimulate or alter chondrocytes to express this protein. Several authors have investigated gene therapy with Hsp70 in OA chondrocytes. Chondrocytes to which Hsp70 gene was transduced had a significantly higher metabolic activity and viability under heat stress (Kubo et al. 2001), showing that Hsp70 expressed in OA chondrocytes protects the cells against excessive stress and accelerates cartilage metabolism with enhanced expression of proteoglycan core protein (Arai et al. 1997). Induction of Hsp70 by the reversible proteasome inhibitor MG132 or by gene transfer, efficiently protects chondrocytes from cytotoxic stress *in vitro*, and so increasing cell survival (Grossin et al. 2004, 2006). *In vivo*, the induction by intra-articular injections of MG132 (Grossin et al. 2004; Etienne et al. 2008) or overexpression via electric gene transfer (Grossin et al. 2006) of Hsp70 decreased the severity of OA lesions in a mono-iodoacetate-induced (Grossin et al. 2004, 2006) or surgically-induced (Etienne et al. 2008) OA rat model. Cartilage hypocellularity due to cell death also contributes to the development of OA, and chondrocyte apoptosis has been described as one possible pathway for OA patophysiology (Blanco et al. 1998). Stress-induced activation of Hsp not only protects cells from stress but also suppresses apoptosis in various cells (Mosser et al. 2000; Park et al. 2000; Ravagnan et al. 2001). Given the fact that stress-induced intracellular Hsp70, through its cytoprotective and anti-apoptotic capacities, could have chondroprotective effects, it has been proposed that increasing intracellular expression of Hsp70 in chondrocytes may be an effective approach to prevent apoptotic cell death and thus extracellular matrix destruction in OA. *In vitro*, overexpression of Hsp70 in cultured articular chondrocytes via gene delivery almost completely suppressed apoptosis through inhibition of the activation of caspase 3 (Terauchi et al. 2003). Additionally, treatment of articular chondrocytes with glutamine has shown anti-apoptotic effects that may be mediated by Hsp70 (Tonomura et al. 2006). In addition, since mechanical stress can, under certain conditions, induce Hsp70 expression, the effects of exercise in OA have also been investigated. In a rat model of OA, it has been reported that light or moderate exercise is beneficial on the severity of chondral lesions, probably due to a reduced level of chondrocyte apoptosis through the anti-apoptotic capacities of stress-induced Hsp70 overexpression, whereas intense exercise abolishes this chondroprotective effect (Galois et al. 2004). This may be due to the intense exercise-induced release of eHsp72, which usually provokes pro-inflammatory effects.

Heat stimulation is another method for inducing Hsp70 expression. Hydrotherapy, balneotherapy, mud therapy, hot packs, paraffin baths, ultrashort waves, microwave, ultrasound, and other heat stimuli are thermal treatments widely used as non-pharmacological strategies for the treatment of OA in clinical settings (Yildirim et al. 2010; McAlindon et al. 2014). Appropriate microwave heat-induced Hsp70 expression might partially mediate an increase in the expression of matrix genes such as proteoglycan core protein and type II collagen in the articular cartilage *in vivo*, thus contributing to promote metabolic activity in the cartilage (Tonomura et al. 2008). Fujita and co-workers (2012) examined the combination of glutamine and microwave-induced heat stimulation therapy *in vivo* to increase Hsp70 expression in OA cartilage. After the treatment, the severity of OA was reduced, thus suggesting that Hsp70, stimulated by the combination of microwave heat and glutamine, may be involved in the suppression of OA progression. Similarly, mild electrical stimulation and heat stimulation have been reported to increase Hsp70 in chondrocytes and promote articular cartilage matrix metabolism (Hiraoka et al. 2013). These studies suggest that heat-induced increase of Hsp70 in OA cartilage may be useful for slowing the progression of OA.

Taking into account that increased systemic levels of eHsp72 have been found in OA patients, most likely a systemic reflection of an inflammatory and stress status in this pathology, the question now is: what about a role for eHsp72 in the management of OA? It seems to be plausible to speculate that strategies focused on diminishing the circulating concentration of eHsp72 could also be effective as a therapeutic aid for OA patients. Nevertheless, the systemic effect of therapies such as hyperthermia on eHsp72 in OA has not been sufficiently investigated in clinical studies. Recently, our research group described a reduction in systemic eHsp72 concentrations in aged OA patients (with closer values to those of age-matched healthy individuals) after a cycle of 10 days of mud therapy with mineral-medicinal water at 38–42 °C. This effect occurred in parallel with a marked decrease in the serum concentration of the pro-inflammatory cytokines IL-1β, TNF-α, IL-8, and IL-6, which strongly suggests a role for eHsp72 in the systemic anti-inflammatory effects induced by the cycle of balneotherapy or, at least, confirms eHsp72 as a good inflammatory and stress systemic biomarker in OA (Ortega et al. 2017). Uzunoglu and co-workers (2017) have also assessed mild heat stress using thermal water baths for 3 weeks in OA patients. Serum eHsp72 levels initially increased after the first session of balneotherapy, but at the end of the 3-week protocol eHsp72 systemic concentration was lower than baseline, implying that an adaptation might occur at the end of the intervention. It could therefore be speculated that the paradoxical decrease of eHsp72 in our study reflects a lower release of eHsp72 after the heat-induced intracellular Hsp70 increase in OA, for example in chondrocytes, with a decrease in expression and release of the inflammatory and stress mediators, such as inflammatory cytokines and stress proteins. Thus, the ratio [iHsp70]/[eHsp70] could be crucial to evaluate the effectiveness of heat therapies, particularly in the joint. In addition, response to eHsp72 also depends on the basal activation status of NF-κB in the cells. In some chronic inflammatory diseases associated with aging, such as diabetes (García et al. 2013) and other rheumatic diseases such as rheumatoid arthritis (RA), the effects of eHsp72 via the activation of TLR-4 can be anti-inflammatory (or not pro-inflammatory) by inhibiting NF-κB (Luo et al. 2008a, 2008b; Martínez de Toda and De la Fuente 2015). The anti-inflammatory effects of eHsp72 under certain physiological or pathophysiological conditions, such as aging or RA, can also be induced by a strong IL-10 release brought about by stress proteins in FLS or macrophages (Luo et al. 2008a; Ortega et al. 2012a). Thus, it would also be important to corroborate that eHsp72 exerts pro-inflammatory responses in OA patients in the same way as in healthy individuals. In this way, post-heat modulation of eHsp72 circulating concentrations could trigger an Hsp-cytokine-HPA-cortisol anti-inflammatory feedback mechanism, leading to anti-inflammatory effects and neuroendocrine-immune regulation (Ortega et al. 2017).

In sum, although the number of studies is small, they demonstrate a role for Hsp70 and thermotherapy. This suggests that hyperthermia can be effectively applied to OA treatments. It is still necessary to determine the optimal intensity, duration, and interval of heat stimulation for clinical application in humans, since the effect of hyperthermia shows significant variations in response to small differences in the applied temperature (Hojo et al. 2003). Given that some treatments for OA are associated with problems including invasiveness, prognosis and high cost of the procedures, it would be interesting to develop alternative conservative therapies like thermotherapy which could effectively prevent or delay the progression of OA (Takahashi et al. 2009; Ortega et al. 2017).

2.1.4 Aging as a Low-Grade Inflammation- and Stress-Related Condition: Extracellular Hsp70 and Aging

As can be easily deduced from reading the previous sections, the pathologies addressed in relation to inflammatory and stress response dysregulation are closely related to the aging process, mainly OA and type 2 diabetes, which are associated with MetS, but also FM, since patients suffering from FM show signs of premature aging (Hassett et al. 2015). Low-grade inflammation is, in turn, closely related to the aging process. Together with an oxidative stress state (Harman 1956), the aging process is also characterized by chronic low-grade inflammation, so-called "inflamm-aging" (Franceschi et al. 2000). This term denotes an upregulation of the inflammatory response that occurs with age, resulting in a low-grade chronic systemic pro-inflammatory state characterized by elevated concentrations of pro-inflammatory cytokines (such as IL-1β, IL-6, and TNF-α) and other inflammatory compounds, all of them involved in the pathogenesis of most age-associated diseases (Vasto et al. 2007). Due to the very interrelated oxidative and inflammatory processes, the term "oxi-inflamm-aging" was proposed to better define what occurs during the aging process, linking chronic oxidative and inflammatory stress and involving the immune system in these interactions (De la Fuente and Miquel 2009). Thus, immunosenescence, or age-related changes in the immune system, is characterized by the impairment of many immune functions but also by the overactivation of others that could, at the same time, contribute to the increase of inflammatory-oxidative stress, and to the acceleration of aging. Therefore, successful aging and longevity depend on the maintenance of efficient anti-inflammatory and antioxidant mechanisms, in which Hsp70 can play a crucial role (Ortega et al. 2012a; Martínez de Toda and De la Fuente 2015). Functional longevity also depends on the preservation of an adequate balance at all physiological levels, especially between the function and the interaction of the homeostatic systems such as the nervous, endocrine and immune systems and their communications. Age-related alterations of these regulatory systems lead to loss of homeostasis and therefore to the increase in morbidity and mortality that appears with age (De la Fuente et al. 2011). eHsp72 can also exert protective actions by affecting the neuro-endocrine-immune network, particularly the interactions between cytokines and neuroendocrine agents (Ortega et al. 2012a).

It is widely accepted that intracellular Hsp70 confers protection against the harmful effects of oxidative stress and inflammatory status, playing an important role as a regulator of the rate of aging. In this context, Hsp70 has been proposed as a promising biomarker of biological age and lifespan. Nevertheless, the different actions of the inducible and basal levels of Hsp70 must be taken into account (reviewed by Martínez de Toda and De la Fuente 2015). Martínez de Toda and De la Fuente conclude that, since Hsp70 induction levels after a stressor are high in adult and long-lived individuals but low in old ones (or prematurely aged subjects), these inducible levels can be proposed as a biomarker of biological age. Together with impaired induction, the decreased basal levels during aging could lead to even more damage accumulation in cells and tissues. However, the age-related changes of Hsp70 basal levels are still controversial, mainly due to differential accumulation of Hsp70 between mitotic or postmitotic tissues (Martínez de Toda and De la Fuente 2015). Recent results have confirmed that aging-associated variations of the levels of Hsp70 followed a different pattern in post-mitotic (heart, renal cortex, cerebral cortex, spleen) and mitotic (liver and renal medulla) tissues, being lower or higher in old mice compared to adults, respectively. It is curious that Hsp70 levels were similar in tissues from long-lived mice to those in adult animals. These results strongly suggest that Hsp70 basal levels show tissue-specific age-associated variations and are preserved in long-lived animals, demonstrating their role as markers of the rate of aging and longevity (Martínez de Toda et al. 2016).

With respect to extracellular Hsp70, results are even more controversial, not only with regard to age-related changes in the circulating concentration of eHsp72 but also with respect to the immunophysiological role of this protein in the adaptive homeostasis during the aging process. What do we expect from the relation between aging and eHsp72? It could be hypothesized that, due to the inflammatory state in aging (inflammaging), systemic concentration of eHsp72 would increase with advancing age and, in turn, these high levels of eHsp72 could increase the inflammatory state in aged individuals in a vicious circle of positive feedback. However, most human studies indicate that eHsp72 decreases with advancing age in the normal population, and low serum eHsp72 levels have been associated with longevity (Rea et al. 2001; Terry et al. 2004, 2006) reflecting an anti-inflammatory status (Terry et al. 2006; Franceschi et al. 2007). Conversely, high eHsp72 levels have been positively correlated with the level of inflammation, being associated with inflammation and frailty in elderly patients (Njemini et al. 2004, 2011). Centenarians and centenarians' offspring have shown lower serum eHsp72 levels compared to an old group and their respective offspring, the authors proposing that serum eHsp72 concentration can predict the development of cardiovascular diseases, and that low eHsp72 levels can be a predictor of longevity (Terry et al. 2004).

However, it has also been proposed that circulating eHsp72 could play a protective role against inflammation-related conditions, such as aging (Bautmans et al. 2010). Results from our laboratory in mice were also contradictory with respect to those reported in humans: circulating concentrations of eHsp72 were only detectable in old (13 months) and very old (26 months) C57BL/6 J mice (no detectable concentration could be determined in 3 month-old mice). Surprisingly, very old mice also had the lowest glucose (and exhibited a better glucose tolerance than old mice) and the highest corticosterone plasma levels. In addition, no significant age-dependent changes were found in constitutive and eHsp72-induced release of IL-1β, but IL-6 release was significantly decreased in very old mice. Taken together, these results suggest that very old animals present a different inflammatory response, which may contribute to endocrine and metabolic alterations. Alternatively, they may indicate an immune-endocrine adaptation that contributes to longevity (Ortega et al. 2011). Furthermore, while immune cells from very old animals produced low levels of IL-6, subsequent results showed that the capacity to produce IL-10 was much higher than in younger mice (unpublished data). Glucocorticoids, which can also induce the production of Hsp72 (Sun et al. 2000), not only inhibit pro-inflammatory cytokines synthesis and release, but also increase IL-10 production, the most important anti-inflammatory cytokine. Thus, eHsp72-glucocorticoid-IL-10 interaction appears to be relevant to keep inflammaging under control (Ortega et al. 2012a). Basal systemic levels of eHsp72 may not reflect the magnitude of the increase and use of this protein that occurs under stress and other danger conditions. Physiological levels of both eHsp72 and corticosterone can stimulate certain aspects of the innate response, and the release of eHsp72 could be a countermeasure against the deleterious effects of inflammatory mediators, particularly during the aging process when disruptions in the connections of the neuro-immune-endocrine network are inexorable. eHsp72 could therefore exert protective actions by affecting the neuro-endocrine-immune network, particularly the interactions between cytokines and neuroendocrine agents. In addition to the immune/inflammatory regulatory responses, this network would be able to mediate neuroendocrine metabolic adjustments (Ortega et al. 2012a).

2.2 Conclusions

The ability of Hsp70 to modulate the inflammatory response in the context of stress responses, with a protective anti-inflammatory activity of intracellular Hsp70 and a pro-inflammatory activity of extracellular Hsp70, is well accepted. Together with other stress stimuli, inflammation also results in the release of eHsp72. Consequently, high circulating levels of eHsp72 can conceptually be considered as a biomarker to at least reinforce the diagnosis of low-grade inflammation- and stress-related pathologies. Nevertheless, this would be a simplistic approach if the fact that possible dissimilarities in immunophysiological activity and relevance of eHsp72 between healthy individuals and people with inflammation- and stress-related diseases is not considered. Participation of stress-induced Hsp70 in response and physiological regulation of these pathologies may be more than a simply protective anti-inflammatory effect of iHsp72 and potentially dangerous pro-inflammatory effect of eHsp72. Not only can the (eHsp70)/(iHsp70) ratio be determinant in triggering chronic pro-inflammatory states, but also the levels of anti-Hsp70 antibodies in response to increased concentrations of eHsp72, and different pro- or anti-inflammatory effects of this extracellular protein, can be crucial in regarding eHsp72 as a good biomarker of inflammation and stress-related conditions and longevity, as well as for approaching strategies focused on Hsp70 as a target for treatment of these diseases. This is particularly important in aged people when disruptions in the connections of the neuro-immune-endocrine network are inexorable. In this situation eHsp72 could exert protective actions by regulating the interactions between cytokines and neuroendocrine agents, and mediating anti-inflammatory and metabolic adjustments.

Acknowledgements This work was partially supported by the Gobierno de Extremadura-FEDER (GR 15041, GR EE-14-0082-4) and Ministerio de Economía y Competitividad (DEP 2015-66093-R), Spain. Isabel Gálvez is recipient of a "Formación del Profesorado Universitario (FPU)" predoctoral contract (FPU15/02395) from the Ministerio de Educación, Cultura y Deporte, Spain.

References

Amorim F, Moseley PL (2010) Heat shock protein and inflammation. In: Asea A, Pedersen BK (eds) Heat shock proteins and whole body physiology. Netherlands, Springer, pp 57–83

Arai Y, Kubo T, Kobayashi K et al (1997) Adenovirus vector-mediated gene transduction to chondrocytes: in vitro evaluation of therapeutic efficacy of transforming growth factor-beta 1 and heat shock protein 70 gene transduction. J Rheumatol 24:1787–1795

Asea A (2006) Initiation of the immune response by extracellular Hsp72: chaperokine activity of Hsp72. Curr Immunol Rev 2:209–215

Asea A, Kraeft SK, Kurt-Jones EA et al (2000) HSP70 stimulates cytokine production through a CD14-dependant pathway, demonstrating its dual role as a chaperone and cytokine. Nat Med 6:435–442

Bassuk SS, Manson JE (2005) Epidemiological evidence for the role of physical activity in reducing risk of type 2 diabetes and cardiovascular disease. J Appl Physiol 99:1193–1204

Bastard JP, Maachi M, Lagathu C et al (2006) Recent advances in the relationship between obesity, inflammation, and insulin resistance. Eur Cytokine Netw 17:4–12

Bautmans I, Njemini R, Mets T (2010) Biochemical changes in response to intensive resistance exercise training in the elderly. In: Asea A, Pedersen BK (eds) Heat shock proteins and whole body physiology. Springer, Netherlands, pp 365–385

Bazzichi L, Rossi A, Massimetti G et al (2007) Cytokine patterns in fibromyalgia and their correlation with clinical manifestations. Clin Exp Rheumatol 25:225–230

Berenbaum F (2013) Osteoarthritis as an inflammatory disease (osteoarthritis is not osteoarthrosis!). Osteoarthr Cartil 21:16–21

Besedovsky HO, Del Rey A (2000) The cytokine-HPA axis feed-back circuit. Z Rheumatol 59:26–30

Besedovsky HO, Del Rey A (2007) Physiology of psychoneuroimmunology: a personal view. Brain Behav Immun 21:34–44

Blanco FJ, Guitian R, Vazquez-Martul E, de Toro FJ, Galdo F (1998) Osteoarthritis chondrocytes die by apoptosis: a possible pathway for osteoarthritis pathology. Arthritis Rheum 41:284–289
Bote ME, García JJ, Hinchado MD, Ortega E (2012) Inflammatory/stress feedback dysregulation in women with fibromyalgia. Neuroimmunomodulation 19:343–351
Bote ME, García JJ, Hinchado MD, Ortega E (2013) Fibromyalgia: anti-inflammatory and stress responses after acute moderate exercise. PLoS One 8:e74524
Bote ME, García JJ, Hinchado MD, Ortega E (2014) An exploratory study of the effect of regular aquatic exercise on the function of neutrophils from women with fibromyalgia: role of IL-8 and noradrenaline. Brain Behav Immun 39:107–112
Bray GA (1985) Complications of obesity. Ann Int Med 103:1052–1062
Bruusgaard H (2005) Physical activity and modulation of systemic low-level inflammation. J Leukoc Biol 78:819–835
Campisi J, Leem TH, Fleshner M (2003) Stress-induced extracellular Hsp72 is a functionally significant danger signal to the immune system. Cell Stress Chaperones 8:272–286
Chung J, Nguyen AK, Henstridge DC et al (2008) Hsp72 protects against obesity-induced insulin resistance. Proc Natl Acad Sci U S A 105:1739–1744
Cruz TF, Kandel RA, Brown IR (1991) Interleukin 1 induces the expression of heat-shock gene in chondrocytes. Biochem J 277:327–330
Das UN (2002) Is metabolic syndrome X an inflammatory condition? Exp Biol Med 227:989–997
Das UN (2004) Metabolic syndrome X: an inflammatory condition? Curr Hypertens Rep 6:66–73
De la Fuente M, Miquel J (2009) An update of the oxidation-inflammation theory of aging: the involvement of the immune system in oxi-inflamm-aging. Curr Pharm Des 15:3003–3026
De la Fuente M, Cruces J, Hernández O, Ortega E (2011) Strategies to improve the function and redox state of the immune system in aged subjects. Curr Pharm Des 17:3966–3993
De Lange-Brokaar BJE, Ioan-Facsinay A, van Osch GJVM et al (2012) Synovial inflammation, immune cells and their cytokines in osteoarthritis: a review. Osteoarthr Cartil 20:1484–1499
Del Rey A, Besedovsky HO (2000) The cytokine-HPA axis circuit contributes to prevent or moderate autoimmune processes. Z Rheumatol 59:31–35
Del Rey A, Welsh CJ, Schwarz MJ, Besedovsky HO (2012) Foreword for neuroimmunomodulation in health and disease. Ann N Y Acad Sci 1262:VII–VIII
Eckel RH, Grundy SM, Zimmet PZ (2005) The metabolic syndrome. Lancet 365:1415–1428
Elenkov IJ (2008) Neurohormonal-cytokine interactions: implications for inflammation, common human diseases and Well-being. Neurochem Int 52:40–51
Elenkov IJ, Chrousos GP (2002) Stress hormones, proinflammatory and anti-inflammatory cytokines, and autoimmunity. Ann N Y Acad Sci 966:290–303
Elenkov IJ, Iezzoni DG, Daly A, Harris AG, Chrousos GP (2005) Cytokine dysregulation, inflammation and well-being. Neuroimmunomodulation 12:255–269
Etienne S, Gaborit N, Henrionnet C et al (2008) Local induction of heat shock protein 70 (Hsp70) by proteasome inhibition confers chondroprotection during surgically induced osteoarthritis in the rat knee. Biomed Mater Eng 18:253–260
Fain JN, Bahouth SW, Madan AK (2004) TNF release by the nonfat cells of human adipose tissue. Int J Obes Relat Metab Disord 28:616–622
Fleshner M, Campisi J, Johnson JD (2003) Can exercise stress facilitate innate immunity? A functional role for stress-induced extracellular Hsp72. Exerc Immunol Rev 9:6–24
Fleshner M, Campisi J, Amiri L, Diamond DM (2004) Cat exposure induces both intra- and extracellular Hsp72: the role of adrenal hormones. Psychoneuroendocrinology 29:1142–1152
Foell D, Wittkowski H, Roth J (2007) Mechanisms of disease: a 'DAMP' view of inflammatory arthritis. Nat Clin Pract Rheumatol 3:382–390
Franceschi C, Bonafe M, Valensin S et al (2000) Inflamm-aging: an evolutionary perspective on immunosenescence. Ann N Y Acad Sci 908:244–254
Franceschi C, Capri M, Monti D et al (2007) Inflammaging and anti-inflammaging: a systemic perspective on aging and longevity emerged from studies in humans. Mech Ageing Dev 128:92–105

Fujita S, Arai Y, Nakagawa S et al (2012) Combined microwave irradiation and intraarticular glutamine administration-induced HSP70 expression therapy prevents cartilage degradation in a rat osteoarthritis model. J Orthop Res 30:401–407
Galois L, Etienne S, Grossin L et al (2004) Dose-response relationship for exercise on severity of experimental osteoarthritis in rats: a pilot study. Osteoarthr Cartil 12:779–786
Gálvez I, Torres-Piles S, Hinchado MD et al (2017) Immune-neuroendocrine dysregulation in patients with osteoarthritis: a revision and a pilot study. Endocr Metab Immune Disord Drug Targets 17:78–85
García JJ, Sáez MC, De la Fuente M, Ortega E (2003) Regulation of phagocytic process of macrophages by noradrenaline and its end metabolito 4- hydroxy-3-metoxyphenyl-glycol. Role of alpha- and betaadrenoreceptors. Mol Cell Biochem 254:299–304
García JJ, Martin-Cordero L, Hinchado MD, Bote ME, Ortega E (2013) Effects of habitual exercise on the eHsp72-induced release of inflammatory cytokines by macrophages from obese Zucker rats. Int J Sports Med 34:559–564
Ghayour-Mobarhan M, Lamb DJ, Lovell DP, Livingstone C, Wang T, Ferns GA (2005) Plasma antibody titres to heat shock proteins-60, −65 and-70: their relationship to coronary risk factors in dyslipidaemic patients and healthy individuals. Scand J Clin Lab Invest 65:601–614
Ghayour-Mobarhan M, Taylor A, Lamb DJ, Ferns GA (2007) Association between indices of body mass and antibody titres to heat-shock protein-60, −65 and −70 in healthy Caucasians. Int J Obes 31:197–200
Giraldo E, Garcia JJ, Hinchado MD, Ortega E (2009) Exercise intensity-dependent changes in the inflammatory response in sedentary women: role of neuroendocrine parameters in the neutrophil phagocytic process and the pro−/anti-inflammatory cytokine balance. Neuroimmunomodulation 16:237–244
Giraldo E, Martin-Cordero L, Garcia JJ, Gehrmann M, Multhoff G, Ortega E (2010a) Exercise-induced extracellular 72 kDa heat shock protein (Hsp72) stimulates neutrophil phagocytic and fungicidal capacities via TLR-2. Eur J Appl Physiol 108:217–225
Giraldo E, Multhoff G, Ortega E (2010b) Noradrenaline increases the expression and release of Hsp72 by human neutrophils. Brain Behav Immun 24:672–677
Glyn-Jones S, Palmer AJ, Agricola R et al (2015) Osteoarthritis. Lancet 386:376–387
Goldring MB, Otero M (2011) Inflammation in osteoarthritis. Curr Opin Rheumatol 23:471–478
Grossin L, Etienne S, Gaborit N et al (2004) Induction of heat shock protein 70 (Hsp70) by proteasome inhibitor MG 132 protects articular chondrocytes from cellular death in vitro and in vivo. Biorheology 41:521–534
Grossin L, Cournil-Henrionnet C, Pinzano A et al (2006) Gene transfer with HSP 70 in rat chondrocytes confers cytoprotection in vitro and during experimental osteoarthritis. FASEB J 20:65–75
Gruden G, Bruno G, Chaturvedi N et al (2009) ANTI-HSP60 and ANTI-HSP70 antibody levels and micro/macrovascular complications in type 1 diabetes: the EURODIAB study. J Intern Med 266:527–536
Gruden G, Barutta F, Pinach S et al (2013) Circulating anti-Hsp70 levels in nascent metabolic syndrome: the Casale Monferrato study. Cell Stress Chaperones 18:353–357
Gupte AA, Bomhoff GL, Swerdlow RH, Geiger PC (2009) Heat treatment improves glucose tolerance and prevents skeletal muscle insulin resistance in rats fed a high-fat diet. Diabetes 58:567–578
Hansen BC (1999) The metabolic syndrome X. Ann N Y Acad Sci 892:1–24
Harman D (1956) Aging: a theory based on free radical andradiation chemistry. J Gerontol 11:298–300
Hassett AL, Clauw DJ, Williams DA (2015) Premature aging in fibromyalgia. Curr Aging Sci 8:178–185
Henstridge DC, Whitham M, Febbraio MA (2014) Chaperoning to the metabolic party: Theemerging therapeutic role of heat-shock proteins in obesity and type 2 diabetes. Mol Metab 3:781–793

Hinchado MD, Bote ME, Ortega E (2011) Expression and release of Hsp72 by neutrophils from fibromyalgia patients. Neuroimmunomodulation 18:378–378
Hiraoka N, Arai Y, Takahashi KA et al (2013) Mild electrical stimulation with heat stimulation increase heat shock protein 70 in articular chondrocyte. J Orthop Res 31:894–900
Hochberg MC, Altman RD, April KT et al (2012) American College of Rheumatology 2012 recommendations for the use of nonpharmacologic and pharmacologic therapies in osteoarthritis of the hand, hip, and knee. Arthritis Care Res (Hoboken) 64:465–474
Hojo T, Fujioka M, Otsuka G, Inoue S, Kim U, Kubo T (2003) Effect of heat stimulation on viability and proteoglycan metabolism of cultured chondrocytes: preliminary report. J Orthop Sci 8:396–399
Hotamisligil GS, Shargill NS, Spiegelman BM (1993) Adipose expression of tumor necrosis factor-alpha: direct role in obesity-linked insulin resistance. Science 259:87–91
Hotamisligil GS, Arner P, Caro JF, Atkinson RL, Spiegelman BM (1995) Increased adipose tissue expression of tumor necrosis factor- alpha in human obesity and insulin resistance. J Clin Investig 95:2409–2415
Hügle T, Geurts J (2016) What drives osteoarthritis?—synovial versus subchondral bone pathology. Rheumatology (Oxford) 56:1461–1471
Iizasa H, Matsushima K (2001) IL-8. In: Oppenheim JJ, Feldman M (eds) Cytokine reference. Academic Press, London, pp 1061–1067
Johnson JD, Fleshner M (2006) Releasing signals, secretory pathways, and immune function of endogenous extracellular heat shock protein 72. J Leukoc Biol 79:425–434
Kapoor M (2015) Pathogenesis of osteoarthritis. In: Kapoor M, Mahomed NN (eds) Osteoarthritis. Pathogenesis, diagnosis, Available treatments, drug safety, regenerative and precision medicine. Springer, Switzerland, pp 1–28
Kapoor M, Martel-Pelletier J, Lajeunesse D, Pelletier JP, Fahmi H (2011) Role of proinflammatory cytokines in the pathophysiology of osteoarthritis. Nat Rev Rheumatol 7:33–42
Karpe PA, Tikoo K (2014) Heat shock prevents insulin resistance-induced vascular complications by augmenting angiotensin-(1-7) signaling. Diabetes 63:1124–1139
Kaur P, Reis MD, Couchman GR, Forjuoh SN, Greene JF, Asea A (2010) Role of heat shock proteins in obesity and type 2 diabetes. In: Asea A, Pedersen B (eds) Heat shock proteins and whole body physiology. Springer, Netherlands, pp 19–29
Knight RJ, Bodian C, Rodriguez-Laiz G, Guy SR, Fishbein TM (2000) Risk factors for intra-abdominal infection after pancreas transplantation. Am J Surg 179:99–102
Krause M, Heck TG, Bittencourt A et al (2015) The chaperone balance hypothesis: the importance of the extracellular to intracellular HSP70 ratio to inflammation-driven type 2 diabetes, the effect of exercise, and the implications for clinical management. Mediators Inflamm 2015:249205
Kubo T, Towle CA, Mankin HJ, Treadwell BV (1985) Stress-induced proteins in chondrocytes from patients with osteoarthritis. Arthritis Rheum 28:1140–1145
Kubo T, Arai Y, Takahashi K et al (2001) Expression of transduced HSP70 gene protects chondrocytes from stress. J Rheumatol 28:330–335
Kumar V, Cotran RS, Robbins S (2003) Robins basic pathology. Saunders, Philadelphia
Lindquist S, Craig EA (1988) The heat shock proteins. Annu Rev Genet 22:631–677
Literati-Nagy B, Kulcsar E, Literati-Nagy Z et al (2009) Improvement of insulin sensitivity by a novel drug, BGP-15, in insulin-resistant patients: a proof of concept randomized double-blind clinical trial. Horm Metab Res 41:374–380
Liu CT, Brooks GA (2012) Mild heat stress induces mitochondrial biogenesis in C2C12 myotubes. J Appl Physiol 112:354–361
Liu-Bryan R (2013) Synovium and the innate inflammatory network in osteoarthritis progression. Curr Rheumatol Rep 15:323
Liu-Bryan R, Terkeltaub R (2015) Emerging regulators of the inflammatory process in osteoarthritis. Nat Rev Rheumatol 11:35–44

Loeser RF, Goldring SR, Scanzello CR, Goldring MB (2012) Osteoarthritis: a disease of the joint as an organ. Arthritis Rheum 64:1697–1707
Luo X, Zuo X, Zhang B et al (2008a) Release of heat shock protein 70 and the effects of extracellular heat shock protein 70 on the production of IL-10 in fibroblast-like synoviocytes. Cell Stress Chaperones 13:365–373
Luo X, Zuo X, Zhou Y et al (2008b) Extracellular heat shock protein 70 inhibits tumour necrosis factor-alpha induced proinflammatory mediator production in fibroblast-like synoviocytes. Arthritis Res Ther 10:R41
Madreperla SA, Louwerenburg B, Mann RW, Towle CA, Mankin HJ, Treadwell BV (1985) Induction of heat shock protein synthesis in chondrocytes at physiological temperatures. J Orthop Res 3:30–35
Martín-Cordero L, García JJ, Hinchado MD, Ortega E (2011) The interleukin-6 and noradrenaline mediated inflammation-stress feedback mechanism is dysregulated in metabolic syndrome: effect of exercise. Cardiovasc Diabetol 10:42
Martínez de Toda I, De la Fuente M (2015) The role of Hsp70 in oxi-inflamma-aging and its use as a potential biomarker of lifespan. Biogerontology 16:709–721
Martínez de Toda I, Vida C, Ortega E, De la Fuente M (2016) Hsp70 levels, a tissue marker of the rate of aging and longevity in mice. Exp Gerontol 84:21–28
Mathur N, Pedersen BK (2008) Exercise as a mean to control low-grade systemic inflammation. Mediat Inflamm 2008:109502
McAlindon TE, Bannuru RR, Sullivan MC et al (2014) OARSI guidelines for the non-surgical management of knee osteoarthritis. Osteoarthr Cartil 22:363–388
McCarty MF (2006) Induction of heat shock proteins may combat insulin resistance. Med Hypotheses 66:527–534
Medzhitov R (2008) Origin and physiological roles of inflammation. Nature 454:428–435
Molina MN, Ferder L, Manucha W (2016) Emerging role of nitric oxide and heat shock proteins in insulin resistance. Curr Hypertens Rep 18(1):1
Morino S, Kondo T, Sasaki K et al (2008) Mild electrical stimulation with heat shock ameliorates insulin resistance via enhanced insulin signaling. PLoS One 3:4068
Mosser DD, Caron AW, Bourget L et al (2000) The chaperone function of Hsp70 is required for protection against stress-induced apoptosis. Mol Cell Biol 20:7146–7159
Musumeci G, Aiello FC, Szychlinska MA, Di Rosa M, Castrogiovanni P, Mobasheri A (2015) Osteoarthritis in the XXIst century: risk factors and behaviours that influence disease onset and progression. Int J Mol Sci 16:6093–6112
Njemini R, Demanet C, Mets T (2004) Inflammatory status as an important determinant of heat shock protein 70 serum concentrations during aging. Biogerontology 5:31–38
Njemini R, Bautmans I, Onyema OO, Van Puyvelde K, Demanet C, Mets T (2011) Circulating heat shock protein 70 in health, aging, and disease. BMC Immunol 12:24
Northoff H, Weinstock C, Berg A (1994) The cytokine response to strenuous exercise. Int J Sports Med 15:167–171
Odrowaz-Sypniewska G (2007) Markers of pro-inflammatory and pro-thrombotic state in the diagnosis of metabolic syndrome. Adv Med Sci 52:246–250
Orlowsky EW, Kraus VB (2015) The role of innate immunity in osteoarthritis: when our first line of defense Goes on the offensive. J Rheumatol 42:363–371
Ortega E (1994) Influence of exercise on phagocytosis. Int J Sports Med 15:172–178
Ortega E (2003) Neuroendocrine mediators in the modulation of phagocytosis by exercise: physiological implications. Exerc Immunol Rev 9:70–93
Ortega E (2016) The "bioregulatory effect of exercise" on the innate/inflammatory responses. J Physiol Biochem 72:361–369
Ortega E, Collazos ME, Maynar M, Barriga C, De la Fuente M (1993) Stimulation of the phagocytic function of neutrophils in sedentary men after acute moderate exercise. Eur J Appl Physiol Occup Physiol 66:60–64
Ortega E, García JJ, Sáez MC, De la Fuente M (2000) Changes with aging in the modulation of macrophages by norepinephrine. Mech Ageing Dev 118:103–114

Ortega E, Giraldo E, Hinchado MD et al (2006) Role of Hsp72 and norepinephrine in the moderate exercise-induced stimulation of neutrophils' microbicide capacity. Eur J Appl Physiol 98:250–255

Ortega E, Giraldo E, Hinchado MD, Martín L, García JJ, De la Fuente M (2007) Neuroinmmunomodulation during exercise: role of catecholamines as 'stress mediator' and/or 'danger signal' for the innate immune response. Neuroinmmunomodulation 98:250–255

Ortega E, Hinchado MD, Martín-Cordero L, Asea A (2009a) The effect of stress-inducible extracellular Hsp72 on human neutrophil chemotaxis: a role during acute intense exercise. Stress 12:240–249

Ortega E, García JJ, Bote ME et al (2009b) Exercise in fibromyalgia and related inflammatory disorders: known effects and unknown chances. Exerc Immunol Rev 15:42–65

Ortega E, Giraldo E, Hinchado MD, Martín-Cordero L, García JJ (2010) 72 kDa extracellular heat shock protein (eHsp72), norepinephrine (NE), and the innate immune response following moderate exercise. In: Asea A, Pedersen BK (eds) Heat shock proteins and whole body physiology. Springer, Netherlands, pp 329–350

Ortega E, Bote ME, Besedovsky HO, Del Rey A (2011) Altered Hsp72-induced release of inflammatory cytokines and circulating levels of Hsp72, glucose, and corticosterone during aging. Neurommunomodulation 18:395

Ortega E, Bote ME, Besedovsky HO, Del Rey A (2012a) Hsp72, inflammation, and aging: causes, consequences, and perspectives. Ann N Y Acad Sci 1261:64–71

Ortega E, Bote ME, Giraldo E, García JJ (2012b) Aquatic exercise improves the monocyte pro- and anti-inflammatory cytokine production balance in fibromyalgia patients. Scand J Med Sci Sports 22:104–112

Ortega E, Gálvez I, Hinchado MD, Guerrero J, Martín-Cordero L, Torres-Piles S (2017) Anti-inflammatory effect as a mechanism of effectiveness underlying the clinical benefits of pelotherapy in osteoarthritis patients: regulation of the altered inflammatory and stress feedback response. Int J Biometeorol 61:1777–1785

Park KC, Kim DS, Choi HO et al (2000) Overexpression of HSP70 prevents ultraviolet B-induced apoptosis of a human melanoma cell line. Arch Dermatol Res 292:482–487

Petersen AM, Pedersen BK (2005) The anti-inflammatory effect of exercise. J Appl Physiol 98:1154–1162

Ploeger HE, Takken T, de Greef MH, Timmons BW (2009) The effects of acute and chronic exercise on inflammatory markers in children and adults with a chronic inflammatory disease: a systematic review. Exerc Immunol Rev 15:6–41

Pockley AG, De Faire U, Kiessling R, Lemne C, Thulin T, Frostegård J (2002) Circulating heat shock protein and heat shock protein antibody levels in established hypertension. J Hypertens 20:1815–1820

Pockley AG, Georgiades A, Thulin T, de Faire U, Frostegård J (2003) Serum heat shock protein 70 levels predict the development of atherosclerosis in subjects with established hypertension. Hypertension 42:235–238

Rahmati M, Mobasheri A, Mozafari M (2016) Inflammatory mediators in osteoarthritis: a critical review of the state-of-the-art, current prospects, and future challenges. Bone 85:81–90

Ravagnan L, Gurbuxani S, Susin SA et al (2001) Heat-shock protein 70 antagonizes apoptosis-inducing factor. Nat Cell Biol 3:839–843

Rea IM, McNerlan S, Pockley AG (2001) Serum heat shock protein and anti-heat shock protein antibody levels in aging. Exp Gerontol 36:341–352

Sanders VM (2006) Interdisciplinary research: noradrenergic regulation of adaptive immunity. Brain Behav Immun 20:1–8

Scanzello C, Goldring S (2012) The role of synovitis in osteoarthritis pathogenesis. Bone 51:249–257

Scanzello CR, Plaas A, Crow MK (2008) Innate immune system activation in osteoarthritis: is osteoarthritis a chronic wound? Curr Opin Rheumatol 20:565–572

Sellam J, Berenbaum F (2010) The role of synovitis in pathophysiology and clinical symptoms of osteoarthritis. Nat Rev Rheumatol 6:625–635

Sohn DH, Sokolove J, Sharpe O et al (2012) Plasma proteins present in osteoarthritic synovial fluid can stimulate cytokine production via toll-like receptor 4. Arthritis Res Ther 14:R7
Sokolove J, Lepus CM (2013) Role of inflammation in the pathogenesis of osteoarthritis: latest findings and interpretations. Ther Adv Musculoskelet Dis 5:77–94
Straznicky NE, Lambert EA, Nestel PJ et al (2010a) Sympathetic neural adaptation to hypocaloric diet with or without exercise training in obese metabolic syndrome subjects. Diabetes 59:71–79
Straznicky NE, Lambert GW, Lambert EA (2010b) Neuroadrenergic dysfunction in obesity: an overview of the effects of weight loss. Curr Opin Lipidol 21:21–30
Sun L, Chang J, Kirchhoff SR, Knowlton AA (2000) Activation of HSF and selective increase in heat-shock proteins by acute dexamethasone treatment. Am J Physiol Heart Circ Physiol 278:H1091–H1097
Suzuki K, Nakaji S, Yamada M, Totsuka M, Sato K, Sugawara K (2002) Systemic inflammatory response to exhaustive exercise. Cytokine kinetics. Exerc Immunol Rev 8:6–48
Takahashi K, Kubo T, Goomer RS et al (1997a) Analysis of heat shock proteins and cytokines expressed during early stages of osteoarthritis in a mouse model. Osteoarthr Cartil 5:321–329
Takahashi K, Kubo T, Arai Y, Imanishi J, Kawata M, Hirasawa Y (1997b) Localization of heat shock protein in osteoarthritic cartilage. Scand J Rheumatol 26:368–375
Takahashi KA, Tonomura H, Arai Y et al (2009) Hyperthermia for the treatment of articular cartilage with osteoarthritis. Int J Hyperth 25:661–667
Terauchi R, Takahashi KA, Arai Y et al (2003) Hsp70 prevents nitric oxide-induced apoptosis in articular chondrocytes. Arthritis Rheum 48:1562–1568
Terry DF, McCormick M, Andersen S et al (2004) Cardiovascular disease delay in centenarian offspring: role of heat shock proteins. Ann N Y Acad Sci 1019:502–505
Terry DF, Wyszynski DF, Nolan VG et al (2006) Serum heat shock protein 70 level as a biomarker of exceptional longevity. Mech Ageing Dev 127:862–868
Tonomura H, Takahashi KA, Mazda O et al (2006) Glutamine protects articular chondrocytes from heat stress and NO-induced apoptosis with HSP70 expression. Osteoarthr Cartil 14:545–553
Tonomura H, Takahashi KA, Mazda O et al (2008) Effects of heat stimulation via microwave applicator on cartilage matrix gene and HSP70 expression in the rabbit knee joint. J Orthop Res 26:34–41
Uzunoglu E, Yentur S, Kayar AH et al (2017) Effect of mild heat stress on heat shock protein 70 in a balneotherapy model. Eur J Integr Med 9:86–90
Van West D, Maes M (2001) Neuroendocrine and immune aspects of fibromyalgia. BioDrugs 15:521–531
Vasto S, Candore G, Balistreri CR et al (2007) Inflammatory networks in ageing, age-related diseases and longevity. Mech Ageing Dev 128:83–91
Wallace DJ, Linker-Israeli M, Hallegua D, Silverman S, Silver D, Weisman MH (2001) Cytokines play an aetiopathogenetic role in fibromyalgia: a hypothesis and pilot study. Rheumatology (Oxford) 40:743–749
Wellen KE, Hotamisligil GS (2005) Inflammation, stress, and diabetes. J Clin Invest 115:1111–1119
Whitham M, Febbraio MA (2010) HSP and diabetes. In: Asea A, Pedersen BK (eds) Heat shock proteins and whole body physiology. Netherlands, Springer, pp 3–18
Wolfe F, Smythe HA, Yunus MB et al (1990) The American College of Rheumatology 1990 criteria for the classification of fibromyalgia. Report of the multicenter criteria committee. Arthritis Rheum 33:160–172
Wu T, Ma J, Chen S et al (2001) Association of plasma antibodies against the inducible Hsp70 with hypertension and harsh working conditions. Cell Stress Chaperones 6:394–401
Yildirim N, Filiz Ulusoy M, Bodur H (2010) The effect of heat application on pain, stiffness, physical function and quality of life in patients with knee osteoarthritis. J Clin Nurs 19:1113–1120
Zimmet P, George K, Alberti MM, Serrano M (2005) Una nueva definición mundial del síndrome metabólico propuesta por la Federación Internacional de Diabetes: fundamento y resultados. Rev Esp Cardiol 58:1371–1376

Chapter 3
Heat Shock Proteins and Endoplasmic Reticulum Stress

Azhwar Raghunath, Lakshmikanthan Panneerselvam, Kiruthika Sundarraj, and Ekambaram Perumal

Abstract The endoplasmic reticulum (ER) is a dynamic as well as an intricate organelle that carries out diverse essential functions within the cell. The efficient functions of the ER regulate the cellular processes and maintain cell homeostasis. When this dynamic equilibrium gets imbalanced due to perturbations in the ER functions, a condition termed ER stress occurs. ER stress accumulates and aggregates unfolded proteins within the ER lumen. An evolutionarily conserved signal transduction mechanism called unfolded protein response (UPR) recuperates the ER function and proteostasis in the ER. A diverse group of heat shock proteins (HSP) has been implicated in this ER stress response. In recent years, studies have revealed the roles of HSP in sensing the ER stress and activating the three axes of UPR. In addition, the HSP are also responsible for the ER associated degradation of unfolded proteins when ER stress persists for a long time. The HSP in ER stress have been implicated in diverse pathophysiological conditions like cardiovascular diseases, cancer, aging, diabetes and obesity. In the present chapter, we have summarized the advances made in the manifold roles of HSP that govern ER stress signaling in health.

Keywords Endoplasmic reticulum stress · Heat shock proteins · Molecular chaperones · Protein folding · Signal transduction · Unfolded protein response

Abbreviations

ADP	Adenosine diphosphate
AgNPs	Silver nanoparticles
ASK1	Apoptosis signal regulating kinase 1

Azhwar Raghunath, Lakshmikanthan Panneerselvam, Kiruthika Sundarraj are contributed equally to this work.

A. Raghunath · L. Panneerselvam · K. Sundarraj · E. Perumal (✉)
Molecular Toxicology Laboratory, Department of Biotechnology, Bharathiar University, Coimbatore, Tamilnadu, India
e-mail: ekas2009@buc.edu.in

A. A. A. Asea, P. Kaur (eds.), *Heat Shock Proteins and Stress*,
Heat Shock Proteins 15, https://doi.org/10.1007/978-3-319-90725-3_3

ATF4	Activating transcription factor 4
ATF6	Activating transcription factor 6
BAG1	BCL2 associated athano-gene 1
BAK	BCL-2 antagonist/killer
BAX	BCL-2-associated X protein
BBG2H7	BBF2 human homolog on chromosome 7
BiP	Immunoglobin heavy chain binding protein
bZIP	Basic leucine zipper
C/EBP	CCAAT/enhancer binding protein
CHIP	C terminus of HSP70 interacting protein
CHOP	C/EBP homologous protein
CREB3	Cyclic AMP-responsive element-binding protein 3
CREB3L1	cAMP responsive element binding protein 3-like 1
CREB3L3	cAMP responsive element binding protein 3-like 3
CREB4	cAMP responsive element binding 4
CREBH	cAMP responsive element binding protein H
CReP	Constitutive repressor of eIF2α phosphorylation
CryαB	AlphaB-crystallin
EIF2S1	Eukaryotic translation initiation factor 2 subunit alpha
EIF2α	Eukaryotic translation-initiation factor 2α
ER	Endoplasmic reticulum
ERAD	Endoplasmic reticulum associated degradation
ERdj3	A mammalian endoplasmic reticulum (ER) Hsp40/DnaJ family member
ERp29	Endoplasmic reticulum protein 29
ERp99	Endoplasmic reticulum resident protein 99
GADD34	Growth arrest and DNA damage-inducible gene 34
GLS	Golgi localization sequences
GRP78	Glucose-regulated protein 78
GRP94	Glucose-regulated protein 94
GSK3	Glycogen synthase kinase 3
HEDJ	Human ER-associated DnaJ
Hip	Hsp70 interacting protein
HSF1	Heat shock factor 1
HSJ1b	Heat shock protein DNAJ-containing protein 1b
HSP	Heat shock proteins
HSP90b1	Heat shock protein 90, beta (Grp94), member 1
HSPA5	Heat shock protein family A (Hsp70) member 5
HSPBP1	HSP70 binding protein 1
IGF	Insulin like growth factor
INM	Inner nuclear membrane
IRE1	Inositol-requiring transmembrane kinase/endonuclease 1
JNK	JUN N-terminal kinase
kDa	Kilo daltons
KDEL	Lys-Asp-Glu-Leu

LcSSc	Limited cutaneous systemic sclerosis
LRP6	Low-density lipoprotein receptor-related protein 6
LRRC32	Leucine-rich repeat-containing protein 32
LZIP	Leucine zipper protein
MCL	Mantle cell lymphoma
MEEVD	Met-Glu-Glu-Val-Asp
mtHSP60	Mitochondrial HSP60
mtHSP70	Mitochondrial HSP70
NCK-1	NCK adaptor protein 1
NE	Nuclear envelope
NLRP-3	Nucleotide-binding oligomerization domain-like receptor containing pyrin domain 3
NRF2	Nuclear factor erythroid 2-related factor 2
OASIS	Old astrocyte specifically induced substance
OHDA	Hydroxydopamine
ONM	Outer nuclear membrane
OS-9	Osteosarcoma-9
PAB	Pulmonary arterial hypertensioned
PBMC	Peripheral blood mononuclear cells
PDI	Protein disulphide isomerase
PERK	Pancreatic endoplasmic reticulum kinase/double-stranded RNA-activated protein kinase-like endoplasmic reticulum kinase
PP1	Protein phosphatase 1
PPDs	Polypeptide-binding domains
RER	Rough endoplasmic reticulum
RING	Really interesting new gene
ROS	Reactive oxygen species
S1P	Site-1 proteases
S2P	Site-2 proteases
SER	Smooth endoplasmic reticulum
SIRT1	Silent mating type information regulation 2 homolog 1
SREBPs	Sterol regulatory element binding proteins
TDP-43	Transactive response DNA binding protein 43
TLR	Toll-like receptors
TPR	Tetratricopeptide repeat
Tra-1	Tumor rejection. antigen 1
TRAF	Tumour-necrosis factor receptor associated factor 2
UBX	Ubiquitin regulatory X
UBXD8	UBX domain-containing protein 8
uORF	Upstream open reading frames
UPR	Unfolded protein response
XBP1	X-box binding protein 1
XBP1s	Spliced XBP1
XBP1u	Unspliced XBP1

3.1 Introduction

The endoplasmic reticulum (ER) is a dynamic three-dimensional membrane-bound organelle contiguous with the nuclear envelope (NE) and an interconnected network of tubules, vesicles, and cisternae extending from the nucleus towards the periphery of the cells. In 1902, Emilio Veratti, a student of Camillo Golgi, discovered ER as a new subcellular structure distinct that of Golgi apparatus (Veratti 1961). The scientific community disregarded the work of Emilio Veratti and abandoned the existence of such organelle. After five decades, in 1953, with advancements in high-resolution imaging technologies, Keith Porter and George Palade captured a complex tubular network within the cytoplasm and proved Emilio Veratti was correct (Porter 1953; Palade and Porter 1954; Porter and Palade 1957). Keith Porter named this complex tubular network as ER. Since then, the diverse functions of the ER are still being discovered.

3.1.1 ER Structure, Function and Stress

Based on the difference in morphologies, ER has three distinct domains, namely NE, peripheral ER cisternae, and interconnected tubules (D'Angelo and Hetzer 2006; Burke and Ellenberg 2002). The nucleus is encircled by NE with the inner and outer nuclear membranes (INM and ONM) connected only at the nuclear pores. Beneath the inner membrane, a network of lamins holds the nucleus in shape (Hetzer et al. 2005). The INM and ONM envelope the nuclear membrane space. The peripheral ER extends and forms an interlinked network of cisternae and tubules throughout the cytoplasm, where the former occupies the region closer to the NE, the latter resides in the periphery (Terasaki et al. 1994). Based on the presence of bound ribosomes/polysomes on the surface of the ER, it has been classified into two types: rough and smooth ER. Mostly cisternae are classified as rough ER (RER) and tubules as smooth ER (SER) (Shibata et al. 2006). The RER serves as the compartment for the synthesis of one third of the cellular proteins, whereas the SER portion stores lipids and xenobiotics. In addition to having more numbers of ribosomes, cisternae possess large luminal volume to surface area unlike tubules. The primary functions of ER are protein synthesis and transport, protein modification, lipid and steroid synthesis, regulation of Ca^{2+} homeostasis and secretion, carbohydrate metabolism, and drug detoxification (Chen et al. 2013; Schwartz and Blower 2016). Perturbations in any of these ER functions derange the folding and secretion of proteins thereby, promote the accumulation of unfolded proteins in its lumen ultimately resulting in a condition called as "ER Stress" (Zhao and Ackerman 2006; Banhegyi et al. 2007). Energy deprivation, redox or lipid imbalance, Ca^{2+} diminution, accumulation of misfolded, unfolded or abundant proteins, and alteration in ionic conditions in ER trigger a response called unfolded protein response (UPR) or

ER stress response (Schroder and Kaufman 2005a, b). The UPR is an evolutionarily conserved complex response, which reestablishes the ER homeostasis.

3.1.2 UPR

The UPR recuperates the cell from ER stress through three major molecular mechanisms: (i) attenuation of protein synthesis and trafficking into the ER (Brostrom and Brostrom 1998; Harding et al. 2002), (ii) augmentation of transcription of ER lumenal chaperone genes to facilitate the protein folding and processing within ER (Ma and Hendershot 2004), and (iii) elimination of unfolded proteins from the ER through proteasome-dependent ER-associated degradation (ERAD) (Plemper and Wolf 1999; Hampton 2002; Tsai et al. 2002; Jarosch et al. 2003). When the UPR fails to restore the ER homeostasis spatially and temporally, then UPR triggers apoptosis - a programmed cell death, to clear the dysfunctional cells (Ferri and Kroemer 2001; Breckenridge et al. 2003; Rao et al. 2004). The dysregulation of the UPR leads to a range of pathological conditions - atherosclerosis, cardiovascular diseases, diabetes, ischemia, lysosomal storage disease, obesity, neurodegeneration, and viral infection (Aridor and Balch 1999; Rutishauser and Spiess 2002; Paschen 2003). The three major molecular mechanisms of UPR are under the control of three prongs of UPR signal-transduction pathway. These three arms are pancreatic ER kinase (PERK), activating transcription factor 6 (ATF6), and inositol-requiring transmembrane kinase/endonuclease 1 (IRE1) get activated sequentially due to ER stress. In unstressed state, all these three ER-localized transmembrane proteins are bound to an ER-resident chaperone called immunoglobulin heavy chain binding protein (BiP; also called glucose-regulated protein 78 (GRP78) and (HSPA5)).

3.1.3 PERK - Translation Impediment

PERK is a type I transmembrane protein, acts as a starting point in the cascade of translational arrest (Harding et al. 1999). PERK not only resembles IRE1's luminal domain, but also in the activation mechanism (Shi et al. 1998; Liu et al. 2000). Under resting conditions, BiP binds PERK and keeps it in an inactivated state (Harding et al. 2000a). During ER stress, the unfolded proteins recruit BiP and dissociate it from PERK (Bertolotti et al. 2000). As a result, PERK homo-oligomerizes and self-activates by trans-autophosphorylation with the cytoplasmic PERK kinase domain. This activated PERK transiently phosphorylates eukaryotic translation-initiation factor 2α (EIF2α aka EIF2S1) at Ser51 and inactivates EIF2α. The transient phosphorylation of EIF2α attenuates translation of the most mRNA, reduces the traffic of nascent polypeptides into the ER, thus lessens the cargo of newly synthesized proteins ultimately diminishes the burden of the stressed ER (Shi et al. 1998). Though phosphorylated EIF2α halts the translation of the most mRNAs, it preferentially translates a few mRNA components in the ER stress (Harding et al. 2000b; Fernandez et al. 2002a). One such mRNA is ATF4, constitutively expressed, but translated inefficiently on association with monoribosomes and low MW polyribosomes under normal physiological conditions. ATF4 mRNA carries upstream open reading frames (uORFs) in the 5' untranslated regions which hinders translation under resting conditions (Lu et al. 2004; Vattem and Wek 2004; Fernandez et al. 2002b). Upon ER stress, these ATF4 mRNAs recruit heavier ribosomal fractions which bypass uORFs in the 5' leader and favour translation of ATF4 mRNAs. The PERK- EIF2α pathway plays dual roles: one halts translation of the most mRNAs and two prefers translation of genes to manage ER stress. ATF4 is a member of the basic leucine zipper (bZIP) family of transcription factors activates the transcription of UPR genes and recuperates ER homeostasis (Luo et al. 2003). ATF4 target genes include C/EBP homologous protein (CHOP), growth arrest and DNA damage-inducible gene 34 (GADD34), and BIP (Fawcett et al. 1999). CHOP - a bZIP transcription factor and its expression is induced transcriptionally by ATF4 and ATF6 and translationally by EIF2α phosphorylation as CHOP mRNA contains small uORFs (Zinszner et al. 1998; Jousse et al. 2001; Ma et al. 2002). The transcriptional component of ER stress is under the control of CHOP (Ubeda et al. 1996). CHOP forms heterodimers with other families of transcription factors: members of C/EBP and fos-jun, thus regulates the genes that are in the UPR (Ubeda et al. 1999). Once the ER stress is abated, the normalcy is brought back with normal protein synthesis. Protein phosphatase 1 (PP1) - a serine/threonine phosphatase dephosphorylates EIF2α and PPI, which requires GADD34 as cofactor to perform its functions (Novoa et al. 2001; Brush et al. 2003). Thus GADD34 removes the translational halt, restores protein synthesis in the ER and completes a feedback loop (Kojima et al. 2003). In attenuation of protein synthesis, EIF2α activates all the three chief players: ATF4, CHOP, and GADD34. This translation control happens sequentially: BiP senses ER stress and activates PERK, PERK phosphorylates EIF2α, EIF2α triggers the transcription of ATF4, ATF3, and CHOP, and these bZIP transcription factors upregulate GADD34 (Jiang et al. 2004). CReP (homolog of GADD34) and Nck-1 are found to dephosphorylate EIF2α where CReP directly unphosphorylates EIF2α in unstressed conditions and Nck-1 directly dephosphorylates PERK (Jousse et al. 2003; Kebache et al. 2004). PERK pathway activates another bZIP transcription factor nuclear factor erythroid 2-related factor 2 (NRF2) through phosphorylation. The activated NRF2 upregulates an array of antioxidant genes and promotes cell survival during ER stress (Itoh et al. 1997; Cullinan et al. 2003).

3.1.4 ATF6 - Proteolytic Activation

Similar to IRE1, mammalian ATF6 has two homologs, namely ATF6α and ATF6β. ATF6α is a type II transmembrane protein (90 kDa) that has two important functional domains: a bZIP transcription factor in the cytoplasmic domain and a Golgi localization sequences (GLS) in the ER lumen (Haze et al. 1999). Under normal physiological conditions, BiP masks the GLS and the same BiP is freed from the GLS during the ER stress (Shen et al. 2002). Once free from BiP, unmasked GLS translocate ATF6α to the Golgi. In the Golgi, a bZIP transcription factor, ATF6α is proteolytically cleaved by site-1 and site-2 proteases (S1P and S2P) and translocates itself and activates the transcription of UPR target genes (Chen et al. 2002; Ye et al. 2000). These UPR target genes include BiP (GRP78), GRP94, protein disulphide isomerase (PDI), and X-box binding protein 1 (XBP1) (Yoshida et al. 2001). In the UPR transcriptional response, there is a temporal difference in the functions of transcription factors: ATF6 and XBP1 wherein ATF6 performs its action first, followed by XBP1 yet they display complementary roles in UPR (Yoshida et al. 2003). Though ATF6β has bZIP domain like that of ATF6α, ATF6β has little part to play in the UPR (Thuerauf et al. 2004). Like ATF6α, there are other ER transmembrane transcription factors activated by S1P and S2P intransmembrane proteolysis namely, ATF6β, BBF2 human homolog on chromosome 7 (BBG2H7 aka CREB3L2), cAMP responsive element binding protein H (CREBH aka CREB3L3), cAMP responsive element binding 4 (CREB4), leucine zipper protein (LZIP aka CREB3), old astrocyte specifically induced substance (OASIS aka CREB3L1), and sterol regulatory element binding proteins (SREBPs) (Bailey and O'Hare 2007; Asada et al. 2011). Except SREBPs, other ER transmembrane transcription factors functions are not studied well. SREBPs play an important role in lipid synthesis (Brown and Goldstein 1997).

3.1.5 IRE1 - Non-Canonical mRNA Processing

IRE1 is a type I ER transmembrane receptor with a molecular weight of ~100 kDa. Both the serine/threonine kinase and C-terminal endoribonuclease domains of IRE1 reside in the cytoplasm (Wang et al. 1998; Tirasophon et al. 2000). In mammals, there are two homologs of IRE1: IRE1α and IRE1β, the former is ubiquitously expressed, while the latter expression is restricted to gut epithelial cells (Bertolotti et al. 2001). Unfolded proteins activate IRE1 in two ways: firstly unfolded proteins dissociate BiP from IRE1 and secondly on direct interaction with IRE1 (Bertolotti et al. 2000). Either way, the IREI oligomerizes and transautophophorylates itself and other IRE1 in the complex (Shamu and Walter 1996). As a result, both the kinase and endoribonuclease functions get activated. Though the kinase activity has not been studied thoroughly, the activated IRE1 can initiate two major events: apoptosis and autophagy. Autophagy is initiated as an adaptive response to ER stress when IRE1 binds tumour-necrosis factor receptor associated factor 2 (TRAF2) and activates JUN N-terminal kinase (JNK) via apoptosis signal regulating kinase 1 (ASK1) (Urano and Wang 2000; Nishitoh et al. 2002; Ogata et al. 2006). When ER stress persists for a long time beyond unsalvageable levels, JNK incites apoptosis through the activation of caspase 12.

The IRE1 endoribonuclease activity performs non-canonical splicing of *XBP1* mRNA called "XBP1 splicing" resulting in spliced *XBP1* (*XBP1s*). Two post-transcriptional products are possible from *XBP1* mRNA, when it undergoes

XBP1 splicing, a shift in the reading frame occurs resulting in the XBP1s protein with 376 amino acids (Yoshida et al. 2001; Lee et al. 2002; Calfon et al. 2002). In this XBP1 splicing mechanism, *XBP1* mRNA serves as the only substrate for activated IRE1. When there is no splicing, unspliced *XBP1* (*XBP1u*) mRNA produces a truncated and short-lived XBP1u protein with 261 amino acids. XBP1u proteins are instantaneously degraded through both ubiquitin-dependent and ubiquitin-independent pathways (Tirosh et al. 2006). Both the XBP1u and XBP1s have a bZIP DNA-binding/dimerization and the nuclear-localization domains in common, but C-terminal transactivation domain is present only in XBP1s and absent in XBP1u, which is essential for transcriptional activation of cytoprotective genes. IRE1 axis regulates UPR with XBP1u, wherein XBP1u inhibits both translocation and transactivation under specific conditions (Yoshida et al. 2006).

Depending on the magnitude of ER stress, IRE1 axial plays a dual role: when the stress is recoverable, the IRE1 offers protection and when irrecoverable, the IRE1 triggers apoptosis. During recoverable stress conditions, the pro-apoptotic proteins - BCL-2-associated X protein (BAX) and BCL-2 antagonist/killer (BAK) remain inert by directly interacting with the cytosolic domains of IRE1, thereby enhances IRE1's kinase and endoribonuclease activities (Hetz et al. 2006). In contrast, during irrecoverable ER stress these pro-apoptotic proteins trigger apoptosis. The ER stress response collectively depends on IRE1, XBP1 and associated adaptor proteins called "UPRosome" (Hetz and Glimcher 2008).

3.1.6 ERAD

Eukaryotes have two ER quality control systems to take care of the misfolded proteins: one is UPR and the other is ERAD (Travers et al. 2000; Smith et al. 2011 - 31-32). ERAD pathway distinct that of the lysosomal degradative pathway was first reported by Lippincott-Schwartz in the late 80s (Lippincott-Schwartz et al. 1988). ERAD machinery identifies the misfolded proteins retained in the ER and translocates them into the cytosol for degradation in the proteasome. ERAD pathway consists of the following steps: ERAD substrate recognition, substrate targeting, retrotranslocation, ubiquitylation and proteasomal degradation. The substrates of ERAD include damaged, misfolded, or unmodified polypeptides and improperly assembled members of multiprotein complexes. Besides targeting the problematic proteins, ERAD earmarks properly folded proteins to maintain protein concentrations, metabolic enzymes, plasma membrane metal transporters, and transcription factors (Rape et al. 2001; Tyler et al. 2012). The proteins that fail to achieve the native conformation expose their hydrophobic patches to the exterior, which are kept interior in the properly folded proteins and maintain the lowest energy state (Jahn and Radford 2005). Aggregation of misfolded proteins happens through the interaction between these exposed hydrophobic patches. ERAD substrates are recognized through these exposed hydrophobic regions. Most nascent proteins that translocate into the ER while being synthesized on ribosomes are co-translationally glycosylated with an N-linked oligosaccharide (GlcNAC2-Man9-Glc3, where N is acetylglucosamine, Glc is glucose, and Man is mannose). This glycosylation happens mostly in an Asn in a consensus Asn-X-Ser/Thr motif (where X is any amino acid) and determines the fate of the protein whether processed, trafficked or degraded (Tannous et al. 2015). ER mannosidases remove the several mannose residues of the misfolded proteins and direct them for ERAD (Thibault and Ng 2012). Both glycosylation dependent and independent signals direct the misfolded proteins to the ERAD pathway (Christianson et al. 2008; Ninagawa et al. 2015). The ER resident chaperones bind to improperly folded proteins and deliver to Sel1L-Hrd1 ERAD complex for translocation into the cytoplasm (Lilley and Ploegh 2005; Mueller et al. 2006). Hrd1 is a six transmembrane domain and possesses a catalytic RING finger domain in the cytosol (Gardner et al. 2000). The transmembrane regions form a retrotranslocation channel to export the misfolded substrates (Carvalho et al. 2010; Baldridge and Rapoport 2016). The ubiquitination of the misfolded proteins happens once they are exposed to the cytosol and Hrd1 catalyzes this ubiquitination. The Cdc48/p97 ATPase complex recognizes the ubiquitinated substrates (Ye et al. 2001). Ubiquitin regulatory X (UBX) domain-containing protein, UBXD8 recruits the ATPase complex to the ER membrane (Schuberth and Buchberger 2005). Thus ubiquitinated substrates are exported into the cytosol by the ATPase activity (Ye et al. 2004). Once in the cytoplasm, the substrates are turned soluble and cytosolic chaperones – HSP direct the substrates to the proteasome for degradation (Wang et al. 2011).

3.1.7 HSP in ER Stress

HSP plays a decisive role in assisting the protein synthesis from the secretory pathways of ER, protecting the nascent polypeptides from various stress stimuli and apparently maintaining proteostasis regulations (Parsell and Lindquist 1993). In addition to this the subcellular localization of specific molecular chaperones (mtHSP60, mtHSP70) coupled with UPR signaling potentiates adaptive mechanism via ERAD system in eukaryotes (Vembar and Brodsky 2008; Maattanen et al. 2010). HSP90 acts as a transcriptional arm of ER stress and regulates the stability and function of UPR signaling kinases during cellular transduction and stressed states (Marcu et al. 2002). On the other hand, mutation or failure in the ER stress sensors (GRP78 and GRP94)/UPR multiple signaling kinases machineries will trigger an apoptotic stimulus. Numerous studies have reported that differential expression of HSP in various ER stressed disease models (Zheng et al. 2017; Meares et al. 2008). However, the underlying mechanism of HSP in ER stress and the molecular insights associated with UPR signaling kinases, ERAD, autophagy, apoptosis and their regulations remain obscure. This chapter discusses the role of HSP in the ER stress (Table 3.1).

Table 3.1 The heat shock proteins and their co-chaperones in ER stress

HSPs, Co-chaperones, sHSPs, Crystallin Family Proteins, and Stress Related Proteins	Aliases	Location	Size (kDa)	Cofactors	Role in ER Stress	References
HSP40	HSPF1, Hdj1, RSPH16B, Sis1	Cytosol	40	Hsc70, Hip, ERdj4	Protect cells from ER stress induced cytotoxicity. Stimulates the activity of ATPase.	Lüders et al. (1998); Kurisu et al. (2002)
HSP70	DnaK, grp78, hsc70, BiP, Kar2, Ssa, Ssb, Ssc	Cytosol, Nucleus Mitochondria, ER	70	Hsp40, DnaJ, Sis1, Hdj1, NEFs, GrpE, HSP110	Inhibits apoptosis and directly binds with ER stress sensor proteins. Unfolding, disaggregation, stabilization of extended chains, targeting substrates for degradation.	Bertolotti et al. (2000); Saibil (2013)
HSP90	EL52, HEL-S-65p, HSP86, HSP89A, HSP90A, HSP90N, HSPC1, HSPCA, HSPCAL1, HSPCAL4, HSPN, Hsp103, Hsp89, LAP-2, LAP2	Cytosol, Nucleus, ER	90	P23, HOP, AHA1, FKPB52, UNC45	Highly conserved cofactor p23 and HSP90 associated ER stress requires caspase-3 and caspase-7 activity. Binding, stabilization and maturation of steroid receptors and protein kinases, delivery of protease, and myosin assembly.	Rao et al. (2006); Saibil (2013)
HSP100	Hsp104, ClpA, ClpB, CipC, ClPX	Cytosol, Nucleus, Nucleolus	80-110	HSP70 system, ClpP, ClpS	Disaggregation of denatured proteins under stress condition.	Buchberger et al. (2010); Saibil (2013)
Aha1	AHA1, C14orf3, hAha1, p38	Cytosol	38	Cpr6	HSP90 ATPase cycle in proteostasis.	Li et al. (2013); Wolmarans et al. (2016); Koulov et al. (2010)

Grp94	ECGP, GP96, GRP94, HEL-S-125m, HEL35, TRA1	Cytosol, ER	94	HSP90	Secretory cells that induce stress response and up-regulate the protein folding. N-terminal domain dimerization, hydrolysis, nucleotide binding.	Hu et al. (2007)
Hop	CAMEO, HOD, HOP, LAGY, NECC1, OB1, SMAP31, TOTO	Cytosol, Nucleus	64	HSP70 and HSP90	HAP70/HSP90 organizing protein, stress induced phosphoprotein 1, binds EEVD motifs at thr C-termini of HSP70 and HSP90.	Baindur-Hudson et al. (2015)
HSP90α	HSP90AA2, HSPCA, HSPCAL3	Cytosol	90	Hop and Sti1	ATPase activity, extracellular HSP90α interact with matrix metalloproteinase 2 induce cell invasiveness and stimulate the activity of endothelial nitric oxide synthase.	Eustace et al. (2004); Synoradzki et al. (2015)
HSP90β	HSP90B, HSPC2, Hspcb	Cytosol	92	cIAP1	Inhibit the endothelial nitric oxide synthase activity.	Synoradzki and Bieganowski (2015)
p23	TEBP, cPGES	Golgi organelle and throughout the cell	23	Gim3	ER stress inhibited by p23 associated with p53-up regulated modulator of apoptosis. Directly binds to HSP90 and hydrolysis ATP and also bind denatured proteins and inhibit aggregation.	Bose et al. (1996); Grenert et al. (1999); Rao et al. (2006)
p50/Cdc37	CDC37	Nucleus	50	Hop	P50 along with NF-kB dimer forming transcription factor in ER stress. Activates protein kinases and directly binds HSP90, kinase-specific co-chaperone.	Li et al. (2012)
TRAP1	HSP 75, HSP75, HSP90L	Mitochondria	75	c-Src	Regulates the opening of mitochodiral permeability transition pore. Inhibits oxidative stress mediated apoptosis and also ATPase activity.	Montesano Gesualdi et al. (2007); Kang et al. (2007)

(continued)

Table 3.1 (continued)

HSPs, Co-chaperones, sHSPs, Crystallin Family Proteins, and Stress Related Proteins	Aliases	Location	Size (kDa)	Cofactors	Role in ER Stress	References
APG-1	DREG, GPR126, LCCS9, PR126, PS1TP2, VIGR	Cytosol	105	P53	Heat stress in germ cells. Inhibits the aggregation of citrate synthase. ATP binding.	Kumagai et al. (2000); Matsumori et al. (2002)
APG-2	HEL-S-5a, HS24/P52, HSPH2, RY, hsp70, hsp70RY	Peripheral membrane	110	APG9	Formation of cytosolic sequestering vesicle. Degrading peroxisome.	Wang et al. (2001)
Bag-1	HAP, RAP46	Nucleus	29	HSP70, Bcl-2, Siah, Raf-1, RAR, CHIP, Hip	Novel chaperone link with cell signaling, cell death and stress response. Potential modulator of molecular chaperones, bind to the ATPase domain. Protects from heat shock induced cell death.	Takayama et al. (1997); Wang et al. (1996)
Bag-2	dJ417I1.2	Cytosol	24	Hsc70, Hip, MAPKAPK2, CHIP	When Bag-2 is overexpressed, mutant cystic fibrosis conductance regulator reaches the plasma membrane. Main component of CHIP protein. Inhibit the ubiquitin ligase activity	Arndt et al. (2005); Arndt et al. (2005)
Bag-3	BIS, CAIR-1, MFM6	Cytosol	54	PLC-γ, p65	Protects cells from heat shock and heavy metal exposure and antistress activity. Forming an trimeric complex with HAP70 provides a link between HSP70 and EGF signaling pathway. Directly binds with antiapoptotic protein Bcl-2, preventing cell death.	Doong et al. (2000); Liao et al. (2001)

Bag-4	SODD	Cytosol	60	TNF-R1, DF3	Binds to TNF-R1 by an ATP dependent mechanism. Negatively regulate TNF-R1 and DR3 to inhibit self aggregation.	Tschopp et al. (1999)
Bag-5	BAG-5	Cytosol	51	Parkin	Involved in induction of ER stress. During stress condition, Bag5 relocates from cytoplasm to ER. Binds HSP70, inhibit parkin E3 ubiquitin ligase and HSP70 chaperone activity.	Kalia et al. (2004); Bruchmann et al. (2013)
CHIP	HSPABP2, NY-CO-7, SCAR16, SDCCAG7, UBOX1	Cytosol	32	HSP70	Intrinsic E3 ubiquitin ligase activity and promotes ubiquitylation.	Jiang et al. (2001)
CSP	CSPA	-	7.7	-	Highly expressed in cold shock condition and regulate the protein expression in low temperature.	Graumann and Marahiel (1996)
CSP-b	BB2249	-	7.23	-	Expressed in cold shock condition and expressed only in the absence of CspV.	Phadtare et al. (1999)
CSP-g	CSP1, ECK0980, JW0974	-	-	-	Expressed in cold shock condition and expressed only in the absence of CspV.	Phadtare et al. (1999)
DRIP78	HDJ3, DnaJC14	ER	78	-	Interacts with Gγ2 subunits.	Bermak et al. (2001)
ERdj3	DnaJB11, ERj3, HEDJ, ABBP-2	ER, Nucleus, Cytosol	40	Bip	ER stress tolerance. Inhibition of distruption of calcium homeostasis mechanism in the region of ER.	Lau et al. (2001)
ERdj5	JPDI, DnaJC10	ER	91	DnaJ, Bip	ERdj5 transcription induced by ER stress. Protein folding.	Cunnea et al. (2003)

(continued)

Table 3.1 (continued)

HSPs, Co-chaperones, sHSPs, Crystallin Family Proteins, and Stress Related Proteins	Aliases	Location	Size (kDa)	Cofactors	Role in ER Stress	References
Grp75/Mortalin	MtHSP70, HSPA9, HSP70-9, MOT, Mot-2	Intracellular	75	FGF-1	Expressed in heat shock condition. Glucose regulation, antigen processing and cell mortality.	Jackson et al. (1992); Tarantini et al. (1998); Mizukoshi et al. (1999)
Grp78/BiP	HSPA5, HSP70-5	-	78	-	Control the transmembrane ER stress sensors and regulator, anti-apoptotic.	Lee (2005)
Hdj2	DnaJA1, DJ2, DJA1, HSDJ, HSJ2	ER	45	Hsc70	Hdj1 binds with HSP70 to fold nascent polypeptides in the cytosolic ribosomes and protect from thermal stress. Suppresses the aggregation of nucleotide binding domain.	Ohtsuka (1993); Frydman et al. (1994); Meacham et al. (1999)
Hip	HSPBP1	-	25		Induces cell death and caspase 12 activation by ER stress. Proteolytic activation.	Song et al. (2008)
HSC40	DnaJB5	-	42	HSP70	Promotes membrane integration.	Kalbfleisch et al. (2007)
HSC70	HSPA8, HSP70-8, HSP73, HSC71	Cytosol	71	HSP40	Tail-anchored protein biogenesis in ER.	Rabu et al. (2008)
HSP110	HSP105, HSPH1	Nucleolus/ Cytosol	110	Hsc70, HSP40	Thermotolerance.	Benjamin and McMillan (1998)
HSP40/Hdj1	DnaJB1, HSPF1	-	40	-	Induces ATP hydrolysis for HSP70 activity.	Hohfeld (1998); Takayama et al. (1999)

HSP70-4	HSP70L1, HSPA14	--	70	-	Binds with hydrophobic region and activity was controlled by ATP binding, hydrolysis and nucleotide exchange.	Minami et al. (1996)
αA-Crystallin	CRYAA, CRYA1, HSPB4	-	20	-	Highly involved in cataract and skeletal muscle desmic related myopathy.	Litt et al. (1998); Kappe et al. (2003); Quraishe et al. (2008)
αB-Crystallin	CRYAB, CRYA2, HSPB5	-	20	-	Stress inducible protein. Involved in autophagy and apoptosis.	Kappe et al. (2003); Quraishe et al. (2008)
Heat Shock 22 KDa protein 8	H11, E2IG1, HSP22, HSPB8	Cytosol, Nucleus	22	HSPB1	Degradation of unfolded proteins, autophagy regulation.	Sun et al. (2003); Irobi et al. (2004); Hedhli et al. (2008)
Heat Shock 27 KDa protein 1	HSP27, HSP25, HSPB1	Cytosol, Nucleus	27	-	Phosphorylation of HSP27 due to oxidative stress, TNF, heat shock and growth factor. Maintains homeostasis response in stress condition.	Huot et al. (1995)
Heat Shock protein α-crystallin related, B6	HSPB6, HSP20	-	20	-	Involved in apoptosis and autophagy signaling pathway. Stress indicible protein.	Taylor and Benjamin (2005); Stegh et al. (2008); Carra et al. (2008)
Calnexin	CNX, P90	ER	67	Calreticulin, UGGT, ERp57	Accumulation of unfolded protein in ER induces proapoptotic response involved in apoptosis.	Coe et al. (2008); Guérin et al. (2008)
Calreticulin	CRTC, ERp60, grp60	ER	60	Calnexin	Calcium depletion. Glycoprotein in the ER binding with calnexin maintain glycoprotein folding and MHC class I pathway.	Rizvi et al. (2004); Gao et al. (2002)

(continued)

Table 3.1 (continued)

HSPs, Co-chaperones, sHSPs, Crystallin Family Proteins, and Stress Related Proteins	Aliases	Location	Size (kDa)	Cofactors	Role in ER Stress	References
CHIP	STUB1, UBOX1, HSPABP2	Cytosol	32	Hsc70	CHIP binds with stress responsive ubiquitin conjugative enzyme family UBCH5.	Jiang et al. (2001)
HO-1	Heme oxygenase (decyclizing) 1, HMOX1, HSP32	Mitochondria	32	-	Highly expressed in oxidative stress condition, and heat stress condition. Mainly targeting heme degradation.	Bauer et al. (2003); Converso et al. (2006)
HO-2	Heme oxygenase (decyclizing) 2, HMOX2	Nucleus	36	-	Cleaves the heme molecule in the neurons.	Ewing and Maines (1991); Maines et al. (1996); Muñoz-Sánchez et al. (2014)
HSF-1	HSTF1	Nucleus	82	-	Regulates the activity of chaperones, transcription of stress response proteins, protein disaggregation activity	Chen and Currie (2006); Panoeski and Chánez-Cárdenas (2007)
HSF-2	HSTF2	Nucleus	79	-	Involved in the transcriptional regulation of heat shock response.	Denegri et al. (2002); Alastalo et al. (2003)
HSP47	SerpinH1, Colligin, Gp46	ER	47	-	Stress condition act as quality control system for secretion of procollagen.	Nagata (1996)

3.1.7.1 GRP78

GRP78 - an ER resident and the most abundant ER chaperone with a molecular weight of 78 kDa and is discovered in transformed chick embryo fibroblasts grown in glucose-deprived medium (Pouyssegur et al. 1977). It is evolutionarily conserved from unicellular yeast to higher eukaryote humans. It is induced during ER stress and commonly called as BiP as it binds to pre B-cells immunoglobulin heavy chains (Haas and Wabi 1983). GRP78 remains as an integral part of ER stress and serves as a signaling master regulator (Lee 2005; Wang et al. 2009). The excess unfolded proteins during ER stress deplete GRP78 and lead to the dissociation of PERK, ATF6, and IREi, thus regulate the UPR (Schroder 2006). It transduces signals in ER stress by regulating three sensors: PERK, IRE1, and ATF6 (Bertolotti et al. 2000). Under ER stress, UPR activates the expression of GRP78, which inhibits apoptosis and makes the cells survive (Rao et al. 2002). GRP78 has two domains: an N-terminal ATPase and a C-terminal peptide-binding domain. The peptide-binding domain of GRP78 binds hydrophobic residues of unfolded proteins (Lee 2001). GRP78 prevents the aggregation of mutant prion proteins and facilitates their degradation in the proteasome (Jin et al. 2000). GRP78 is determined to reside in the perinuclear ER with its C-terminal retention signal Lys-Asp-Glu-Leu (KDEL) domain (Zhang and Zhang 2010) and a novel isoform (GRP78va) that lacks this signal is retained in the cytosol (Ni et al. 2011). GRP78 transiently binds the nascent polypeptides inside the ER and promotes the formation of multimeric protein complexes (Gething 1999). When the interaction of GRP78 with misfolded or mutant proteins is persistent and this extended association directs the bound protein for ERAD (Sörgjerd et al. 2006; Petrova et al. 2008). Under ischemia, GRP78 translocates from ER to the neuronal cell surface. The binding of the serine protease tissue plasminogen activator to the cell surface GRP78 fails to activate the PERK/EIF2α axis in the UPR and thus protects neuronal cells from cerebral ischemia mediated ER stress (Louessard et al. 2017). This negative feedback of the PERK/EIF2α axis blocks the ATF4 leading to the non-activation of CHOP - an inducer of death signals during ER stress hyperactivation.

3.1.7.2 GRP94

GRP94 (aka gp96, endoplasmin, ERp99, HSP90b1, HSP108 or Tra-1) is a ubiquitously expressed HSP90 family ER luminal chaperone and gets upregulated during ER stress (Argon and Simen 1999). GRP94 has four domains: an N-terminal ATPase domain, a middle domain, an acidic charged linker domain, and a C-terminal dimerization domain. Similar to GRP78, GRP94 becomes an ER resident with its C-terminal KDEL sequence - a retrieval signal for the KDEL receptor (Munro and Pelham 1987). It functions in the proper folding and assembly of membrane and secretory proteins. A long-awaited enigma of the ATPase activity of GRP94 was proved from the GRP94 structure at atomic resolution. The GRP94 clients include heavy and light chains of immunoglobulin family, integrins family, TLRs, LRP6,

LRRC32 (GARP), platelet glycoprotein Ib/IX/V and insulin like proteins (Melnick et al. 1994; Ostrovsky et al. 2009; Morales et al. 2009; Liu et al. 2010; Staron et al. 2010; Staron et al. 2011; Liu et al. 2013; Zhang et al. 2015). The only common feature used for the GRP94 selectivity is the presence of disulfide bonds in proteins. The misfolded proteins are detected and directed for ERAD with the physiological interaction of both GRP94 and OS-9 - an ERAD sensor protein (Christianson et al. 2008). Insulin like growth factor (IGF) signaling plays an important role in ER stress response by increasing the chaperone reserves within the ER. UPR controls IGF production post-translationally, whereas GRP94 is essential for the generation of mature IGFs (Ostrovsky et al. 2009). In the ER functional network, GRP94 remains as the chief player. As more and more proteins are synthesized into the ER, the functional burden of ER increases, this will co-regulate the transcription of GRP94 and other chaperones essential for folding efficacy and restrict the export of misfolded proteins out of ER. GRP94 forms a large complex with GRP78 and an array of ER proteins, particularly itinerant proteins and these complexes are abundant (Melnick et al. 1994; Meunier et al. 2002). Immunoglobulins and thyroglobulins are properly folded in this fashion (Muresan and Arvan 1997). Both GRP94 and GRP78 are co-induced during ER stress and they are the distinctive features of the UPR response.

3.1.7.3 HSP90

HSP90 dimerizes to perform its functions *in vivo* (Wayne and Bolon 2007; Mayer and Le Breton 2015). HSP90 monomer has three domains: an N-terminal ATP binding domain, a central hydrophobic patch binds the clients and is essential for ATP hydrolysis, and a C-terminal helical motif for dimerization (Prodromou et al. 1997; Meyer et al. 2003; Harris et al. 2004). The C-terminal Met-Glu-Glu-Val-Asp (MEEVD) motif is required for its interaction with co-chaperones that carry tetratricopeptide repeat (TPR) domains (Buchner 1999). The interesting feature of HSP90 is that heat shock factor 1 (HSF1) is an HSP90 client. HSF1 is a transcription factor that regulates the expression of heat shock genes during stress. HSF1 is held in an indolent state on forming a complex with HSP70 and HSP90 (Voellmy and Boellmann 2007; Solis et al. 2016). Thus HSP90 is the key component in the stress status. HSP90 interacts with a broad spectrum of client proteins and the selectivity and specificity of these client proteins are determined by a group of co-chaperones (Röhl et al. 2013). In the UPR, HSP90 associates with both transmembrane kinase IRE1α and PERK and stabilizes both, which is essential for the functioning of these kinases (Marcu et al. 2002). Similar to HSP70, the interaction partners decide the fate of the clients either towards folding or proteolytic degradation. Thus HSP90 has the ability to switch the chaperone function between pro-folding and pro-degradative modes (Brodsky 2007).

3.1.7.4 HSP70

HSP70 (MW 70 kDa) has three domains: ATPase domain in the N-terminal region, substrate-binding domain in the centre, and a lid at the C-terminal region (Sousa and Lafer 2006). The N-terminal ATPase domain binds and hydrolyzes ATP. The substrate domain interacts with different protein substrates with hydrophobic area clouded with positive charges (Jordan and McMacken 1995). The lid of the C-domain opens and allows the substrate to hold in place by forming hydrogen and ionic bonds with the substrate (Han and Christen 2003a). Eleven different HSP70 chaperones are found in mammals and exhibit an array of proteostasis functions: regulation of stress-responsive signaling pathways, protein degradation, protein disaggregation, protein folding, and translocation of proteins across membranes (Kampinga and Craig 2010; Nillegoda and Bukau 2015). This vast array of HSP70 functions is chiefly operated by the diverse HSP40 co-chaperones. ATPase activity, specificity of protein binding, disaggregation, and localization functions of HSP70 are influenced by HSP40 co-chaperones. In addition, other players regulate the functions of HSP70 include Bag-1, Hip, and HSPBP1. Hip retains the substrate within HSP70, whereas HSPBP1 and Bag-1 facilitate the ADP dissociation from the HSP70 and release the protein substrate. The damaged and irreversibly denatured proteins are recognized by HSP70 with its cochaperone HSP40 and directs them for degradation in the ubiquitin-proteasome system with another group of co-chaperones - Bag-1, C terminus of HSP70 interacting protein (CHIP), HSJ1 (Cyr et al. 2002; Petrucelli et al. 2004; Lüders et al. 2006). CHIP first forms a HSP70-HSP40-substrate protein-CHIP-E2 complex and inhibits the protein substrate refolding by blocking the ATPase activity of HSP70. CHIP - an E3 ligase tags the substrate protein with ubiquitin and destines it for proteasomal degradation (Glover and Lindquist 1998; Ballinger et al. 1999; Connell et al. 2001; Meacham et al. 2001). Bag-1 has a dual role, inhibits proteasomal degradation of substrate proteins not taggged with ubiquitin and favours the normal folding of susbtrate proteins (Elliott et al. 2007). On the other hand, Bag-1 carries the ubiquitinated substrate protein for degradation in the proteosome (Alberti et al. 2003). HSJ1b takes care of non-hydrolysis of ubiquitin chain from the protein substrate by ubiquitin hydrolases. HSJ1b has affinity towards ubiquitin-binding sites, which make it to bind at the sites of ubiquitin protein substrate and prevent it from ubiquitin hydrolases and targets the ubiquitinated proteins for degradation (Westhoff et al. 2005). The HSP70 primarily favours the protein folding, this function will be shifted based on the stress experienced by the cell, and determined by the ratio between the folding-stimulating co-chaperones and the degradation stimulating factors (Connell et al. 2001).

3.1.7.5 HSP40

HSP40 (MW 40 kDa) - a co-chaperone for HSP70. HSP40 is called as J-proteins due to the presence of a J-domain, which is conserved with *Escherichia coli* DnaJ (Georgopoulos et al. 1980). This J-domain triggers the HSP70 ATPase activity (Wall et al. 1994) and is present in all the members of HSP40. The complex formation between polypeptides and HSP70 is controlled through three mechanisms: specific polypeptide-binding domains (PPDs) of HSP40 bind and deliver distinct polypeptides to HSP70 (Cheetham and Caplan 1998), HSP40 stabilizes the polypeptide-HSP70 interaction by the hydrolysis of HSP70 ATP to ADP (Liberek et al. 1991), and unique HSP40 members at different sites with the cellular compartment determines the binding of specific polypeptides with HSP70-HSP40 pairs (Cyr and Neupert 1996). The substrate specificity and binding affinity differ for HSP40 and HSP70 as they interact with same protein at different regions (Han and Christen 2003 b). In the ER alone, seven HSP40 co-chaperones regulate the protein folding and trafficking (Kampinga and Craig 2010; Otero et al. 2010; Melnyk et al. 2015). ERdj3 - a new member of the HSP40 family glycoprotein resides within the ER. It is a stress-inducible ER DnaJ homologue that functions as a cochaperone of BiP. The histidine, proline, and aspartic acid residues (HPD motif) are essential for the ERdj3 function. ERdj3 first binds nascent unfolded and mutant secretory pathway proteins and then recruits BiP onto them for folding and assembly. The interaction of BiP with ERdj3 after recruitment releases the ERdj3 from the protein substrate:BiP complex (Shen and Hendershot 2005). The human ER-associated DnaJ (HEDJ) - a novel HSP40 chaperone is expressed in the human cell and localized to the ER. ERdj3 assembles into a native tetramer and this tetramer assembly coordinates the secretory proteostasis in ER stress (Chen et al. 2017). Similar to that of ERdj3, HEDJ interacts with BiP in an ATP-dependent manner and stimulates its activity (Yu et al. 2000). ERdj3 not only maintains proteostasis within the cells, but also involved in the regulation of extracellular proteostasis (Genereux et al. 2015). UPR is the major regulator of extracellular proteostasis network through the increased expression and secretion of ERdj3. Thus ERdj3 takes care of the improper proteins in the secretory pathway and extracellular chaperoning potential. The secreted ERdj3 binds misfolded protein clients extracellularly. In addition, ERdj3 has the potential to co-secrete with aggregation proteins, thus controls the extracellular chaperoning activity of misfolded proteotoxic clients that elude the ER quality control. ERdj3 co-chaperone is capable of extracellular proteostasis under ER stress.

3.1.8 HSP and ER Stress in Health

3.1.8.1 Cardiovascular Diseases

The role of HSP on ER stress mediated ischemic heart diseases was extensively studied in the chronic inflammatory mice models. The increased cardiac expression of both mRNA and protein levels of HSP25, HSP40 and HSP70 confers cardio-protection by refolding the misfolded and aggregated proteins against adverse effects of ER stress response in monocyte chemoattractant protein-1 treated mice (Azfer et al. 2006). Chien et al. (2014) documented that treatment with progressive thermal preconditioning regulates cardiac function through induction of myocardial HSP32, HSP70 and preserves the subcellular organelles - mitochondrial and ER function and its integrity in the rat hearts of ischemia/reperfusion injury. Aberrant protein folding affects the ER quality control, which leads to the accumulation of misfolded proteins and causes ER stress in diseased conditions. Impairment of Lys-Asp-Glu-Leu (KDEL) receptor - a retrieval receptor for ER chaperones induces the expression of cytosolic HSP70 and Bip chaperone due to mechanical stress response during nascent protein synthesis. The dissociation of HSP70 and Bip chaperone form ER kinases initiates the UPR in the pressure overload heart models (Hamada et al. 2004). Mutation in small HSP alphaB-Crystallin (CryαB) causes desmin related cardiomyopathies in the heart and skeleton muscles. The overexpression of CryαB R120G mutant impairs the function of Ca^{2+} handling proteins, sarcoplasmic/endoplasmic reticulum calcium ATPase 2, phospholamban, ryanodine receptor 2 and calsequestrin 2, heart rate variability, increased the protein expression of ER degradation enhancing α-mannosidase-like protein, inositol requirement 1 and XBP1 and enhances the susceptibility of cardiac arrhythmia (Jiao et al. 2014). However, the underlying mechanism in the induction of ER stress associated with the mutation of CryαB R120G chaperone remains unclear. A proteomics study demonstrated that CryαB offers protection against cardiac specific overexpression of calcineurin, a calmodulin dependent serine/threonine protein phosphatase induces ER stress mediated apoptosis. This result suggests that CryαB could be used for the potential therapeutic target for the treatment of cardiac ER stress (Bousette et al. 2010). In addition, systemic transgenic overexpression of cardiac HO-1 (HSP32) showed that decreased calcineurin level. Besides under pressure overload condition, elevated calcineurin is associated with cardiac hypertrophy (Chen et al. 2011a). The implication of HSP, ER stress and activation of UPR signaling was associated in the pathophysiology of limited cutaneous systemic sclerosis (lcSSc) and pulmonary arterial hypertensioned (PAH) patients. In this study, upregulation of HSP and its cochaperones genes (HSP70, HSP40, HSP90 and HSP105/110) were observed in the peripheral blood mononuclear cells (PBMC) from patients with LcSSc. Among these genes, HSP40 was found to be positively correlated with BiP. IL-6 was associated with ER stress mediated inflammation in the patients of IcSSc with PAH (Lenna et al. 2013). The excessive ingestion of fluoride induces ER stress and eventually intervened protein synthesis and secretion (Sharma et al. 2008). The differential expression of HSP in the heart of fluoride treated animals was reported (Panneerselvam et al. 2017). HSP27, HSP32, HSP60, and HSP70 were upregulated, whereas HSP40 and HSP90 were downregulated in the myocardium of rats treated with fluoride.

3.1.8.2 Cancer

Extensive literatures suggest that ER stress plays double-edged sword mechanism in the enhancing (Allavena et al. 2008) and controlling the cell proliferation (Nakagawa et al. 2000). In recent years, emerging evidence shows that the implication of HSP and its molecular inhibitors were closely linked to its therapeutic strategies in the mechanism of cancer therapy. In general, increased accumulation of protein folding and their aggravations were commonly seen in the cancer cells due to their uncontrollable proliferation. Moreover, most of the currently available anticancer drugs cause ER stress through activating UPR mediated apoptosis and autophagy (Rutkowski and Kaufman 2007). It is well acknowledged that knockdown of HSP27 using antisense oligonucleotide targeting or downregulating HSP27 with small interfering (si) RNA increases the sensitivity to paclitaxel in UMUC-3 cells and enhances the resistance to *in vivo* tumor growth in the bladder cancer model (Kamada et al. 2007). Recent evidence suggest that combinational co-targeting of HSP27 and autophagy induces ER stress in prostate cancer. Inhibition of HSP27 suppresses the ERAD machinery in proteostasis and activates the UPR kinases molecules resulting in induction of autopaghy flux in LNCaP and C4-2 cells. Combined inhibition of HSP27 and autophagy induces apoptotic cell death through caspase activation (Kumano et al. 2012). Further, it has been reported that ER stress induces the phosphorylation of HSP27 along with intracellular inclusions called aggresomes in the U251MG cells (Ito et al. 2005). These aggresomes were later found to be pericentriolar membrane-free, cytoplasmic inclusions, which possess cytoprotection against the improper toxic misfolded protein aggregation in the cells (Kopito 2000). The ER protein 29 (ERp29) plays a decisive role in the trafficking protein folding and confers tumor suppressor activity through upregulation of HSP27 and downregulating the expression of EIF2α in doxorubicin induced breast cancer cells. In addition, knockdown of HSP27 decreases the cell viability and induces apoptotic signaling in the ERp29 overexpressing MDA-MB-231 cells and parental MCF-7 cells. These results confirm that HSP27 plays a critical role in the regulation of ER stress in cancer cells (Zhang and Putti 2010).

In addition, it is well known that pharmacological inhibition of CK2, a serine/theronine protein kinase abundantly expressed in the cancer cells induces apoptotic cell death in cancer therapy (Wang et al. 2005). Hessenauer et al. (2011) reported that inhibition of CK2 induces apoptosis via ER stress response in prostate cancer cells (LNCaP and PC-3). In this study, the protein expression of HSP70 and HSP90 were found to remain unchanged after CK2 inhibition, whereas HSP27 expression was found to decrease in both LNCaP and PC-3 cells. Further, it has been demonstrated that HSP27 is mainly responsible for the regulation of ER stress mediated autophagy activation in hepatocellular carcinoma (Chen et al. 2011b). Inhibition of HSP90 deregulates BiP/Grp78 accumulation and enhances apoptosis thereby diminishes the growth of bortezomib-resistant tumors in a mouse model of MCL xenotransplantation (Roue et al. 2011). Lamoureux et al. (2014) documented that inhibitors of HSP90 regulate the HSP27 expression through increased ER stress mediated apoptosis and synergistic cases the castration-resistant prostate cancer model. Combined inhibition of HSP90 and HSP32 induces ROS dependent ER stress and activates apoptotic-signaling pathway in human melanoma (A375) cell. The inhibition of these chaperones also reduces the phosphorylation of AKT signaling which is an essential marker of cell survival mechanism and drastically elevates the ER stress functional markers of ER oxidoreductin and protein disulfide isomerase expression in A375 cells (Barbagallo et al. 2015).

3.1.8.3 Neurodegenerative Diseases and Aging

Impairment in the ubiquitin-proteasome system in the eukaryotic cells has been associated with the mechanism of neuronal cell death in various neurodegenerative disease models. Ryu et al. (2002) identified that set of ER stress and UPR signaling genes, including molecular chaperones HSP27, HSP40, HSP70, HSP90β, and HSP110 were upregulated in the pathogenesis of 6-OHDA-induced neuronal cell death in Parkinson's disease model. A time course microarray study confirmed that pharmacological inhibition of potassium degradation pathway induces ER stress related genes in the primary cultures of mouse neocortical neurons. Incidentally, the upregulation of HSP genes (HSP27, HSP40, HSP47, and HSP70) in these conditions elicits the neuroprotective response against the proteasomal inhibitor lactacystin-induced cultured cortical neurons (Choy et al. 2011). Further, the increased neuronal expression of HSP27 and HSP40 were directed correctly with adaptive neuronal defense, survival mechanism against methamphetamine induced dopamine D1 receptor dependent ER stress mediated apoptosis in rat striatum (Jayanthi et al. 2009). Alternatively, the overexpression of ATF3 potentiates HSP27 and AKT activation and inhibits c-Jun N-terminal kinase-induced neuronal death in PC12 cells and superior nerve ganglion neurons (Nakagomi et al. 2003). It has been reported that binding of HSP72 regulates the XBP1 signaling via increased RNase activity of IRE1a/XBP1 binding and inhibits ER stress-induced apoptosis in PC12 cells. Further, the physical interaction between cytosolic chaperone HSP72 and XBP1 confer prosurvival mechanism in the ER stress (Gupta et al. 2010). A recent study revealed that sustained expression of HSP25 decreases the astroglial apoptosis and exacerbates the clasmatodendrosis (autophagic astroglial death) through prolonged ER stress in the rat hippocampus induced by status epilepticus (Kim et al. 2017). Under peroxidase stress condition, the expression of HSP70 was increased drastically with decreased ER stress chaperones (GRP78, GRP94 and GADD153) in primary neuronal cell cultures. The mechanism underlying the deregulation of ER stress chaperone was due to increased oxidative insults from peroxidase stress and enhances the sensitivity of the neurons to nonlethal state (Paschen et al. 2001). Impairment in the ER Ca^{2+} homeostasis causes neuronal injury called transient cerebral ischemia. Increased mRNA expression of HSP32 was observed in the cortical and hippocampal neurons during a disturbance in the ER Ca^{2+}-ATPase pool. Activation of HO1 was believed to enhance the resistance against cerebral ischemic and posses adaptive survival mechanism against ER stress (Gissel et al. 1997). HSP105 is one of the members of larger HSP family members, found to be associated with HSP70, HSP40 and assisted in maintaining stable protein folding and prevents aggregation during ER stress. Meares et al. (2008) revealed that knockdown of HSP105 did not influence the UPR signaling in ER stress, rather it maintains its chaperonic activity via interacting with GRP78 and glycogen synthase kinase 3 (GSK3) during ER stress and blocks caspase 3 activation and proceeds alternate death pathway in SH-SY5Y human neuroblastoma cells. Previous studies also revealed that inhibition of HSP90 causes ER stress mediated apoptosis in rat histiocytoma (Taiyab et al. 2009) and SH-SY5Y cell (Xie et al. 2012). Vaccaro et al. (2013) reported that pharmacological inhibition of ER stress protects against TDP-43 neuronal toxicity in *Caenorhabditis elegans* and *Danio rerio* mutants. In this study, transgenic *C. elegans* reporter strains express a chaperone like HSP-4 under ER stress.

3.1.8.4 Obesity and Diabetes

ER stress has been considered to play a vital role in the pathology of obesity, insulin resistances and diabetic disease models. It has been reported that the dysregulation of HSP70 and HSP40 homologs were found to be associated with the ER stress related factor in the diabetic retinopathy (Yan et al. 2012). Over expression of HSPA5 (heat shock 70 kDa protein 5) found to attenuate ER stress by eliminating the deleterious misfolded proteins and regulates the UPR signaling kinases in the lipid-induced apoptosis in MIN6 cells (Laybutt et al. 2007). Further, the upregulation of mitochondrial HSP60, HSP20 and HSP27 were augmented in the mitochondrial stress, which transports the nascent proteins from cytosol to mitochondria during ER stress in the adipose tissue of obese volunteers (Boden et al. 2008). A recent study showed that exenatide (exendin-4), a glucagon-like peptide-1 receptor agonist activates the HSP1/SIRT1-mediated deacetylation and enhances the expression of HSP40 and HSP70 in the treatment of hepatic ER stress in both *in vivo* and *in vitro* models (Zheng et al. 2017). Hagiwara et al. (2009) reported that hyperthermia induces the expression of HSP72 and attenuates the lipopolysaccharides induced systemic inflammation mediated ER stress in the pancreatic β cells. In this study, the induction HSP72 believed to prevent the formation of misfolded proteins during the hyperthermic condition and aids cytoprotection to pancreatic β cells.

3.1.9 HSP and ER Stress in Xenobiotics Toxicity

In recent years, there have been enormous reports that suggest occupational and accidental exposure of heavy metals and nanoparticles contaminants in the environment either natural or anthropogenic causes adverse effects in human health. Lin et al. (2007) demonstrated that ER stress mediated apoptosis plays a crucial role in the arsenite-induced neurodegeneration in rats. The induction of HSP 70 and HO1 was correlated with oxidative stress response mechanism in the substantia nigra of rat brain (Lin et al. 2007). It has been reported that chronic exposure of lithium increases the expression of the phosphorylated form of HSP27 and GRP94 in SH-SY5Y Cells. In this study, GRP94, phosphorylation of HSP27 and GSK-3β/α (ser21/9) were believed to regulate the post translational modification of molecular chaperones and confers neurogenesis in SH-SY5Y cells exposed to lithium (Allagui et al. 2009). Increased calcineurin, calcium and activation of caspase 3, 4, 8, 9 regulates ER stress mediated apoptosis in human neutrophils (Binet et al. 2010). In a recent study, Shirriff and Heikkila (2017) revealed that cadmium chloride increased the expression of BiP, HSP70 and HO1 in a dose and time dependent manner in *Xenopus laevis* A6 kidney epithelial cells. Induction of BiP, HSP70 and HO1 were possibly attenuated due to the aggregation of unfolded/misfolded protein though UPR, HSR and NRF2 activation. In contrary, titanium dioxide nanoparticles exposure did not alter the expression of HSP70 and GRP78 rather it influence the cytotoxic effects in the human bronchial epithelial A549 cells (Aueviriyavit et al. 2012).

It has been denoted that silver nanoparticles inhibit ER stress sensor - ATF6 and cause pyroptosis mediated cell death through induction of NLRP-3 inflammasome in human THP-1 monocytes. Interestingly, lower concentration of silver nanoparticles induces the HSP70, and at higher concentration it inhibits the synthesis of HSP90. Thus, the differential expression of these HSP70 and HSP90 were implicated in the ER stress response events in human THP-1 monocytes (Simard et al. 2015). In another study, the co-localization of high HSP70 expression was observed in the ER of human bronchial epithelial cells (16HBE) treated with AgNPs (Huo et al. 2015). High throughput screening analysis portrayed that silicon dioxide nanoparticles alter the levels of E3 ubiquitin-protein ligase UBR5, and HSP90α isoform 1 in the assessment of ubiquitin-proteasome pathway and induces ER stress in A549 cells (Okoturo-Evans et al. 2013). Molecular chaperones HSP4, HSP70, and HSP16.1 regulates the activation of canonical UPR signaling mediated ER stress in the toxicogenomic responses of *C.elegans* exposed to gold nanoparticles (Tsyusko et al. 2012). The exposure of urban particulate matter induces UPR associated post-translational modification of HSP27, S6 ribosomal protein, and protein kinases and cleaves ATF-6 in cultured human bronchial epithelial (BEAS-2B) cells for 24 h. The elevated expression of HSP27, HSP70 and HSP90 were believed to enhance the cellular resistance and prevents the protein misfolding against ER stress (Watterson et al. 2009). The modulation in the levels of HSP60, HSP70, and HSP90 was observed in the seminal vesicle and prostate glands of mice exposed to iron oxide nanoparticles (Sundarraj et al. 2017).

3.2 Conclusions

In the last few decades, our understanding of the pathophysiological roles of HSP in ER stress has vastly increased. Nevertheless, the roles of HSP still remain an enigma in the ER stress. Research studies addressing these questions will lead to novel treatments and preventive strategies to promote the human health targeting these

HSP. HSP associated with ER stress have been implicated in several diseases. To able to intervene these diseases, a thorough understanding of the signal transduction mechanisms is warranted on the roles of HSP in ER stress. Currently, it appears that diverse HSP involve in orchestrating the normal functions of the ER particularly proteostasis. The investigations so far revealed that each of these HSP has manifold functions in the ER. Hence, relentless research in these HSP is mandatory in order to unravel the complexities in the signal transduction mechanisms within the ER.

Acknowledgements Azhwar Raghunath is the recipient of a University Grants Commission - Basic Scientific Research Senior Research Fellowship (UGC-BSR-SRF - No.F.7-25/2007) funded by UGC-BSR, New Delhi, India. We thank the Department of Science and Technology, Science and Engineering Board, New Delhi, for financial assistance (No.: SB/EMEQ-246/2014 and EMR/2014/000600). This work was also supported by the University Grants Commission – Special Assistance Programme (UGC-SAP-II:F-3-20/2013) and Department of Science and Technology, Fund for Improvement of Science and Technology infrastructure in universities and higher educational institutions (DST-FIST:SR/FST/LSI-618/2014), New Delhi, India.

References

Alastalo TP, Hellesuo M, Sandqvist A, Hietakangas V, Kallio M, Sistonen L (2003) Formation of nuclear stress granules involves HSF2 and coincides with the nucleolar localization of Hsp70. J Cell Sci 116:3557–3570

Alberti S, Esser C, Höhfeld J (2003) BAG-1–a nucleotide exchange factor of Hsc70 with multiple cellular functions. Cell Stress Chaperones 8:225–231

Allagui MS, Nciri R, Rouhaud MF, Murat JC, El Feki A, Croute F, Vincent C (2009) Long-term exposure to low lithium concentrations stimulates proliferation, modifies stress protein expression pattern and enhances resistance to oxidative stress in SH-SY5Y cells. Neurochem Res 34:453–462

Allavena P, Sica A, Garlanda C, Mantovani A (2008) The Yin-Yang of tumor-associated macrophages in neoplastic progression and immune surveillance. Immunol Rev 222:155–161

Argon Y, Simen BB (1999) GRP94, an ER chaperone with protein and peptide binding properties. Semin Cell Dev Bio 10:495–505

Aridor M, Balch WE (1999) Integration of endoplasmic reticulum signaling in health and disease. Nat Med 5:745–751

Arndt V, Daniel C, Nastainczyk W, Alberti S, Höhfeld J (2005) BAG-2 Acts as an Inhibitor of the Chaperone-associated Ubiquitin Ligase CHIP. Mol Boil cell 16:5891–5900

Asada R, Kanemoto S, Kondo S, Saito A, Imaizumi K (2011) The signalling from endoplasmic reticulum-resident bZIP transcription factors involved in diverse cellular physiology. J Biochem 149:507–518

Aueviriyavit S, Phummiratch D, Kulthong K, Maniratanachote R (2012) Titanium dioxide nanoparticles-mediated in vitro cytotoxicity does not induce Hsp70 and Grp78 expression in human bronchial epithelial A549 cells. Biol Trace Elem Res 149:123–132

Azfer A, Niu J, Rogers LM, Adamski FM, Kolattukudy PE (2006) Activation of endoplasmic reticulum stress response during the development of ischemic heart disease. Am J Physiol Heart Circ Physiol 291:H1411–H1420

Bailey D, O'Hare P (2007) Transmembrane bZIP transcription factors in ER stress signaling and the unfolded protein response. Antioxid Redox Signal 9:2305–2321

Baindur-Hudson S, Edkins AL, Blatch GL (2015) Hsp70/Hsp90 organising protein (hop): beyond interactions with chaperones and prion proteins. Suncell Biochem 78:69–90

Baldridge RD, Rapoport TA (2016) Autoubiquitination of the Hrd1 ligase triggers protein retro translocation in ERAD. Cell 166:394–407

Ballinger CA, Connell P, Wu Y, Hu Z, Thompson LJ, Yin LY, Patterson C (1999) Identification of CHIP, a novel tetratricopeptide repeat-containing protein that interacts with heat shock proteins and negatively regulates chaperone functions. Mol Cell Biol 19:4535–4545

Banhegyi G, Baumeister P, Benedetti A, Dong D, Fu Y, Lee AS, Li J, Mao C, Margittai E, Ni M, Paschen W, Piccirella S, Senesi S, Sitia R, Wang M, Yang W (2007) Endoplasmic reticulum stress. Ann N Y Acad Sci 1113:58–71

Barbagallo I, Parenti R, Zappalà A, Vanella L, Tibullo D, Pepe F, Onni T, Li Volti G (2015) Combined inhibition of Hsp90 and heme oxygenase-1 induces apoptosis and endoplasmic reticulum stress in melanoma. Acta Histochem 117:705–711

Bauer I, Rensing H, Florax A, Ulrich C, Pistorius G, Red H, Bauer M (2003) Expression pattern and regulation of heme oxygenase-1/heat shock protein 32 in human liver cells. Shock 20:116–122

Benjamin IJ, McMillan DR (1998) Stress (Heat Shock) proteins molecular chaperones in cardiovascular biology and disease. Circ Res 83:117–132

Bermak JC, Li M, Bullock C, Zhou QY (2001) Regulation of transport of the dopamine D1 receptor by a new membrane-associated ER protein. Nat Cell Biol 3:492–498

Bertolotti A, Zhang Y, Hendershot LM, Harding HP, Ron D (2000) Dynamic interaction of BiP and ER stress transducers in the unfolded-protein response. Nature Cell Biol 2:326–332

Bertolotti A, Wang X, Novoa I, Schlessinger K, Cho JH, West AB, Ron D (2001) Increased sensitivity to dextran sodium sulfate colitis in IRE1 β-deficient mice. J Clin Invest 107:585–593

Binet F, Chiasson S, Girard D (2010) Arsenic trioxide induces endoplasmic reticulum stress-related events in neutrophils. Int Immunopharmacol 10:508–512

Boden G, Duan X, Homko C, Molina EJ, Song W, Perez O, Cheung P (2008) Merali S.Increase in endoplasmic reticulum stress-related proteins and genes in adipose tissue of obese, insulin-resistant individuals. Diabetes 57:2438–2444

Bose S, Weikl T, Bugl H, Buchner J (1996) Chaperone function of Hsp90-associated proteins. Science 274:1715–1717

Bousette N, Chugh S, Fong V, Isserlin R, Kim KH, Volchuk A, Backx PH, Liu P, Kislinger T, Maclennan DH, Emili A, Gramolini AO (2010) Constitutively active calcineurin induces cardiac endoplasmic reticulum stress and protects against apoptosis that is mediated by a-crystallin-B. Proc Natl Acad Sci U S A 107:18481–18486

Breckenridge DG, Germain M, Mathai JP, Nguyen M, Shore GC (2003) Regulation of apoptosis by endoplasmic reticulum pathways. Oncogene 22:8608–8618

Brostrom CO, Brostrom MA (1998) Regulation of translational initiation during cellular responses to stress. Prog Nucleic Acid Res Mol Biol 58:79–125

Brown MS, Goldstein JL (1997) The SREBP pathway: regulation of cholesterol metabolism by proteolysis of a membrane-bound transcription factor. Cell 89:331–340

Bruchmann A, Roller C, Walther TV, Schäfer G, Lehmusvaara S, Visakorpi T, Klocker H, Cato ACB, Maddalo D (2013) Bcl-2 associated athanogene 5 (Bag5) is overexpressed in prostate cancer and inhibits ER-stress induced apoptosis. BMC Cancer 13(1):96

Brush MH, Weiser DC, Shenolikar S (2003) Growth arrest and DNA damage-inducible protein GADD34 targets protein phosphatase 1 alpha to the endoplasmic reticulum and promotes dephosphorylation of the alpha subunit of eukaryotic translation initiation factor 2. Mol Cell Biol 23:1292–1303

Brodsky JL (2007) The protective and destructive roles played by molecular chaperones during ERAD (endoplasmic-reticulum-associated degradation). Biochem J 404(3):353–363

Buchberger A, Bukau B, Sommer T (2010) Protein quality control in the cytosol and the endoplasmic reticuum: Brothers in Arms. Molecular Cell. 40:238–252

Burke B, Ellenberg J (2002) Remodelling the walls of the nucleus. Nat Rev Mol Cell Biol 3:487–497

Buchner J (1999) Hsp90 & Co. – a holding for folding. Trends Biochem Sci 24:136–141

Calfon M, Zeng H, Urano F, Till JH, Hubbard SR, Harding HP, Clark SG, Ron D (2002) IRE1 couples endoplasmic reticulum load to secretory capacity by processing the XBP-1 mRNA. Nature 415:92–96
Carra S, Seguin SJ, Lambert H, Landry J (2008) HspB8 chaperone activity toward poly(Q)-containing proteins depends on its association with Bag3, a stimulator of macroautophagy. J Biol Chem 283:1437–1444
Carvalho P, Stanley AM, Rapoport TA (2010) Retrotranslocation of a misfolded luminal ER protein by the ubiquitin-ligase Hrd1p. Cell 143:579–591
Cheetham ME, Caplan AJ (1998) Structure function and evolution of DnaJ: conservation and adaptation of chaperone function. Cell Stress Chaperones 3:28–36
Chen Y, Currie RW (2006) Small interfering RNA knocks down heat shock factor-1 (HSF-1) and exacerbates pro-inflammatory activation of NFkappaB and AP-1 in vascular smooth muscle cells. Cardiovasc Res 69:66–75
Chen X, Shen J, Prywes R (2002) The luminal domain of ATF6 senses endoplasmic reticulum (ER) stress and causes translocation of ATF6 from the ER to the Golgi. J Biol Chem 277:13045–13052
Chen C, Huo R, Tong Y, Sheng Y, Liu HB, Gao X, Nakajima O, Yang BF, Dong DL (2011a) Systemic heme oxygenase-1 transgenic overexpression aggravates pressure overload-induced cardiac hypertrophy in mice. Cell Physiol Biochem 28:25–32
Chen R, Dai RY, Duan CY, Liu YP, Chen SK, Yan DM, Chen CN, Wei M, Li H (2011b) Unfolded protein response suppresses cisplatin-induced apoptosis via autophagy regulation in human hepatocellular carcinoma cells. Folia Biol (Praha) 57:87–95
Chen S, Novick P, Ferro-Novick S (2013) ER structure and function. Curr Opin Cell Biol 25:428–433
Chen KC, Qu S, Chowdhury S, Noxon IC, Schonhoft JD, Plate L, Powers ET, Kelly JW, Lander GC, Wiseman RL (2017) The endoplasmic reticulum HSP40 co-chaperon ERdj3/DNAJB11 assembles and functions as a tetramer. EMBO J 36:2296–2309
Chien CY, Chien CT, Wang SS (2014) Progressive thermo pre conditioning attenuates rat cardiac ischemia/reperfusion injury by mitochondria-mediated antioxidant and anti-apoptotic mechanisms. J Thorac Cardiovas Surg 148:705–713
Choy MS, Chen MJ, Manikandan J, Peng ZF, Jenner AM, Melendez AJ, Cheung NS (2011) Up-regulation of endoplasmic reticulum stress-related genes during the early phase of treatment of cultured cortical neurons by the proteasomal inhibitor lactacystin. J Cell Physiol 226:494–510
Christianson JC, Shaler TA, Tyler RE, Kopito RR (2008) OS-9 and GRP94 deliver mutant alpha1-antitrypsin to the Hrd1-SEL1L ubiquitin ligase complex for ERAD. Nat Cell Biol 10:272–282
Coe H, Bedard K, Groenendyk J, Jung J, Michalak M (2008) Endoplasmic reticulum stress in the absence of calnexin. Cell Stress and Chaperones 13:497–507
Connell P, Ballinger CA, Jiang J, Thompson LJ, Hohfeld T, Patterson C (2001) The co chaperone CHIP regulates protein triage decisions mediated by heat-shock proteins. Nat Cell Biol 3:93–96
Converso DP, Taillé C, Carreras MC, Jaitovich A, Poderoso JJ, Boczkowski J (2006) HO-1 is located in liver mitochondria and modulates mitochondrial heme content and metabolism. FASEB J 20(8):1236–1238
Cullinan SB, Zhang D, Hannink M, Arvisais E, Kaufman RJ, Diehl JA (2003) Nrf2 is a direct PERK substrate and effector of PERK-dependent cell survival. Mol Cell Biol 23:7198–7209
Cunnea PM, Miranda-Vizuete A, Bertoli G, Simmen T, Damdimopoulos AE, Hermann S, Leinonen S, Huikko MP, Gustafsson JA, Sitia R, Spyrou G (2003) ERdj5, an Endoplasmic Reticulum (ER)-resident Protein Containing DnaJ and Thioredoxin Domains, Is Expressed in Secretory Cells or following ER Stress. J Biol Chem 278:1059–1066
Cyr DM, Neupert W (1996) Roles for hsp70 in protein translocation across membranes of organelles. EXS 77:25–40

Cyr DM, Höhfeld J, Patterson C (2002) Protein quality control: U-box-containing E3 ubiquitin ligases join the fold. Trends Biochem Sci 27:368–375
D'Angelo MA, Hetzer MW (2006) The role of the nuclear envelope in cellular organization. Cell Mol Life Sci 63:316–332
Denegri M, Moralli D, Rocchi M, Biggiogera M, Raimondi E, Cobianchi F, De Carli L, Riva S, Biamonti G (2002) Human chromosomes 9, 12, and 15 contain the nucleation sites of stress-induced nuclear bodies. Mol Biol Cell 13:2069–2079
Doong H, Price J, Kim YS, Gasbarre C, Probst J, Liotta LA, Blanchette J, Rizzo K, Kohn E (2000) CAIR-1/ BAG-3 forms an EGF-regulated ternary complex with phospholipase C-gamma and Hsp70/Hsc70. Oncogene 19:4385–4395
Elliott E, Tsvetkov P, Ginzburg I (2007) BAG-1 associates with Hsc70.Tau complex and regulates the proteasomal degradation of Tau protein. J Biol Chem 51:37276–37284
Eustace BK, Sakurai T, Stewart JK, Yimlamai D, Unger C, Zehetmeier C, Lain B, Torella C, Henning SW, Beste G, Scroggins BT, Neckers L, llag LL, Jay DG (2004) Functional proteomic screens reveal an essential extracellular role for hsp90aplha in cancer cell invasiveness. Nat Cell Biol 6:507–514
Ewing JF, Maines MD (1991) Rapid induction of heme oxygenase 1 mRNA and protein by hyperthermia in rat brain: Heme oxygenase 2 is not a heat shock protein. Proc Natl Acad Sci U S A 88:5364–5368
Fawcett TW, Martindale JL, Guyton KZ, Hai T, Holbrook NJ (1999) Complexes containing activating transcription factor (ATF)/cAMP-responsive-element-binding protein (CREB) interact with the CCAAT/enhancer-binding protein (C/EBP)-ATF composite site to regulate Gadd153 expression during the stress response. Biochem J 339:135–141
Fernandez J, Yaman I, Merrick WC, Koromilas A, Wek RC, Sood R, Hensold J, Hatzoglou M (2002a) Regulation of internal ribosome entry site-mediated translation by eukaryotic initiation factor-2alpha phosphorylation and translation of a small upstream open reading frame. J Biol Chem 277:2050–2058
Fernandez J, Yaman I, Sarnow P, Snider MD, Hatzoglou M (2002b) Regulation of internal ribosomal entry site-mediated translation by phosphorylation of the translation initiation factor eIF2alpha. J Biol Chem 277:19198–19205
Ferri KF, Kroemer G (2001) Organelle-specific initiation of cell death pathways. Nat Cell Biol 3:E255–E263
Frydman J, Nimmesgern E, Ohtsuka K, Hartl FU (1994) Folding of nascent polypeptide chains in a high molecular mass assembly with molecular chaperones. Nature 370:111–117
Gao B, Adhikari R, Howarth M, Nakamura K, Gold MC, Hill AB, Knee R, Michalak M, Elliott T (2002) Assembly and antigen-presenting function of MHC class I molecules in cells lacking the ER chaperone calreticulin. Immunity 16:99–109
Gardner RG, Swarbrick GM, Bays NW, Cronin SR, Wilhovsky S, Seeliq L, Kim C, Hampton RY (2000) Endoplasmic reticulum degradation requires lumen to cytosol signaling. Transmembrane control of Hrd1p by Hrd3p. J Cell Biol 151:69–82
Genereux JC, Qu S, Zhou M, Ryno LM, Wang S, Shoulders MD, Kaufman RJ, Lasmezas CI, Kelly JW, Wiseman RL (2015) Unfolded protein response-induced ERdj3 secretion links ER stress to extracellular proteostasis. EMBO J 34:4–19
Georgopoulos CP, Lundquist-Heil A, Yochem J, Feiss M (1980) Identification of the E. coli dnaJ gene product. Mol Gen Genet 178:583–588
Gething MJ (1999) Role and regulation of the ER chaperone BiP. Semin Cell Dev Biol 10:465–472
Gissel C, Doutheil J, Paschen W (1997) Activation of heme oxygenase-1 expression by disturbance of endoplasmic reticulum calcium homeostasis in rat neuronal cell culture. Neurosci Lett 231:75–78
Glover JR, Lindquist S (1998) Hsp104, Hsp70, and Hsp40: a novel chaperone system that rescues previously aggregated proteins. Cell 94:73–82
Graumann P, Marahiel MA (1996) Some like it cold: response of microorganisms to cold shock. Arch Microbiol 166:293–300

Grenert JP, Johnson BD, Toft DO (1999) The importance of ATP binding and hydrolysis by Hsp90 in formation and function of protein heterocomplexes. J Biol Chem 274:17525–17533
Guérin R, Arseneault G, Dumont S, Rokeach LA (2008) Calnexin Is Involved in Apoptosis Induced by Endoplasmic Reticulum Stress in the Fission Yeast. Mol Biol Cell 19:4404–4420
Gupta S, Deepti A, Deegan S, Lisbona F, Hetz C, Samali A (2010) HSP72 protects cells from ER stress-induced apoptosis via enhancement of IRE1alpha-XBP1 signaling through a physical interaction. PLoS Biol 8:e1000410
Haas IG, Wabi M (1983) Immunoglobulin heavy chain binding protein. Nature 306:387–389
Hagiwara S, Iwasaka H, Shingu C, Matsumoto S, Hasegawa A, Asai N, Noguchi T (2009) Heat shock protein 72 protects insulin-secreting beta cells from lipopolysaccharide-induced endoplasmic reticulum stress. Int J Hyperthermia 25:626–633
Hamada H, Suzuki M, Yuasa S, Mimura N, Shinozuka N, Takada Y, Suzuki M, Nishino T, Nakaya H, Koseki H, Aoe T (2004) Dilated cardiomyopathy caused by aberrant endoplasmic reticulum quality control in mutant KDEL receptor transgenic mice. Mol Cell Biol 24:8007–8017
Hampton RY (2002) ER-associated degradation in protein quality control and cellular regulation. Curr Opin Cell Biol 14:476–482
Han W, Christen P (2003a) Interdomain communication in the molecular chaperone DnaK. Biochem J 369:627–634
Han W, Christen P (2003b) Mechanism of the targeting action of DnaJ in the DnaK molecular chaperone system. J Biol Chem 278:19038–19043
Harding HP, Zhang Y, Ron D (1999) Protein translation and folding are coupled by an endoplasmic-reticulum-resident kinase. Nature 397:271–274
Harding HP, Zhang Y, Bertolotti A, Zeng H, Ron D (2000a) Perk is essential for translational regulation and cell survival during the unfolded protein response. Mol Cell 5:897–904
Harding HP, Novoa I, Zhang Y, Zeng H, Wek R, Schapira M, Ron D (2000b) Regulated translation initiation controls stress-induced gene expression in mammalian cells. Mol Cell 6:1099–1108
Harding HP, Calfon M, Urano F, Novoa I, Ron D (2002) Transcriptional and translational control in the Mammalian unfolded protein response. Annu Rev Cell Dev Biol 18:575–599
Harris SF, Shiau AK, Agard DA (2004) The crystal structure of the carboxy-terminal dimerization domain of htpG, the Escherichia coli Hsp90, reveals a potential substrate binding site. Structure 12(6):1087–1097
Haze K, Yoshida H, Yanagi H, Yura T, Mori K (1999) Mammalian transcription factor ATF6 is synthesized as a transmembrane protein and activated by proteolysis in response to endoplasmic reticulum stress. Mol Biol Cell 10:3787–3799
Hedhli N, Wang L, Wang Q, Rashed E, Tian Y, Sui X, Madura K, Depre C (2008) Proteasome activation during cardiac hypertrophy by the chaperone H11 Kinase/Hsp22. Cardiovasc Res 77:497–505
Hessenauer A, Schneider CC, Götz C, Montenarh M (2011) CK2 inhibition induces apoptosis via the ER stress response. Cell Signal 23(1):145–151
Hetz C, Glimcher L (2008) The daily job of night killers: alternative roles of the BCL-2 family in organelle physiology. Trends Cell Biol 18:38–44
Hetz C, Bernasconi P, Fisher J, Lee AH, Bassik MC, Antonsson B, Brandt GS, Iwakoshi NN, Schinzel A, Glimcher LH, Korsmeyer SJ (2006) Proapoptotic BAX and BAK modulate the unfolded protein response by a direct interaction with IRE1α. Science 312:572–576
Hetzer MW, Walther TC, Mattaj IW (2005) Pushing the envelope: structure, function, and dynamics of the nuclear periphery. Annu Rev Cell Dev Biol 21:347–380
Hohfeld J (1998) Regulation of the heat shock conjugate Hsc70 in the mammalian cell: the characterization of the antiapoptotic protein BAG-1 provides novel insights. Biol Chem 379:269–274
Hu MC, Gong HY, Lin GH, Hu SY, Chen MH, Huang SJ, Liao CF, Wu JL (2007) XBP-1, a key regulator of unfolded protein response, activates transcription of IGF1 and Akt phosphorylation in zebrafish embryonic cell line. Biochem Biophys Res Commun 359:778–783

Huo L, Chen R, Zhao L, Shi X, Bai R, Long D, Chen F, Zhao Y, Chang YZ, Chen C (2015) Silver nanoparticles activate endoplasmic reticulum stress signaling pathway in cell and mouse models: The role in toxicity evaluation. Biomaterials 61:307–315

Huot J, Lambert H, Lavoie JN, Guimond A, Houle F, Landry J (1995) Characterization of 45-kDd54-kDa HSP27 kinase, a stress-sensitive kinase which may activate the phosphorylation-dependent protective function of mammalian 27-kDa heat-shock protein HSP27. Eur J Biochem 227:416–427

Irobi J, Van Impe K, Seeman P, Jordanova A, Dierick I, Verpoorten N, Michalik A, De Vriendt E, Jacobs A, Van Gerwen V, Vennekens K, Mazanec R, Tournev I, Hilton-Jones D, Talbot K, Kremensky I, Van Den Bosch L, Robberecht W, Van Vandekerckhove J, Van Broeckhoven C, Gettemans J, De Jonghe P, Timmerman V (2004) Hot-spot residue in small heat-shock protein 22 causes distal motor neuropathy. Nature genetics 36:597–601

Ito H, Iwamoto I, Inaguma Y, Takizawa T, Nagata K, Asano T, Kato K (2005) Endoplasmic reticulum stress induces the phosphorylation of small heat shock protein, Hsp27. J Cell Biochem 95:932–941

Itoh K, Chiba T, Takahashi S, Ishii T, Igarashi K, Katoh Y, Oyake T, Hayashi N, Satoh K, Hatayama I, Yamamoto M, Nabeshima Y (1997) An Nrf2/small Maf heterodimer mediates the induction of phase II detoxifying enzyme genes through antioxidant response elements. Biochem Biophys Res Commun 236:313–322

Jackson A, Friedman S, Zhan X, Engleka KA, Forough R, Maciag T (1992) Heat shock induces the release of fibroblast growth factor 1 from NIH 3T3 cells. Proc Natl Acad Sci U S A 89:10691–10695

Jahn TR, Radford SE (2005) The yin and yang of protein folding. FEBS J 272:5962–5970

Jarosch E, Lenk U, Sommer T (2003) Endoplasmic reticulum-associated protein degradation. Int Rev Cytol 223:39–81

Jayanthi S, McCoy MT, Beauvais G, Ladenheim B, Gilmore K, Wood W, Becker K, Cadet JL (2009) Methamphetamine induces dopamine D1 receptor-dependent endoplasmic reticulum stress-related molecular events in the rat striatum. PLoS One 4:e6092

Jiang J, Ballinger CA, Wu Y, Dai Q, Cyr DM, Höhfeld J, Patterson C (2001) CHIP is a U-box-dependent E3 ubiquitin ligase: Identification of Hsc70 as a target for ubiquitylation. J Biol Chem 276:42938–42944

Jiang HY, Wek SA, McGrath BC, Lu D, Hai T, Harding HP, Wang X, Ron D, Cavener DR, Wek RC (2004) Activating transcription factor 3 is integral to the eukaryotic initiation factor 2 kinase stress response. Mol Cell Biol 24:1365–1377

Jiao Q, Sanbe A, Zhang X, Liu J-P, Minamisawa S (2014) aB-Crystallin R120G variant causes cardiac arrhythmias and alterations in the expression of Ca2+ handling proteins and ER stress in mice. Clin Exp Pharmacol Physiol 41(8):589–599

Jin T, Gu Y, Zanusso G, Sy M, Kumar A, Cohen P, Gambetti P, Singh N (2000) The chaperone protein BiP binds to a mutant prion protein and mediates its degradation by the proteasome. J Biol Chem 275:38699–38704

Jordan R, McMacken R (1995) Modulation of the ATPase activity of the molecular chaperone DnaK by peptides and the DnaJ and GrpE heat shock proteins. J Biol Chem 270:4563–4569

Jousse C, Bruhat A, Carraro V, Urano F, Ferrara M, Ron D, Fafournoux P (2001) Inhibition of CHOP translation by a peptide encoded by an open reading frame localized in the chop 50UTR. Nucleic Acids Res 29:4341–4351

Jousse C, Oyadomari S, Novoa I, Lu P, Zhang Y, Harding HP, Ron D (2003) Inhibition of a constitutive translation initiation factor 2alpha phosphatase, CReP, promotes survival of stressed cells. J Cell Biol 163:767–775

Kalbfleisch T, Cambon A, Wattenberg BW (2007) A bioinformatics approach to identifying tail-anchored proteins in the human genome. Traffic 8:1687–1694

Kalia SK, Lee S, Smith PD, Liu L, Crocker SJ, Thorarinsdottir TE, Glover JR, Fon EA, Park DS, Lozano AM (2004) BAG5 inhibits parkin and enhances dopaminergic neuron degeneration. Neuron 44:931–945

Kamada M, So A, Muramaki M, Rocchi P, Beraldi E, Gleave M (2007) Hsp27 knockdown using nucleotide-based therapies inhibit tumor growth and enhance chemotherapy in human bladder cancer cells. Mol Cancer Ther 6:299–308

Kampinga HH, Craig EA (2010) The HSP70 chaperone machinery: J proteins as drivers of functional specificity. Nat Rev Mol Cell Biol 11:579–592

Kang BH, Plescia J, Dohi T, Rosa J, Doxsey SJ, Altieri DC (2007) Regulation of tumor cell mitochondrial homeostasis by an organelle-specific Hsp90 chaperone network. Cell 131:257–270

Kappe G, Franck E, Verschuure P, Boelens WC, Leunissen JAM, de Jong WW (2003) The human genome encodes 10 alpha-crystallin-related small heat shock proteins: HspB1–10. Cell Stress Chaperones 8:53–61

Kebache S, Cardin E, Nguyen DT, Chevet E, Larose L (2004) Nck-1 antagonizes the endoplasmic reticulum stress-induced inhibition of translation. J Biol Chem 279:9662–9671

Kim JE, Hyun HW, Min SJ, Kang TC (2017) Sustained HSP25 Expression Induces Clasmatodendrosis via ER Stress in the Rat Hippocampus. Front Cell Neurosci 11:47

Kojima E, Takeuch A, Haneda M, Yagi A, Hasegawa T, Yamaki K, Takeda K, Akira S, Shimokata K, Isobe K (2003) The function of GADD34 is a recovery from a shutoff of protein synthesis induced by ER stress: elucidation by GADD34-deficient mice. FASEB J 17:1573–1575

Kopito RR (2000) Aggresomes, inclusion bodies and protein aggregation. Trends Cell Biol 10:524–530

Koulov AV, LaPointe P, Lu B, Razvi A, Coppinger J, Dong MQ, Matteson J, Laister R, Arrowsmith C, Yates JR, Balch WE (2010) Biological and Structural Basis for Aha1 Regulation of Hsp90 ATPase Activity in Maintaining Proteostasis in the Human Disease Cystic Fibrosis. Mol Biol Cell 21:871–884

Kumagai J, Fukuda J, Kodama H, Murata M, Kawamura K, Itoh H, Tanaka T (2000) Germ cell-specific heat shock protein 105 binds to p53 in a temperature-sensitive manner in rat testis. Eur J Biochem 267:3073–3078

Kumano M, Furukawa J, Shiota M, Zardan A, Zhang F, Beraldi E, Wiedmann RM, Fazli L, Zoubeidi A, Gleave ME (2012) Co targeting stress-activated Hsp27 and autophagy as a combinatorial strategy to amplify endoplasmic reticular stress in prostate cancer. Mol Cancer Ther 11:1661–1671

Kurisu J, Honma A, Miyajima H, Kondo S, Okumura M, Imaizumi K (2002) MDG1/ERdj4, an ER-resident DnaJ family member, suppresses cell death induced by ER stress. Genes Cells 8:189–202

Lamoureux F, Thomas C, Yin MJ, Fazli L, Zoubeidi A, Gleave ME (2014) Suppression of heat shock protein 27 using OGX-427 induces endoplasmic reticulum stress and potentiates heat shock protein 90 inhibitors to delay castrate-resistant prostate cancer. Eur Urol 66:145–155

Lau PP, Villanueva H, Kobayashi K, Nakamuta M, Chang BH, Chan L (2001) A DnaJ protein, Apobec-1-binding protein-2, modulates apolipoprotein B mRNA editing. J Biol Chem 276:46445–46452

Laybutt DR, Preston AM, Akerfeldt MC, Kench JG, Busch AK, Biankin AV, Biden TJ (2007) Endoplasmic reticulum stress contributes to beta cell apoptosis in type 2 diabetes. Diabetologia 50:752–763

Lee AS (2001) The glucose-regulated proteins: stress induction and clinical applications. Trends Biochem Sci 26:504–510

Lee AS (2005) The ER chaperone and signaling regulator GRP78/BiP as a monitor of endoplasmic reticulum stress. Methods 35:373–381

Lee K, Tirasophon W, Shen X, Michalak M, Prywes R, Okada T, Yoshida H, Mori K, Kaufman RJ (2002) IRE1-mediated unconventional mRNA splicing and S2P-mediated ATF6 cleavage merge to regulate XBP1 in signaling the unfolded protein response. Genes Dev 16:452–466

Lenna S, Farina AG, Martyanov V, Christmann RB, Wood TA, Farber HW, Scorza HW, Scorza R, Whitfield ML, Lafyatis R, Trojanowska M (2013) Increased expression of endoplasmic reticulum stress genes in patients with limited cutaneous Systemic Sclerosis and Pulmonary Arterial Hypertension. Arthritis Rheum 65:1357–1366

Li J, Richter K, Reinstein J, Buchner J (2013) Integration of the accelerator Aha1 in the Hsp90 co-chaperone cycle. Nat Struct Mol Biol 20(3):326–331

Li J, Soroka J, Buchner J (2012) The Hap90 chaperone machinery- conformational dynamics and regulation by co-chaperones. Biochim Biophys Acta 1823:624–635

Liao Q, Ozawa F, Friess H, Zimmermann A, Takayama S, Reed JC, Kleeff J, Büchler MW (2001) The anti-apoptotic protein BAG-3 is overexpressed in pancreatic cancer and induced by heat stress in pancreatic cancer cell lines. FEBS Lett 503:151–157

Liberek K, Marszalek J, Ang D, Georgopoulos C, Zylicz M (1991) Escherichia coli DnaJ and GrpE heat shock proteins jointly stimulate ATPase activity of DnaK. Proc Natl Acad Sci U S A 88:2874–2878

Lilley BN, Ploegh HL (2005) Multiprotein complexes that link dislocation, ubiquitination, and extraction of misfolded proteins from the endoplasmic reticulum membrane. Proc Natl Acad Sci U S A 102:14296–14301

Lin AM, Chao PL, Fang SF, Chi CW, Yang CH (2007) Endoplasmic reticulum stress is involved in arsenite-induced oxidative injury in rat brain. Toxicol Appl Pharmacol 224:138–146

Lippincott-Schwartz J, Bonifacino JS, Yuan LC, Klausner RD (1988) Degradation from the endoplasmic reticulum: disposing of newly synthesized proteins. Cell 54:209–220

Litt M, Kramer P, la Morticella DM, Murphey W, Lovrien EW, Weleber RG (1998) Autosomal dominant congenital cataract associated with a missense mutation in the human alpha crystalline gene CRYAA. Hum Mol Genet 7:471–474

Liu CY, Schroder M, Kaufman RJ (2000) Ligand-independent dimerization activates the stress response kinases IRE1 and PERK in the lumen of the endoplasmic reticulum. J Biol Chem 275:24881–24885

Liu B, Yang Y, Qiu Z, Staron M, Hong F, Li Y, Wu S, Li Y, Hao B, Bona R, Han D, Li Z (2010) Folding of Toll-like receptors by the HSP90 paralogue gp96 requires a substrate-specific cochaperone. Nat Commun 1:79

Liu B, Staron M, Hong F, Wu BX, Sun S, Morales C, Crosson CE, Tomlinson S, Kim I, Wu D, Li Z (2013) Essential roles of grp94 in gut homeostasis via chaperoning canonical Wnt pathway. Proc Natl Acad Sci U S A 110:6877–6882

Louessard M, Bardou I, Lemarchand E, Thiebaut AM, Parcq J, Leprince J, Terrisse A, Carraro V, Fafournoux P, Bruhat A, Orset C, Vivien D, Ali C, Roussel BD (2017) Activation of cell surface GRP78 decreases endoplasmic reticulum stress and neuronal death. Cell Death Differ 24:1518–1529

Lu PD, Harding HP, Ron D (2004) Translation reinitiation at alternative open reading frames regulates gene expression in an integrated stress response. J Cell Biol 167:27–33

Lüders J, Demand J, Schönfelder S, Frien M, Zimmermann R, Höhfeld J (1998) Cofactor-induced modulation of the functional specificity of the molecular chaperone Hsc70. Biol Chem 379:1217–1226

Lüders J, Demand J, Höhfeld J (2006) The ubiquitin-related BAG-1 provides a link between the molecular chaperones Hsc70/Hsp70 and the proteasome. J Biol Chem 275:4613–4617

Luo S, Baumeister P, Yang S, Abcouwer SF, Lee AS (2003) Induction of Grp78/BiP by translational block: activation of the Grp78 promoter by ATF4 through and upstream ATF/CRE site independent of the endoplasmic reticulum stress elements. J Biol Chem 278:37375–37385

Ma Y, Hendershot LM (2004) ER chaperone functions during normal and stress conditions. J Chem Neuroanal 28:51–65

Ma Y, Brewer JW, Diehl JA, Hendershot LM (2002) Two distinct stress signaling pathways converge upon the CHOP promoter during the mammalian unfolded protein response. J Mol Biol 318:1351–1365

Maattanen P, Gehring K, Bergeron JJ, Thomas DY (2010) Protein quality control in the ER: the recognition of misfolded proteins. Semin Cell Dev Biol 21:500–511

Maines MD, Eke BC, Zhao X (1996) Corticosterone promotes increased heme oxygenase-2 protein and transcript expression in the newborn rat brain. Brain Research 722:83–94

Marcu MG, Doyle M, Bertolotti A, Ron D, Hendershot L, Neckers L (2002) Heat shock protein 90 modulates the unfolded protein response by stabilizing IRE1alpha. Mol Cell Biol 22:8506–8513
Martins VR, Graner E, Garcia-Abreu J, de Souza SJ, Mercadante AF, Veiga SS, Zanata SM, Neto VM, Brentani RR (1997) Complementary hydropathy identifies a cellular prion protein receptor. Nature Medicine 3:1376–1382
Mayer MP, Le Breton L (2015) Hsp90: breaking the symmetry. Mol Cell 58(1):8–20
Matsumori M, Itoh H, Toyoshima I, Komatsuda A, Sawada K, Fukuda J, Tanaka T, Okubo A, Kinouchi H, Mizoi K, Hama T, Suzuki A, Hamada F, Otaka M, Shoji Y, Takada G (2002) Characterization of the 105-kDa molecular chaperone Identification, biochemical properties, and localization. Eur J Biochem:1–10
Meacham GC, Lu Z, King S, Sorscher E, Tousson A, Cyr DM (1999) The Hdj-2/Hsc70 chaperone pair facilitates early steps in CFTR biogenesis. The EMBO Journal 18:1492–1505
Meacham GC, Patterson C, Zhang W, Younger JM, Cyr DM (2001) The Hsc70 co-chaperone CHIP targets immature CFTR for proteasomal degradation. Nat Cell Biol 3:100–105
Meares GP, Zmijewska AA, Jope RS (2008) HSP105 interacts with GRP78 and GSK3 and promotes ER stress-induced caspase-3 activation. Cell Signal 20(2):347–358
Melnick J, Dul JL, Argon Y (1994) Sequential interaction of the chaperones BiP and GRP94 with immunoglobulin chains in the endoplasmic reticulum. Nature 370:373–375
Melnyk A, Rieger H, Zimmerman R (2015) Co-chaperones of the mammalian endoplasmic reticulum. Subcell Biochem 78:179–200
Meunier L, Usherwood YK, Chung KT, Hendershot LM (2002) A subset of chaperones and folding enzymes form multiprotein complexes in endoplasmic reticulum to bind nascent proteins. Mol Biol Cell 13:4456–4469
Meyer P, Prodromou C, Hu B, Vaughan C, Roe SM, Panaretou B, Piper PW, Pearl LH (2003) Structural and functional analysis of the middle segment of hsp90: implications for ATP hydrolysis and client protein and cochaperone interactions. Mol Cell 11:647–658
Minami Y, Hohfeld J, Ohtsuka K, Hartl FU (1996) Regulation of the heat-shock protein 70 reaction cycle by the mammalian DnaJ homolog, Hsp40. J Biol Chem 271:19617–19624
Mizukoshi E, Suzuki M, Loupatov A, Uruno T, Hayashi H, Misono T, Kaul SC, Wadhwa R, Imamura T (1999) Fibroblast growth factor-1 interacts with the glucose-regulated protein GRP75/mortalin. Biochem J 2:461–466
Montesano Gesualdi N, Chirico G, Pirozzi G, Costantino E, Landriscina M, Esposito F (2007) Tumor necrosis factor-associated protein 1 (TRAP- 1) protects cells from oxidative stress and apoptosis. Stress 10:342–350
Morales C, Wu S, Yang Y, Hao B, Li Z (2009) Drosophila glycoprotein 93 Is an ortholog of mammalian heat shock protein gp96 (grp94, HSP90b1, HSPC4) and retains disulfide bond independent chaperone function for TLRs and integrins. J Immunol 183:5121–5128
Mueller B, Lilley BN, Ploegh HL (2006) SEL1L, the homologue of yeast Hrd3p, is involved in protein dislocation from the mammalian ER. J Cell Biol 175:261–270
Muñoz-Sánchez J, Chánez-Cárdenas ME (2014) A review on hemeoxygenase-2 focus on cellular protection and oxygen response. Oxid Med Cell Longev 2014:604981
Munro S, Pelham HR (1987) A C-terminal signal prevents secretion of luminal ER proteins. Cell 48:899–907
Muresan Z, Arvan P (1997) Thyroglobulin transport along the secretory pathway. Investigation of the role of molecular chaperone, GRP94, in protein export from the endoplasmic reticulum. J Biol Chem 272:26095–26102
Nakagawa T, Zhu H, Morishima N, Li E, Xu J, Yankner BA (2000) Caspase-12 mediates endoplasmic-reticulum-specific apoptosis and cytotoxicity by amyloid-beta. Nature 403:98–103
Nakagomi S, Suzuki Y, Namikawa K, Kiryu-Seo S, Kiyama H (2003) Expression of the activating transcription factor 3 prevents c-Jun N-terminal kinase-induced neuronal death by promoting heat shock protein 27 expression and Akt activation. J Neurosci 23:5187–5196

Wayne N, Bolon DN (2007) Dimerization of Hsp90 is required for function. J Biol Chem 282(48):35386–35395
Nagata K (1996) Hsp47: a collagen-specific molecular chaperone. Trends Biochem Sci 21(1):23–26
Ni M, Zhang Y, Lee AS (2011) Beyond the endoplasmic reticulum: atypical GRP78 in cell viability, signalling and therapeutic targeting. Biochem J 434:181–188
Nillegoda NB, Bukau B (2015) Metazoan Hsp70-based protein disaggregases: emergence and mechanisms. Front Mol Biosci 2:57
Ninagawa S, Okada T, Sumitomo Y, Horimoto S, Suqimoto T, Ishikawa T, Takeda S, Yamamoto T, Suzuki T, Kamiya Y, Kato K, Mori L (2015) Forcible destruction of severely misfolded mammalian glycoproteins by the non-glycoprotein ERAD pathway. J Cell Biol 211:775–784
Nishitoh H, Matsuzawa A, Tobiume K, Saequsa K, Takeda K, Inoue K, Hori S, Kalkizuka A, Ichijo H (2002) ASK1 is essential for endoplasmic reticulum stress-induced neuronal cell death triggered by expanded polyglutamine repeats. Genes Dev 16:1345–1355
Novoa I, Zeng H, Harding HP, Ron D (2001) Feedback inhibition of the unfolded protein response by GADD34-mediated dephosphorylation of eIF2alpha. J Cell Biol 153:1011–1022
Ogata M, Hino S, Saito A, Morikawa K, Kondo S, Kanemoto S, Murakami T, Taniquchi M, Tanii I, Yoshinaga K, Shiosaka S, Hammarback JA, Urano F, Imaizumi K (2006) Autophagy is activated for cell survival after endoplasmic reticulum stress. Mol Cell Biol 26:9220–9231
Okoturo-Evans O, Dybowska A, Valsami-Jones E, Cupitt J, Gierula M, Boobis AR, Edwards RJ (2013) Elucidation of toxicity pathways in lung epithelial cells induced by silicon dioxide nanoparticles. PLoS One 8:e72363
Ostrovsky O, Ahmed NT, Argon Y (2009) The chaperone activity of GRP94 towards of insulin-like growth factor II is necessary for the stress response to serum deprivation. Mol Biol Cell 20:1855–1864
Palade GE, Porter KR (1954) Studies on the endoplasmic reticulum. I. Its identification in cells in situ. J Exp Med 100:641–656
Panneerselvam L, Raghunath A, Perumal E (2017) Differential expression of myocardial heat shock proteins in rats acutely exposed to fluoride. Cell Stress Chaperones 22:743–750
Panowski SH, Wolff S, Aguilaniu H, Durieux J, Dillin A (2007) PHA-4/Foxa mediates diet-restriction-induced longevity of C. elegans. Nature 447:550–555
Parsell DA, Lindquist S (1993) The function of heat shock proteins in stress tolerance; degradation and reactivation of damaged proteins. Annu Rev Genet 27:437–496
Paschen W (2003) Endoplasmic reticulum: a primary target in various acute disorders and degenerative diseases of the brain. Cell Calcium 34:365–383
Paschen W, Mengesdorf T, Althausen S, Hotop S (2001) Peroxidative stress selectively down-regulates the neuronal stress response activated under conditions of endoplasmic reticulum dysfunction. J Neurochem 76:1916–1924
Phadtare S, Alsina J, Inouye M (1999) Cold-shock response and cold-shock proteins. Curr Opin Microbiol 2(2):175–180
Petrova K, Oyadomari S, Hendershot LM, Ron D (2008) Regulated association of misfolded endoplasmic reticulum lumenal proteins with P58/DNAJc3. EMBO J 27:2862–2872
Petrucelli L, Dickson D, Kehoe K, Taylor J, Snyder H, Grover A, De Lucia M, McGowan E, Lewis J, Prihar G, Kim J, Dilmann WH, Browne WH, Hall A, Voellmy R, Tsuboi Y, Dawson TM, Wolozin B, Hardy J, Hutton M (2004) CHIP and Hsp70 regulate tau ubiquitination, degradation and aggregation. Hum Mol Genet 13:703–714
Plemper RK, Wolf DH (1999) Retrograde protein translocation: ERADication of secretory proteins in health and disease. Trends Biochem Sci 24:266–270
Prodromou C, Roe SM, O'Brien R, Ladbury JE, Piper PW, Pearl LH (1997) Identification and structural characterization of the ATP/ADP-Binding Site in the Hsp90 molecular chaperone. Cell 90(1):65–75
Porter KR (1953) Observations on a submicroscopic basophilic component of cytoplasm. J Exp Med 97:727–750

Porter KR, Palade GE (1957) Studies on the endoplasmic reticulum. III. Its form and distribution in striated muscle cells. J Biophys Biochem Cytol 3:269–300
Pouyssegur J, Shiu RP, Pastan I (1977) Induction of two transformation-sensitive membrane polypeptides in normal fibroblasts by a block in glycoprotein synthesis or glucose deprivation. Cell 11:941–947
Ohtsuka K (1993) Cloning of a cDNA for Heat-Shock Protein hsp40, a Human Homolog of Bacterial DnaJ. Biochem Biophys Res Commun 197(1):235–240
Otero JH, Lizák B, Hendershot LM (2010) Life and death of a BiP substrate. Semin Cell Dev Biol 21(5):472–478
Quraishe S, Asuni A, Boelens WC, O'Connoro V, Wyttenbach A (2008) Expression of the small heat shock protein family in the mouse CNS: differential anatomical and biochemical compartmentalization. Neuroscience 153:483–491
Rabu C, Wipf P, Brodsky JL, High S (2008) A Precursor-specific Role for Hsp40/Hsc70 during Tail-anchored Protein Integration at the Endoplasmic Reticulum. J Biol Chem 283:27504–27513
Rao RV, Peel A, Logvinova A, del Rio G, Hermel E, Yokota T, Goldsmith PC, Ellerby LM, Ellerby HM, Bredesen DE (2002) Coupling endoplasmic reticulum stress to the cell death program: role of the ER chaperone GRP78. FEBS Letters 514:122–128
Rao RV, Ellerby HM, Bredesen DE (2004) Coupling endoplasmic reticulum stress to the cell death program. Cell Death Differ 11:372–380
Rao RV, Niazi K, Mollahan P, Mao X, Crippen D, Poksay KS, Chen DE (2006) Bredesen. Coupling endoplasmic reticulum stress to the cell-death program a novel HSP90-independent role for the small chaperone protein p23. Cell Death Differ 13:415–425
Rape M, Hoppe T, Gorr I, Kalocay M, Richly H, Jentsch S (2001) Mobilization of processed, membrane-tethered SPT23 transcription factor by CDC48 (UFD1/NPL4), a ubiquitin -selective chaperone. Cell 107:667–677
Rizvi SM, Mancino L, Thammavongsa V, Cantley RL, Raghavan MA (2004) Polypeptide binding conformation of calreticulin is induced by heat shock, calcium depletion, or by deletion of the C-terminal acidic region. Mol Cell 15:913–923
Röhl A, Rohrberg J, Buchner J (2013) The chaperone Hsp90: changing partners for demanding clients. Trends Biochem Sci 38(5):253–262
Roue G, Perez-Galan P, Mozos A, Lopez-Guerra M, Xargay-Torrent S, Rosich L, Saborit-Villarroya I, Normant E, Campo E, Colomer D (2011) The Hsp90 inhibitor IPI-504 overcomes bortezomib resistance in mantle cell lymphoma in vitro and in vivo by down-regulation of the prosurvival ER chaperone BiP/Grp78. Blood 117(4):1270–1279
Rutishauser J, Spiess M (2002) Endoplasmic reticulum storage diseases. Swiss Med Wkly 132:211–222
Rutkowski DT, Kaufman RJ (2007) That which does not kill me makes me stronger: adapting to chronic ER stress. Trends Biochem Sci 32:469–476
Ryu EJ, Harding HP, Angelastro JM, Vitolo OV, Ron D, Greene LA (2002) Endoplasmic reticulum stress and the unfolded protein response in cellular models of Parkinson's disease. J Neurosci 22:10690–10698
Saibil H (2013) Chaperone machines for protein folding unfolding and disaggregation. Nat Rev Mol Cell Biol 14:630–642
Schroder M (2006) The unfolded protein response. Mol Biotechnol 34:279–290
Schroder M, Kaufman RJ (2005a) ER stress and the unfolded protein response. Mutat Res 569:29–63
Schroder M, Kaufman RJ (2005b) The Mammalian unfolded protein response. Annu Rev Biochem 74:739–789
Schuberth C, Buchberger A (2005) Membrane-bound Ubx2 recruits Cdc48 to ubiquitin ligases and their substrates to ensure efficient ERassociated protein degradation. Nat Cell Biol 7:999–1006
Schwartz DS, Blower MD (2016) The endoplasmic reticulum: structure, function and response to cellular signaling. Cell Mol Life Sci 73:79–94

Shamu CE, Walter P (1996) Oligomerization and phosphorylation of the Ire1p kinase during intracellular signaling from the endoplasmic reticulum to the nucleus. EMBO J. 15:3028–3039
Sharma R, Tsuchiya M, Bartlett JD (2008) Fluoride induces endoplasmic reticulum stress and inhibits protein synthesis and secretion. Environ Health Perspect 116:1142–1146
Shen Y, Hendershot LM (2005) ERdj3, a stress-inducible endoplasmic reticulum DnaJ homologue, serves as a cofactor for BiP's interactions with unfolded sustrates. Mol Biol Cell 16:40–50
Shen J, Chen X, Hendershot L, Prywes R (2002) ER stress regulation of ATF6 localization by dissociation of BiP/GRP78 binding and unmasking of Golgi localization signals. Dev Cell 3:99–111
Shi Y, Vattem KM, Sood R, An J, Lianq J, Stramm L, Wek RC (1998) Identification and characterization of pancreatic eukaryotic initiation factor 2 α-subunit kinase, PEK, involved in translational control. Mol Biol Cell 18:7499–7509
Shibata Y, Voeltz GK, Rapoport TA (2006) Rough sheets and smooth tubules. Cell 126:435–439
Shirriff CS, Heikkila JJ (2017) Characterization of cadmium chloride-induced BiP accumulation in Xenopus laevis A6 kidney epithelial cells. Comp Biochem Physiol C Toxicol Pharmacol 191:117–128
Simard JC, Vallieres F, de Liz R, Lavastre V, Girard D (2015) Silver nanoparticles induce degradation of the endoplasmic reticulum stress sensor activating transcription factor-6 leading to activation of the NLRP-3 inflammasome. J Biol Chem 290:5926–5939
Smith MH, Ploegh HL, Weissman JS (2011) Road to ruin: targeting proteins for degradation in the endoplasmic reticulum. Science 334:1086–1090
Solís EJ, Pandey JP, Zheng X, Jin DX, Gupta PB, Airoldi EM, Pincus D, Denic V (2016) Defining the essential function of yeast Hsf1 reveals a compact transcriptional program for maintaining Eukaryotic Proteostasis. Mol Cell 63(1):60–71
Song S, Lee H, Kam TI, Tai ML, Lee JY, Noh JY, Shim SM, Seo SJ, Kong YY, Nakagawa T, Chung CW, Choi DY, Oubrahim H, Jung YK (2008) E2-25K/Hip-2 regulates caspase-12 in ER stress – mediated A neurotoxicity. J Cell Biol 182:675–684
Sörgjerd K, Ghafouri B, Jonsson B-H, Kelly JW, Blond SY, Ammarström P (2006) Retention of misfolded mutant transthyretin by the caperone BiP/GRP78 mitigates amyloidogenesis. J Mol Biol 356:469–482
Sousa R, Lafer EM (2006) Keep the traffic moving: mechanism of the Hsp70 motor. Traffic 7:1596–1603
Staron M, Yang Y, Liu B, Li J, Shen Y, Zuniga-Pflucker JC, Aguila I (2010) gp96, an endoplasmic reticulum master chaperone for integrins and Toll-like receptors, selectively regulates early T and B lymphopoiesis. Blood 115:2380–2390
Staron M, Wu S, Feng H, Stojanovic A, Du X, Bona R, Liu B, Li Z (2011) Heat shock protein gp96/grp94 is an essential chaperone for the platelet Ib–IX–V complex. Blood 117:7136–7144
Stegh AH, Kesari S, Mahoney JE, Jenq HT, Forloney KL, Protopopov A, Louis DN, Chin L, DePinho RA (2008) Bcl2L12- mediated inhibition of effector caspase-3 and caspase- 7 via distinct mechanisms in glioblastoma. Proc Natl Acad Sci U S A 105:10703–10708
Sun X, Fontaine JM, Rest JS, Shelden EA, Welsh MJ, Benndorf R (2003) Interaction of human HSPB8 (HSPB8) with other small heat shock proteins. J Biol Chem 279:2394–2402
Sundarraj K, Raghunath A, Panneerselvam L, Perumal E (2017) Iron oxide nanoparticles modulate heat shock proteins and organ specific markers expression in mice male accessory organs. Toxicol Appl Pharmacol 317:12–24
Synoradzki K, Bieganowski P (2015) Middle domain of human Hsp90 isoforms differentially binds Aha1 in human cells and alters Hsp90 activity in yeast. Biochimica et Biophysica Acta (BBA) – Mol Cell Res 1853(2):445–452
Taiyab A, Sreedhar AS, Rao CM (2009) Hsp90 inhibitors, GA and 17AAG, lead to ER stress-induced apoptosis in rat histiocytoma. Biochem Pharmacol 78:142–152

Takayama S, Bimston DN, Matsuzawa S, Freeman BC, Aime-Sempe C, Xie Z, Morimoto RI, Reed JC (1997) BAG-1 modulates the chaperone activity of Hsp70/Hsc70. The EMBO J 16:4887–4896
Takayama S, Xie Z, Reed JC (1999) An evolutionarily conserved family of Hsp70/Hsc70 molecular chaperone regulators. J Biol Chem 274:781–786
Tannous A, Pisoni GB, Hebert DN, Molinari M (2015) N-linked sugar-regulated protein folding and quality control in the ER. Semin Cell Dev Biol 41:79–89
Tarantini F, LaVallee T, Jackson A, Gamble S, Carreira CM, Garfinkel S, Burgess WH, Maciag T (1998) The extravesicular domain of synaptotagmin-1 is released with the latent fibroblast growth factor-1 homodimer in response to heat shock. J Biol Chem 273:22209–22216
Taylor RP, Benjamin IJ (2005) Small heat shock proteins: a new classification scheme in mammals. J Mol Cell Cardiol 38:433–444
Terasaki M, Slater NT, Fein A, Schmidek A, Reese TS (1994) Continuous network of endoplasmic reticulum in cerebellar Purkinje neurons. PNAS 91:7510–7514
Thibault G, Ng DT (2012) The endoplasmic reticulum associated degradation pathways of budding yeast. Cold Spring Harb Perspect Biol 4:a013193
Thuerauf DJ, Morrison L, Glembotski CC (2004) Opposing roles for ATF6α and ATF6β in endoplasmic reticulum stress response gene induction. J Biol Chem 279:21078–21084
Tirasophon W, Lee K, Callaghan B, Welihinda A, Kaufman RJ (2000) The endoribonuclease activity of mammalian IRE1 autoregulates its mRNA and is required for the unfolded protein response. Genes Dev 14:2725–2736
Tirosh B, Iwakoshi NN, Glimcher LH, Ploegh HL (2006) Rapid turnover of unspliced Xbp-1 as a factor that modulates the unfolded protein response. J Biol Chem 281:5852–5860
Travers KJ, Patil CK, Wodicka L, Lockhart DJ, Weissman JS, Walter P (2000) Functional and genomic analyses reveal an essential coordination between the unfolded protein response and ER-associated degradation. Cell 101:249–258
Tsai B, Ye Y, Rapoport TA (2002) Retro-translocation of proteins from the endoplasmic reticulum into the cytosol. Nat Rev Mol Cell Biol 3:246–255
Tschopp J, Martinon F, Hofmann K (1999) Apoptosis: silencing the death receptors. Curr Biol 9:R381–R384
Tsyusko OV, Unrine JM, Spurgeon D, Blalock E, Starnes D, Tseng M, Joice G, Bertsch PM (2012) Toxicogenomic responses of the model organism Caenorhabditis elegans to gold nanoparticles. Environ Sci Technol 46:4115–4124
Tyler RE, Pearce MMP, Shaler TA, Olzmann JA, Greenblatt EJ, Kopito RR (2012) Unassembled CD147 is an endogenous endoplasmic reticulum-associated degradation substrate. Mol Biol Cell 23:4668–4678
Ubeda M, Wang XZ, Zinszner H, Wu I, Habener JF, Ron D (1996) Stress induced binding of the transcriptional factor CHOP to a novel DNA control element. Mol Cell Biol 16:1479–1489
Ubeda M, Vallejo M, Habener JF (1999) CHOP enhancement of gene transcription by interactions with Jun/Fos AP- 1 complex proteins. Mol Cell Biol 19:7589–7599
Urano F, Wang X (2000) Coupling of stress in the ER to activation of JNK protein kinases by transmembrane protein kinase. Science 287:664–666
Vaccaro A, Patten SA, Aggad D, Julien C, Maios C, Kabashi E, Drapeau P, Parker JA (2013) Pharmacological reduction of ER stress protects against TDP-43 neuronal toxicity in vivo. Neurobiol Dis 55:64–75
Vattem KM, Wek RC (2004) Reinitiation involving upstream ORFs regulates ATF4 mRNA translation in mammalian cells. Proc Natl Acad Sci U S A 101:11269–11274
Vembar SS, Brodsky JL (2008) One step at a time: endoplasmic reticulum associated degradation. Nat Rev Mol Cell Biol 9:944–957
Veratti E (1961) Investigations on the fine structure of striated muscle fiber read before the Reale Istituto Lombardo. 13 March 1902. J Biophys Biochem Cytol 10:1–59
Voellmy R, Boellmann F (2007) Chaperone regulation of the heat shock protein response. Adv Exp Med Biol 594:89–99

Wall D, Zylicz M, Georgopoulos C (1994) The NH2-terminal 108 amino acids of the Escherichia coli DnaJ protein stimulate the ATPase activity of DnaK and are sufficient for lambda replication. J Biol Chem 269:5446–5451
Wang HG, Takayama S, Rapp UR, Reed JC (1996) Bcl-2 interacting protein, BAG-1, binds to and activates the kinase Raf-1. Proc Natl Acad Sci U S A 93:7063–7068
Wang XZ, Harding HP, Zhang Y, Jolicoeur EM, Kuroda M, Ron D (1998) Cloning of mammalian Ire1 reveals diversity in the ER stress responses. EMBO J 17:5708–5717
Wang CW, Kim J, Huang WP, Abeliovich H, Stromhaug PE, Dunn WA Jr, Klionsky DJ (2001) Apg2 Is a novel protein required for the cytoplasm to vacuole targeting, autophagy, and pexophagy pathways. J Biol Chem 276:30442–30451
Wang G, Unger G, Ahmad KA, Slaton JW, Ahmed K (2005) Downregulation of CK2 induces apoptosis in cancer cells a potential approach to cancer therapy. Mol Cell Biochem 274:77–84
Wang M, Wey S, Zhang Y, Ye R, Lee AS (2009) Role of the unfolded protein response regulator GRP78/BiP in development, cancer, and neurological disorders. Antioxid Redox Signal 11:2307–2316
Wang Q, Liu Y, Soetandyo N, Baek K, Hegde R, Ye Y (2011) A ubiquitin ligase-associated chaperone holdase maintains polypeptides in soluble states for proteasome degradation. Mol Cell 42:758–770
Watterson TL, Hamilton B, Martin R, Coulombe RA (2009) Urban particulate matter causes ER stress and the unfolded protein response in human lung cells. Toxicol Sci 112:111–122
Westhoff B, Chapple JP, van der Spuy J, Hohfeld J, Cheetham ME (2005) HSJ1 is a neuronal shuttling factor for the sorting of chaperone clients to the proteasome. Curr Biol 15:1058–1064
Wolmarans A, Lee B, Spyracopoulos L, LaPointe P (2016) The Mechanism of Hsp90 ATPase Stimulation by Aha1. Scientific Reports 6:33179
Xie L, Tiong CX, Bian JS (2012) Hydrogen sulfide protects SH-SY5Y cells against 6-hydroxydopamine-induced endoplasmic reticulum stress. Am J Physiol Cell Physiol 303:C81–C91
Yan S, Zheng C, Zhi-qi C, Liu R, Li G, Hu W, Pei H, Li B (2012) Expression of Endoplasmic Reticulum Stress-Related Factors in the Retinas of Diabetic Rats. Exp Diabetes Res 2012:743780
Ye J, Rawson RB, Komuro R, Chen X, Dave UP, Prywes R, Brown MS, Goldstein JL (2000) ER stress induces cleavage of membrane bound ATF6 by the same proteases that process SREBPs. Mol Cell 6:1355–1364
Ye Y, Meyer HH, Rapoport TA (2001) The AAA ATPase Cdc48/p97 and its partners transport proteins from the ER into the cytosol. Nature 414:652–656
Ye Y, Shibata Y, Yun C, Ron D, Rapoport TA (2004) A membrane protein complex mediates retro-translocation from the ER lumen into the cytosol. Nature 429:841–847
Yoshida H, Matsui T, Yamamoto A, Okada T, Mori K (2001) XBP1 mRNA is induced by ATF6 and spliced by IRE1 in response to ER stress to produce a highly active transcription factor. Cell 107:881–891
Yoshida H, Matsui T, Hosokawa N, Kaufman RJ, Nagata K, Mori K (2003) A time-dependent phase shift in the mammalian unfolded protein response. Dev Cell 4:265–271
Yoshida H, Oku M, Suzuki M, Mori K (2006) pXBP1(U) encoded in XBP1 pre-mRNA negatively regulates unfolded protein response activator pXBP1(S) in mammalian ER stress response. J Cell Biol 172:565–575
Yu M, Haslam RH, Haslam DB (2000) HEDJ, an Hsp40 co-chaperone localized to the endoplasmic reticulum of human cells. J Biol Chem 272:24984–24992
Zhang D, Putti TC (2010) Over-expression of ERp29 attenuates doxorubicin-induced cell apoptosis through up-regulation of Hsp27 in breast cancer cells. Exp Cell Res 316:3522–3531
Zhang LH, Zhang X (2010) Roles of GRP78 in physiology and cancer. J Cell Biochem 110:1299–1305

Zhang Y, Wu BX, Metelli A, Thaxton JE, Hong F, Rachidi S, Ansa-Addo E, Sun S, Vasu C, Yang Y, Liu B, Li Z (2015) GP96 is a GARP chaperone and controls regulatory T cell functions. J Clin Invest 125:859–869

Zhao L, Ackerman SL (2006) Endoplasmic reticulum stress in health and disease. Curr Opin Cell Biol 18:444–452

Zheng X, Xu F, Liang H, Cao H, Cai M, Xu W, Weng J (2017) SIRT1/HSF1/HSP pathway is essential for exenatide-alleviated, lipid-induced hepatic endoplasmic reticulum stress. Hepatology 66:809–824

Zinszner H, Kuroda M, Wang X, Batchvarova N, Lightfoot RT, Remotti H, Stevens JL, Ron D (1998) CHOP is implicated in programmed cell death in response to impaired function of the endoplasmic reticulum. Genes Dev 12:982–995

Chapter 4
Features of Stress-Induced Changes of HSP70 Expression in Populations of Immunocompetent Cells

Anna A. Boyko, Natalya I. Troyanova, Julia D. Teterina, Tatyana L. Azhikina, Sergey S. Vetchinin, Elena I. Kovalenko, and Alexander M. Sapozhnikov

Abstract The results of intracellular protein measurement carrying out with the use of monoclonal antibodies, strongly depend on the localization of antibody binding epitopes in the structure of the molecular target. Our previous study has shown pronounced differences in the pattern of cellular stress response detected by monitoring of intracellular HSP70 content, which was assessed using flow cytometry and monoclonal antibodies interacting with various domains of these molecules. In the present work we investigate the features of stress-induced changes in the expression of HSP70 in populations of mononuclear and granulocytic cells from human peripheral blood using standard methods: PCR, Western-blot and flow-cytometric analysis. A particular feature of this study is an application of a number of antibodies including monoclonal antibodies specific for the total pool of HSP70, separately for the inducible or constitutive form of this protein, as well as for the different domains of the HSP70 molecule. The results provide new additional information on HSP70-related stress-induced processes in populations of immunocompetent cells.

Keywords Anti-HSP70 antibodies · Heat shock · Heat shock proteins · HSP70 expression · Human leukocytes

---

A. A. Boyko · N. I. Troyanova · J. D. Teterina · T. L. Azhikina · E. I. Kovalenko
A. M. Sapozhnikov (✉)
Laboratory of Cell Interactions, Department of Immunology, Shemyakin - Ovchinnikov Institute of Bioorganic Chemistry, Moscow, Russia
e-mail: amsap@mx.ibch.ru

S. S. Vetchinin
The State Research Center for Applied Microbiology and Biotechnology, Obolensk, Russia

A. A. A. Asea, P. Kaur (eds.), *Heat Shock Proteins and Stress*,
Heat Shock Proteins 15, https://doi.org/10.1007/978-3-319-90725-3_4

Abbreviations

HS	Heat shock
Hsc70	Constitutive form of HSP70
HSP	Heat shock proteins
HSP70	Heat shock protein 70 kDa
Hsp70	Inducible form of HSP70
MFI	Means of fluorescence intensity
PBMC	Peripheral blood mononuclear cells
PMN	Polymorph nuclear leukocytes

4.1 Introduction

Heat shock proteins (HSP) belong to the group of intracellular proteins, the levels of expression of which significantly increase under the influence of various stress factors. One of the important families of HSP is HSP70 composed of the protective proteins with a molecular mass of 70 kDa, possessing evident chaperone activity. These proteins are expressed by almost all types of cells, including cells of the immune system. The life cycle of the majority of immunocompetent cells is associated with migration and circulation, which causes their exposure to endogenous and exogenous stress factors that affect the intensity of expression and intracellular content of HSP70. Accordingly, these proteins are the involved in immune homeostasis (McLeod et al. 2012; Muralidharan and Mandrekar 2013; Ferat-Osorio et al. 2014). The essential role of HSP70 in immune system is confirmed by evidences on the relationship of the level of HSP70 expression in immunocompetent cells with a number of pathological processes, especially with diseases accompanied by inflammation of certain tissues (Cuanalo-Contreras et al. 2013; Tan et al. 2007). Along with this, age-related changes in the content of HSP70 in various types of immune cells have been demonstrated (Njemini et al. 2002; Calderwood et al. 2009; Kovalenko et al. 2014). However, despite the numerous evidence of the disease-dependent modulation of the intracellular HSP70 level, by now the HSP70 analysis has not found yet a real application for immunodiagnostics and immunotherapy. Additional investigations are needed to expand existing knowledge of relationship between normal and pathological immune reactions and HSP70 expression. It is important to note that for such studies the most informative methodological approach is an integrated analysis using, in particular, RT-PCR, Western-blot and flow cytometry. At the same time it should be taken into account that the results of flow-cytometric analysis can depend on the specificity of used antibodies to certain epitopes of the molecular targets. Indeed, in the structure of HSP70, which form strong complexes with a large number of intracellular proteins, many fragments of the molecule may not be available for detecting antibodies. In the present study we used a number of commercial antibodies, and elaborated by our group a panel of monoclonal antibodies, directed to different domains of HSP70. Fractions of mononuclear and granulocyte cells isolated from human peripheral blood were used for the experiments. Incubation of cell samples in hyperthermia conditions (heat shock, HS) was used as the stress factor.

4.1.1 Materials and Methods

4.1.1.1 Participants

Healthy adult persons (aged 20–65) were recruited to the study. Inclusion criteria for the participation were the absence of active pathologies, of treatment with corticosteroids or high doses of nonsteroid anti-inflammatory drugs for all subjects. The study has been approved by the local research ethics committee. All participants gave their informed consent prior to the study.

4.1.1.2 Granulocyte and Mononuclear Cell Isolation

Cells were isolated from peripheral blood within 0.5 h after blood sampling by centrifugation at 500 g for 30 min at RT in density gradient using PolymorphPrep separation medium (Axis-Shield, Sweden). Fractions containing polymorph nuclear leukocytes (PMN) and mononuclear cells (PBMC) were then collected. Cells were washed twice (400 *g*, 15 min) in Dulbecco's phosphate buffer saline (DPBS), resuspended in RPMI-1640 media (Sigma-Aldrich, USA) supplemented with 2 mM *L*-glutamine, 15 mM HEPES and 2% fetal calf serum (HyClone, Thermo Scientific, USA) (referred hereafter to as assay media) at concentration of 2×10^6 cells/ml and left for 30 min prior to use in assays. PMN fraction purity assessed by flow cytometry analysis with CD66b as a marker for neutrophils was routinely ≥95%. Population of monocytes determined as CD14-positive cells in PBMC fractions amounted as a rule to not more than 10%. Cell viability determined by trypan blue staining was no less than 97%.

4.1.1.3 Heat Treatment

PMN and PBMC in assay media were dispensed into polypropylene tubes (10^6 cells in 500 ml) and heated (heat shock, HS) in a constant-temperature water bath at 40 °C for 1 h, or at 41 °C for 1 h, or at 43 °C for 10 or 20 min. Then the treated cell samples and control samples (without HS) were incubated for a necessary time intervals (5 min, 30 min, 60 min, etc.) at 37 °C in CO_2 incubator (recover period). In some experiments, preliminary incubation of PMN with cycloheximide (100 μM, 30 min) was performed before HS.

4.1.1.4 HSP70 Immunolabeling

Intracellular HSP70 levels were determined by indirect immunofluorescent staining and flow cytometry analysis. For intracellular labeling the cells were fixed and permeabilized in DPBS containing 2% paraformaldehyde (Riedel-de Haen, Germany), 0.05% BSA and 0.05% Triton X-100 (Sigma-Aldrich, USA) at 37 °C for 15 min. The permeabilized cells were incubated with primary HSP70-specific monoclonal antibody BRM22 (Sigma-Aldrich, USA) recognizing both inducible Hsp70 and constitutive Hsc70, or Hsp70-specific antibody C92F3A-5 (Stressgen, Enzo Biochem, USA), or Hsc70-specific antibody 1B5 (Enzo Life Sciences, USA) for 30 min at RT and then stained with secondary sheep anti-mouse or anti-rat IgG Fab-fragments conjugated with PE (Sigma-Aldrich, USA) for 30 min at RT. Each stage of labeling was followed by two washes with DPBS containing 0.2% BSA and 0.1% Triton X-100. HSP70 intracellular levels were determined as means of fluorescence intensity (MFI) corrected for background fluorescence of the negative controls by the formula: $\left(\frac{MFI_{sample}}{MFI_{control}}\right)-1$, where MFI_{sample} is the mean fluorescence intensity of cells in samples, and $MFI_{control}$ is the fluorescence intensity of cells in samples treated only with secondary polyclonal antibodies.

In a number of experiments cell samples were stained for detection of cell surface HSP70. Indirect immunofluorescence staining of PMN and PBMC samples with primary antibodies was performed at 4 °C for 30 min in DPBS supplemented with 0.2% BSA and 0.05% NaN3. Different commercial antibodies (indicated above monoclonal antibodies to Hsp70 + Hsc70, Hsp70, Hsc70, and polyclonal anti-Hsp70/Hsc70 (Santa-Cruz Biotechnology), and elaborated by our group a panel of anti-HSP70 monoclonal antibodies (clones 2E4, 2E11, 2F3, 3C5, 6G2, 2E5) interacting with different epitopes of HSP70 molecule (Boyko et al. 2014)) were used as the primary antibodies. After double washing secondary fluorescent-labeled antibodies staining of the cell samples was performed in the same solution (30 min at 4 °C) followed by washing and flow cytometric analysis. The level of surface expression of HSP70 was calculated in MFI units, similar to the intracellular level of HSP70.

4.1.1.5 Flow Cytometry

Flow cytometry analysis was carried out on a FACSCalibur flow cytometer (BD Biosciences, USA) equipped with 488 and 640 nm lasers and an appropriate set of detectors and filters. Cytofluorimetric analysis was performed using separated PMN и PBMC fractions. In the case of mononuclear cell population, lymphocyte and monocyte subsets were discriminated by gating the cells using forward and side scatter cytogram. A minimum of 10,000 gated events was collected for each sample. Data were analyzed using CellQuest ver. 3.4 (BD Biosciences) and FlowJo version 7.6.5 flow cytometry analysis programs.

4.1.1.6 Western Blot Analysis

PMN and PBMC 6×10^6 cells per sample were harvested by centrifugation, washed one time with cold DPBS and treated with lysis buffer (20 mM Tris-HCl (pH 7.5), 150 mM NaCl, 1 mM Na_2EDTA, 1% Triton X-100, protease inhibitors cocktail (Thermo Scientific, USA)) in 200 μl for 5 min on ice. After the lysed cell sample centrifugation (10,000 g, 10 min) the supernatants were collected and protein concentration was determined in colorimetric assay using Bradford reagent. The protein samples obtained from equal number of cells were loaded into sodium dodecyl sulfate-polyacrylamide gels (10%), separated by electrophoresis and blotted to nitrocellulose membranes for the HSP70 Western blot analysis according the standard protocol (http://www.cellsignal.com/contents/resources-protocols/western-blotting-protocol/western). After the sample transference nitrocellulose membranes were stained with primary antibody BRM22 or C92F3A, anti-β-actin antibody AC-15 (Sigma-Aldrich, USA) and secondary HRP-conjugated antibody (Sigma-Aldrich, USA). The results were visualized using enhanced chemiluminescence western blotting reagents (Immun-Star HRP, Biorad, USA). The data were analyzed using ImageJ software (ver.1.38x Wayne Rasband, NIH, USA).

4.1.1.7 Real-Time PCR Procedure

Total RNA for each time point was extracted from $6x10^6$ cells of PMN or PBMC using TRIzol reagent (MRC, UK) according manufacture's instruction. Before procedure of reverse transcription all RNA samples were treated with DNaseI (Thermo Scientific, USA) to remove residual DNA and quantified (NanoVue Plus spectrophotometer, GE Healthcare Life Science, UK). Complementary DNA synthesis was performed using MINT Reverse Transcriptase (Evrogen, Russia) following the manufacturer's instructions. Real-time PCR primers (5′-3′) are listed below (gene: forward primer, reverse primer).

HSPA1A/B: AGGTGCAGGTGAGCTACAAG, CTCGGCGATCTCCTTCATC;
HSPA6: ACCCAGGTGTATGAGGGTGA, TCTATCTGGGGGACTCCACG;
HSPA8: TGCTGCTCTTGGATGTCACT, AAGGTCTGTGTCTGCTTGGT;
β-actin: CACCACACCTTCTACAATGAG, GTCTCAAACATGATCTGGGTC.

Each real-time PCR sample were prepared using qPCRmix-HS SYBR reaction mixture (Evrogen, Russia) according manufacture's instruction. The reactions repeated in triplicates were carried out using a LightCycler 480 Real-time PCR detection system (Roche Diagnostics, Deutschland GmbH), a detailed amplification protocol is described (Boyko et al. 2017). The fold-increase gene expression ratio was estimated as a ratio of normalized to b-actin *HSPA* mRNA levels in HS-treated cell samples to that in the intact samples.

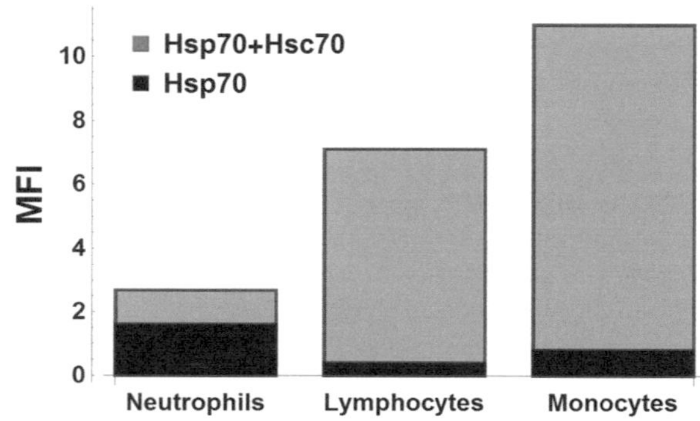

Fig. 4.1 The results of analysis of basal intracellular HSP70 (Hsc70 + Hsp70) and inducible Hsp70 levels in intact human granulocytes, lymphocytes, and monocytes measured by flow cytometry. The results are presented as an average of means of fluorescence intensity (MFI) of several independent experiments

4.1.1.8 Statistical Analysis

Statistical analysis was performed using Microsoft office Excel 2007 and SigmaPlot (ver. 11.0, Systat Software Inc.). Data are expressed as mean ± standard deviation. For data that followed a normal distribution t-tests were performed, for non-normally distributed data a Mann-Whitney *U* test was used. Results were considered statistically significant as * for $p < 0.05$.

4.1.2 Results

4.1.2.1 Intracellular HSP70 Level in Intact Mononuclear Cells and Granulocytes

In the first series of our experiments intracellular HSP70 levels and transcriptional activity of the *HSPA* genes were analyzed in intact human PMN (polymorphonuclear leukocytes, or granulocytes) and PBMC (peripheral blood mononuclear cells containing lymphocytes and monocytes). The results of cytofluorimetric analysis demonstrated that the total pool of intracellular HSP70 detected in granulocytes was more than two times lower than in mononuclear cells; there was no significant difference in the HSP70 content between lymphocytes and monocytes (Fig. 4.1). In contrast, the levels of stress-inducible Hsp70 were considerably higher in granulocytes compared to lymphocytes and monocytes. Thus, intact human PMN (predominantly neutrophils), despite the low total content of HSP70, express more stress-inducible Hsp70 compared with PBMC.

In next series of experiments we analyzed the basal transcriptional activity of *HSPA1A/B*, and *HSPA6* genes encoding stress inducible proteins Hsp70 and Hsp70B, respectively, as well as gene *HSPA8* of constitutive protein Hsc70. The basal *HSPA1A/B* expression that was found using qRT-PCR was higher in PMN compared with PBMC. Similar to *HSPA1A/B*, transcriptional activity of *HSPA6*, most strictly stress-inducible gene, was higher in PMN than in PBMC (Fig. 4.2).

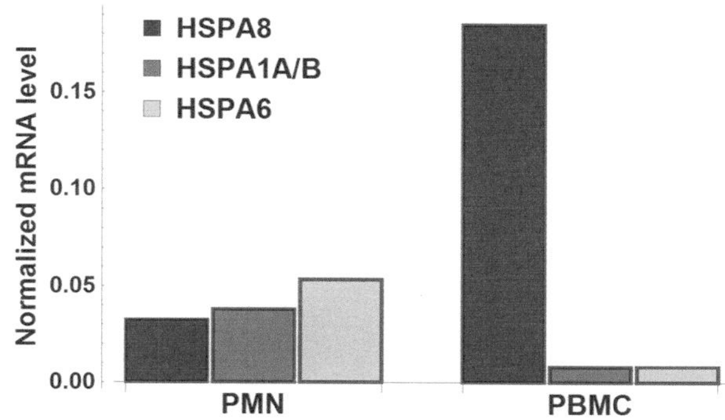

Fig. 4.2 Transcriptional activity of *HSPA* genes in human granulocytes and mononuclear cells analyzed by qRT-PCR; mRNA levels normalized by β-actin. Data were averaged for PMN and PBMC fractions isolated from three different donors

HSPA8 expression was higher in mononuclear cells than in granulocytes; furthermore, in PBMC *HSPA8* expression was considerably more intensive than *HSPA1A/B* or *HSPA6* expression. These results conform to the data described above on the intracellular content of total HSP70 and stress inducible Hsp70 in granulocytes and mononuclear cells (Fig. 4.1). Collectively, these data indicate significant differences in HSP70 expression between the leukocyte populations in healthy humans.

4.1.2.2 Effect of Heat Shock on Transcriptional Activity of *HSPA* Genes in PMN and PBMC

In these experiments two protocols of cell hyperthermia *in vitro* were applied: mild prolong HS (40 °C, 60 min) and sever short-term HS (43 °C, 10 min). Transcriptional activity of *HSPA* genes in the cells subjected to hyperthermia was evaluated for stress induced *HSPA1A/B* and *HSPA6* genes of Hsp70 and Hsp70B proteins, respectively, and for the *HSPA8* gene of the constitutive Hsc70 protein. The accumulated mRNA level of the targeted *HSPA* gene was normalized against house-keeping gene β-actin to correct the sample-to-sample variation.

The results demonstrate that for PMN induction of the *HSPA* mRNA synthesis occurs in response to sever short-term HS (43 °C, 10 min) and practically is absent for the heat treatment at 40 °C, 60 min. The mRNA accumulation arises for both stress-induced genes *HSPA1A/B*, *HSPA6* and the gene of constitutive protein *HSPA8* with a maximum level between 0.5 and 3 h after the end of heating and with the decrease of transcriptional activity to baseline values in about 4 h after the end of the exposure (Fig. 4.3).

The similar protocols were used to analyze the effect of HS on transcriptional activity of *HSPA* genes in mononuclear cells. Like in PMN, in PBMC more pronounced induction of the transcriptional activity was observed under sever short-term HS (43 °C, 10 min). However, heating of PBMC samples at 40 °C for 60 min induced the activation of the *HSPA1A/B* transcription, which was not observed in PMN (Fig. 4.4).

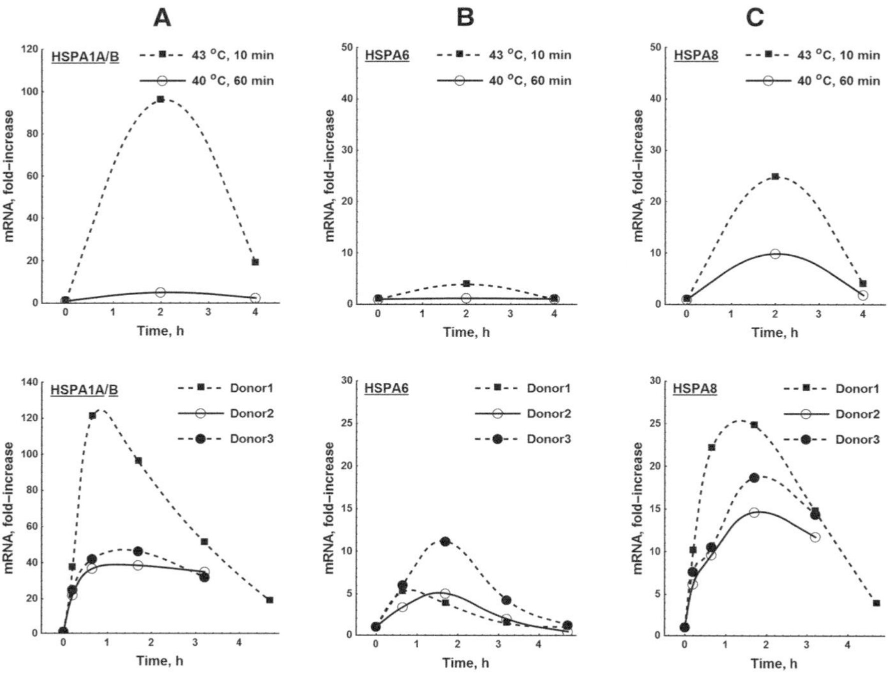

Fig. 4.3 Transcriptional activity of *HSPA* genes in human granulocytes in conditions of hyperthermia were determined by qPCR using β-actin as a reference. Transcription levels of *HSPA1A/B* (**a**), *HSPA6* (**b**) and *HSPA8* (**c**) mRNA synthesis were compared after the cell treatment with two protocols of hyperthermia (40 °C, 60 min, and 43 °C, 10 min) in dynamics (before HS (0 h), 0.2, 0.5, 2, 3 and 5 h after termination of HS). The data were obtained with cell samples from three different donors in independent experiments

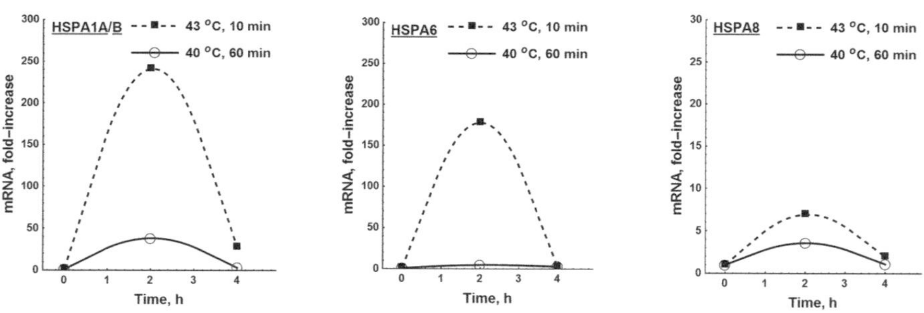

Fig. 4.4 Dynamics of transcriptional activity of *HSPA* genes (*HSPA1A/B*, *HSPA6*, and *HSPA8*) in human mononuclear cells induced by HS (40 °C, 60 min, and 43 °C, 10 min); 0 h - before HS, 2 h and 4 h - after termination of HS. The data were obtained in cell fractions isolated from three different donors in independent experiments

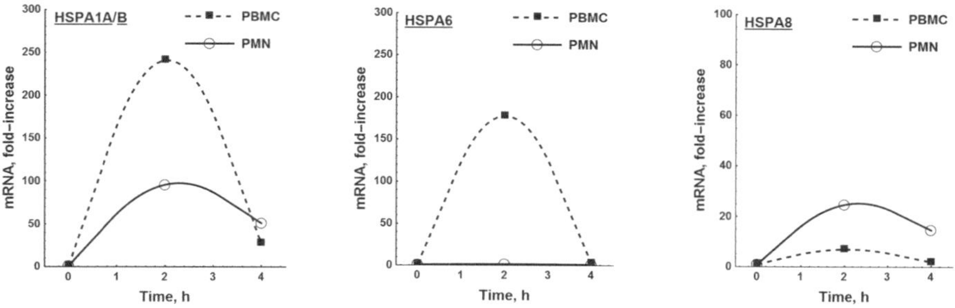

Fig. 4.5 The comparison of effect of HS (43 °C, 10 min) on dynamics (2 h and 4 h after HS) of transcriptional activity of *HSPA* genes in human granulocytes and mononuclear cells

The comparison of effect of HS on HSP70 expression in PMN and PBMC demonstrated a considerable difference in hyperthermia induced patterns of *HSPA* genes transcriptional activity in these types of cells. The response of stress-induced *HSPA1A/B* and *HSPA6* genes for sever short-term HS (43 °C, 10 min) was significantly stronger in mononuclear cells than in granulocytes (Fig. 4.5) despite of higher level of this gene expression in intact PMN in comparison with PBMC (Fig. 4.1). In contrast to the inducible HSP70, the HS-induced accumulation of the mRNA of the *HSPA8* gene was pronounced greater in PMN than in PBMC (Fig. 4.5), in spite of the prevalence of constitutive Hsc70 expression in intact mononuclear cells comparing to granulocytes (Fig. 4.1).

4.1.2.3 Dynamics of Intracellular Level of HSP70 in Granulocytes Subjected to Hyperthermia

It is known that stress-induced increase of intracellular HSP70 level occurs during the cell recovery period, the duration of which can vary from several minutes to several hours, depending on the cells and treatment conditions (Eid et al. 1987; Wang et al. 2003; Polla et al. 1995; Schneider et al. 2002). In our study the dynamics of HS-induced (43 °C, 10 min) alterations of intracellular level of HSP70 in granulocytes was analyzed (before HS, then every hour up to 7 h, and 15–17 h after the termination of HS) by flow cytometry and Western-blot using monoclonal antibodies to inducible and constitutive form of this protein.

As the results of this series of experiments, we did not register a significant HS-induced accumulation of total pool of intracellular HSP70 (Hsp70 + Hsc70), as well as of stress induced Hsp70, during 1–7 h and in 15 h after the termination of the HS (Fig. 4.6a, b). The obtained data did not conform to a considerable induction of *HSPA* transcription in PMN under the same protocol of HS demonstrated in this study (Fig. 4.3). However, immediately after the thermal treatment a significant increase of intracellular HSP70 level was detected by flow cytometry in granulocytes followed by a phase of this level decrease in 15–30 mins after HS (typical patterns of the dynamics are presented in Fig. 4.6**c**).

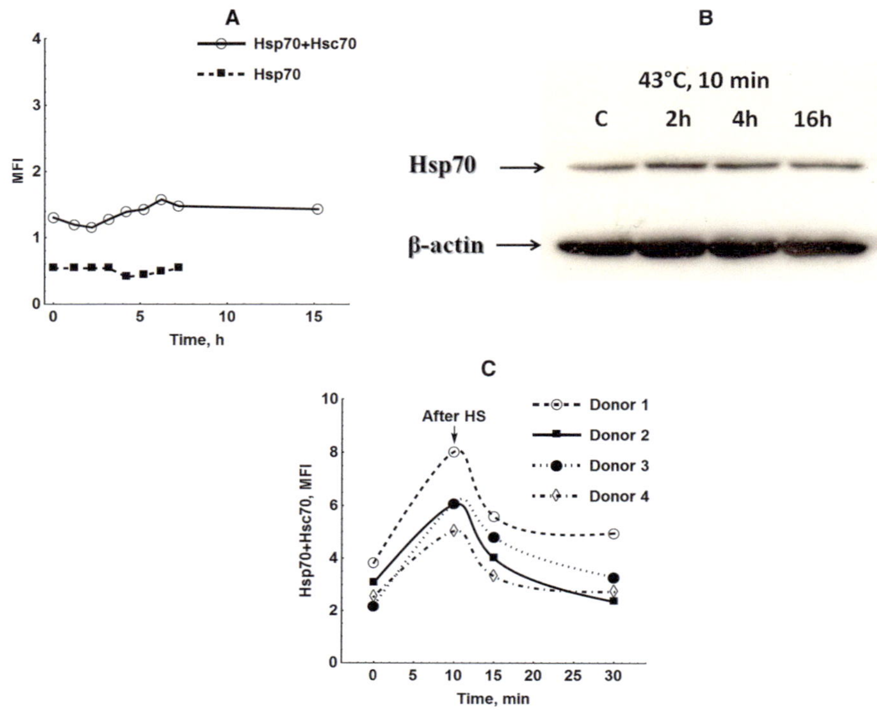

Fig. 4.6 Dynamics of HSP70 response to hyperthermia (43 °C, 10 min) in human neutrophils estimated by flow cytometry and Western-blot methods. (**a**) flow cytometric analysis of intracellular HSP70 (Hsp70 + Hsc70) and Hsp70 (inducible protein) levels before HS (point 0 h) and in response to heat treatment during 7 h, and in 15 h after HS termination. (**b**) Western-blot analysis of intracellular Hsp70 content in granulocytes; time intervals were counted from the start of the HS; (**c**) basal level of Hsp70 without hyperthermia. (**c**) flow cytometric analysis of kinetics of intracellular HSP70 (Hsp70 + Hsc70) before HS (0 min), immediately after the end of HS (10 min) and in 15–30 min after the HS; the data obtained in experiments with neutrophils from four donors

To find out which protein of HSP70 family (Hsp70 or Hsc70) can contribute to the HS-mediated two-phase dynamics of the intracellular protein level in neutrophils detected by flow cytometry analysis, the inducible (Hsp70) and constitutive (Hsc70) protein levels were measured separately in additional experiments. It was revealed that antibodies recognizing both stress induced Hsp70 and constitutive Hsc70, and antibodies specific to Hsp70 or Hsc70 detected the similar profile of the intracellular protein content immediately after HS (Fig. 4.7). Thus, the registered by flow cytometry HS-induced two-phase dynamics of HSP70 intracellular level was mediated by both constitutive Hsc70 and stress induced Hsp70.

It is possible to assume that the HS-induced rapid increase of intracellular HSP70 level could be associated with a real synthesis of these proteins. To check this assumption, a series of experiments was performed with application of cycloheximide as an inhibitor of protein synthesis. Pre-incubation of neutrophils with cycloheximide

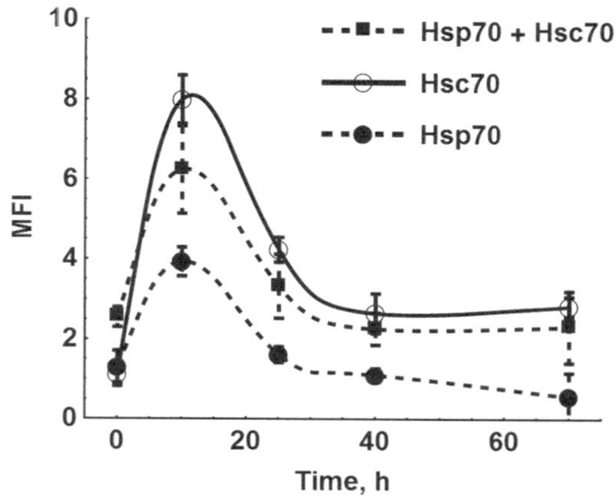

Fig. 4.7 Intracellular level of HSP70 in neutrophils before (point 0 min) and after HS (43 °C, 10 min), estimated by flow cytometry. Detection of the protein was carried out using monoclonal antibodies to Hsp70 (C92F3A-5) and Hsc70 (1B5), as well as with antibodies that recognize both these proteins (BRM-22)

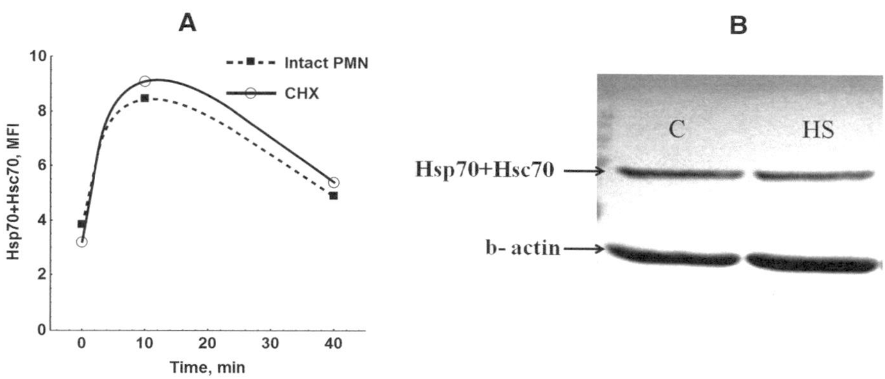

Fig. 4.8 Effect of the protein synthesis inhibitor cycloheximide (100 μM, 30 min) on intracellular level of HSP70 (Hsp70 + Hsc70), evaluated by flow cytometry (**a**) and Western-blot analysis (**b**) in human neutrophils. The representative data of two independent experiments

did not lead to inhibition of the increase of detectable intracellular HSP70 level after HS (Fig. 4.8a). The HS-induced changes of intracellular HSP70 content in neutrophils immediately after hyperthermia were analyzed also by Western blotting. In these experiments, we could not register an increase of intracellular HSP70 content after HS treatment (Fig. 4.8b). Thus, the increase of HSP70 level in neutrophils observed with flow cytometry immediately after thermal exposure is not associated with a real intracellular accumulation of these proteins due to protein synthesis *de novo*.

Another explanation of the HS-induced increase of intracellular HSP70 content in the neutrophils detected by flow cytometry immediately after HS might be connected with HS-caused modification of the protein conformation. To investigate a possible changes of conformation of intracellular HSP70 under HS-conditions we used a panel of six monoclonal anti-HSP70 antibodies elaborated earlier in our

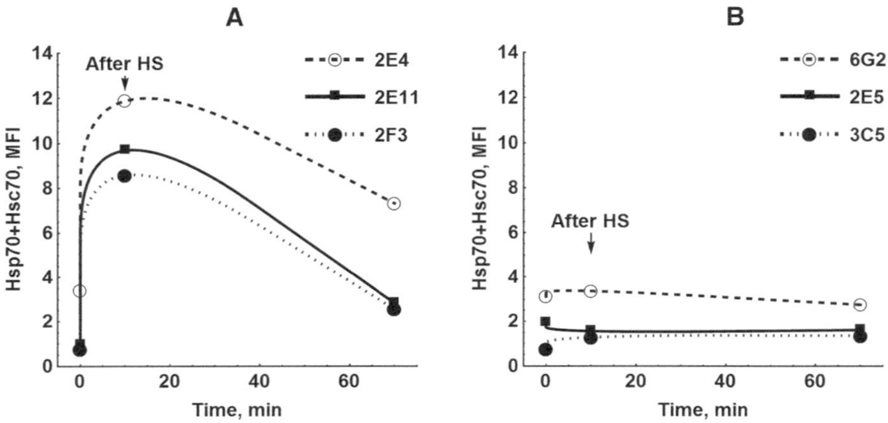

Fig. 4.9 Dynamics of intracellular level of HSP70 after heat shock (43 °C, 10 min) analyzed by flow cytometry using different HSP70-specific antibodies. Immunostaining was performed with antibodies interacting with C-terminal domain of HSP70 (**a**) and with N-terminal domain of HSP70 (**b**). The results are presented from three independent experiments

group. This panel included three antibodies (clones 2E4, 2E11, 2F3) specific to epitopes in the substrate-binding COOH-terminal domain of HSP70 molecule; another three clones (6G2, 2E5, 3C5) are specific to the sites of the conserved NH2-terminal nucleotide-binding domain of the protein (Boyko et al. 2014).

The dynamics of the intracellular level of HSP70 in neutrophils in response to HS, detected by these two groups of antibodies was different. The antibodies 6G2, 2E5, 3C5 did not allow to detect the increase in intracellular HSP70 level immediately after the HS (Fig. 4.9a), while antibodies 2E4, 2E11, 2F3 did display this increase (Fig. 4.9b). The results of these experiments indicate that the increase of intracellular HSP70 level in neutrophils immediately after the thermal stress displayed by flow cytometry may be connected with HS-caused increase of accessibility of epitopes within the substrate-binding domain of HSP70 molecule for the detecting antibodies.

4.1.2.4 Evaluation of HS-Induced Surface Membrane-Associated HSP70 in Human Leukocytes

It is known that stress conditions can induce translocation of intracellular HSP70 to the cell surface (Multhoff 2007). To evaluate a possible effect of HS (43°C, 10 min) on cell surface HSP70 expression in our model we used flow cytometry and all types of the antibodies applied to the study. As a result, all types of monoclonal antibodies used in our study failed to detect HS-induced surface localization of HSP70 in neutrophils. However, a significant effect of HS on surface expression of HSP70 on neutrophils was demonstrated in our model using anti-HSP70 polyclonal antibodies (Fig. 4.10a). In contrast to neutrophils, the most types of monoclonal

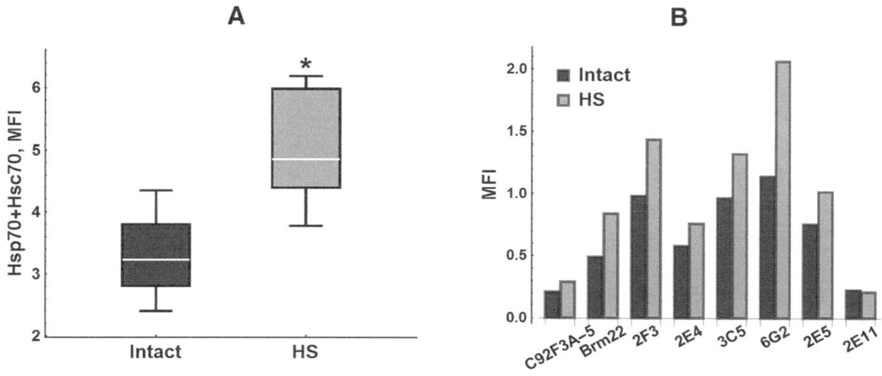

Fig. 4.10 HS-induced (43°C, 10 min) cell-surface expression of HSP70 on neutrophils and monocytes detected by flow cytometry using different antibodies. (**a**) staining of neutrophils with polyclonal antibodies to HSP70 (anti-Hsp70/Hsc70, Santa-Cruz Biotechnology). (**b**) staining of monocytes with different monoclonal antibodies to HSP70

antibodies applied in our study allowed to detect HS-induced cell surface HSP70 expression in population of monocytes (Fig. 4.10b). No effect of HS on surface expression of HSP70 was observed in population of lymphocytes.

4.1.3 Discussion

Intracellular HSP70 proteins are involved in the normal functioning of the cell and assist its survival in adverse conditions, performing some protective functions. The amount of HSP70 in the cell reflects the effectiveness of cell protection, and is increased with accumulation of intracellular damaged proteins. On the other hand, in some pathological conditions associated with stress, accumulation of damaged, improperly folded and aggregated proteins can occur due to impaired cell reparation system. It indicates a possibility that analysis of stress-induced HSP70 expression in cell samples, including immunocompetent cells, from patients may give some considerable diagnostic information. In the present study we investigate the features of stress-induced changes in the expression of HSP70 in populations of mononuclear and granulocytic cells from human peripheral blood using PCR, Western-blot and flow-cytometric analysis.

It is known that in the site of inflammation, where immunocompetent cells migrate, these cells exposed to oxidative stress and hyperthermia. To analyze effects of hyperthermia on the expression of HSP70 in mononuclear cells and granulocytes, we subjected the samples of these cells isolated from peripheral human blood to heat shock (HS). An increase in the transcriptional activity of the *HSPA* genes was observed upon exposure the cells to short-term HS (43 °C, 10 min). In these experiments the mRNA level was increased after HS not only for the induced (Hsp70), but also for the constitutive (Hsc70) form of HSP70 (Figs. 4.3, 4.4). The ability of neutrophils to rapidly change the expression of the protein genes of both Hsp70 and Hsc70 in response to inflammatory stimuli was previously shown (Zhang et al. 2004). Our results demonstrated similar ability for HS response of neutrophils. At the same time, despite an essential increase of transcriptional activity of the *HSPA* genes in response to cell heating we could not detect significant induction of HSP70 synthesis in our model. It is possible that for such short-lived cell population as neutrophils, a significant accumulation of HSP70 proteins is not typical. It is known that the post-transcriptional activity of Hsp70 expression in response to HS is regulated by the stability of mRNA (Theodorakis and Morimoto 1987). A phenomenon has also been described where HS-induced activation of Hsp70 transcription factor HSF1 did not initiate the expression of this protein (Mathur et al. 1994).

In our model of HS-induced cell stress a two-phase dynamics of intracellular level of HSP70 was observed. The dynamics was detected using flow cytometry and was characterized by an increase in the HSP70 level immediately after HS and a subsequent decrease in this level within 15–30 min after the completion of the thermal stress (Fig. 4.6). It was also shown that both constitutive and inducible forms of the protein contributed to the two-phase dynamics of HSP70 (Fig. 4.7). Using cycloheximide as inhibitor of protein synthesis we demonstrated that registered by flow cytometry HS-induced rapid increase of intracellular HSP70 level was not associated with a real intracellular accumulation of these proteins due to protein synthesis *de novo* (Fig. 4.8).

Additionally, using antibodies that recognize either the N-terminal or the C-terminal domain of the HSP70 molecule, we have displayed different profiles of the stress-induced dynamics of intracellular HSP70 detected with these antibodies. In particular, the two-phase dynamics was registered by antibodies specific to C-terminal domain of the HSP70, which was not observed in the case of the antibodies to N-terminal domain (Fig. 4.9). The data demonstrated a connection of the registered two-phase dynamics with C-terminal region of HSP70 molecule indicate HS-induced conformational changes of intracellular HSP70, which lead to an increase in the accessibility of certain epitopes of the substrate-binding domain of this protein for the detecting antibodies. The conformational changes observed in HSP70 molecule during its interaction with other proteins have been described (Zhuravleva et al. 2012). HSP70 proteins practically do not exist *in vivo* in the substrate-free state; they are in interaction with proteins, nucleotides, membranes, and also form auto-aggregates (dimers, trimers, oligomers). All types of interactions mentioned above are realized mainly by substrate-binding domain localized in C-terminal region of HSP70. Auto-aggregates of HSP70 do not possess chaperone activity, but with the transition of the proteins to the non-aggregated state, this activity is restored. The ratio of HSP70 oligomers/monomers in the cell depends on the temperature and is determined by the presence of denatured proteins (Angelidis et al. 1999). Concerning the results of our study, it can be assumed that the increase in the level of intracellular HSP70 detected immediately after HS (Figs. 4.6, 4.7) is due to dissociation of auto-aggregates of these proteins, or is related to amount of monomers of HSP70 released from the complexes with other proteins.

4.2 Conclusions

Thus, the results of this study indicate the importance of application a set of methods for the analysis of stress-induced changes in the intracellular content of heat shock proteins. In particular, the data obtained in our work with this approach show that the heat shock induced short-term increase in the intracellular level of HSP70 in neutrophils measured by flow cytometry is not associated with *de novo* protein synthesis, but is due probably to temporal conformational changes in the molecule, accompanied by augmentation accessibility epitopes of the HSP70 substrate-binding domain to specific antibodies. Our results also demonstrated that the use of monoclonal antibodies directed to different domains of HSP70 for flow-cytometric analysis of the dynamics of intracellular level of this protein might allow evaluation of the features of stress-induced modification of intermolecular interactions of HSP70 molecular chaperones.

Acknowledgements This work was supported by Russian Science Foundation, grant # 16-15-10404.

References

Angelidis CE, Lazaridis I, Pagoulatos GN (1999) Aggregation of hsp70 and hsc70 in vivo is distinct and temperature- dependent and their chaperone function is directly related to non-aggregated forms. Eur J Biochem 259:505–512

Boyko AA, Vetchinin SS, Sapozhnikov AM, Kovalenko EI (2014) Changes in the heat shock 70 kDa protein level in human neutrophils induced by heat shock. Russ J Bioorg Chem 40:488–498

Boyko AA, Azhikina TL, Streltsova MA, Sapozhnikov AM, Kovalenko EI (2017) HSP70 in human polymorphonuclear and mononuclear leukocytes: comparison of the protein content and transcriptional activity of HSPA genes. Cell Stress and Chaperones 22:67–76

Calderwood SK, Murshid A, Prince T (2009) The shock of aging: molecular chaperones and the heat shock response in longevity and aging. Gerontology 55:550–558

Cuanalo-Contreras K, Mukherjee A, Soto C (2013) Role of protein misfolding and proteostasis deficiency in protein misfolding diseases and aging. Int J Cell Biol 2013:638083

Eid NS, Kravath RE, Lanks KW (1987) Heat-shock protein synthesis by human polymorphonuclear cells. J Exp Med 165:1448–1452

Ferat-Osorio E, Sánchez-Anaya A, Gutiérrez-Mendoza M, Boscó-Gárate I, Wong-Baeza I, Pastelin-Palacios R, Pedraza-Alva G, Bonifaz LC, Cortés-Reynosa P, Pérez-Salazar E, Arriaga-Pizano L, López-Macías C, Rosenstein Y, Isibasi A (2014) Heat shock protein 70 down-regulates the production of toll-like receptor-induced pro-inflammatory cytokines by a heat shock factor-1/constitutive heat shock element-binding factor-dependent mechanism. J Inflamm (Lond) 11:19

Kovalenko EI, Boyko AA, Semenkov VF, Lutsenko GV, Grechikhina MV, Kanevskiy LM, Azhikina TL, Telford WG, Sapozhnikov AM (2014) ROS production, intracellular HSP70 levels and their relationship in human neutrophils: effects of age. Oncotarget 5:11800–11812

Mathur SK, Sistonen L, Brown IR, Murphy SP, Sarge KD, Morimoto RI (1994) Deficient induction of human hsp70 heat shock gene transcription in Y79 retinoblastoma cells despite activation of heat shock factor 1. Proc Natl Acad Sci U S A 91:8695–8699

McLeod IX, Jia W, He YW (2012) The contribution of autophagy to lymphocyte survival and homeostasis. Immunol Rev 249:195–204

Multhoff G (2007) Heat shock protein 70 (Hsp70): membrane location, export and immunological relevance. Methods 43:229–237
Muralidharan S, Mandrekar P (2013) Cellular stress response and innate immune signaling: integrating pathways in host defense and inflammation. J Leukoc Biol 94:1167–1184
Njemini R, Abeele MV, Demanet C, Lambert M, Vandebosch S, Mets T (2002) Age-related decrease in the inducibility of heat-shock protein 70 in human peripheral blood mononuclear cells. J Clin Immunol 22:195–205
Polla BS, Stubbe H, Kantengwa S, Maridonneau-Parini I, Jacquier-Sarlin MR (1995) Differential induction of stress proteins and functional effects of heat shock in human phagocytes. Inflammation 19:363–378
Schneider EM, Niess AM, Lorenze I, Northoff H, Fehrenbach E (2002) Inducible hsp70 expression analysis after heat and physical exercise: transcriptional, protein expression, and subcellular localization. Ann N Y Acad Sci 973:8–12
Tan H, Xu Y, Xu J, Wang F, Nie S, Yang M, Yuan J, Tanguay RM, Wu T (2007) Association of increased heat shock protein 70 levels in the lymphocyte with high risk of adverse pregnancy outcomes in early pregnancy: a nested case-control study. Cell Stress Chaperones 12:230–236
Theodorakis NG, Morimoto RI (1987) Posttranscriptional regulation of hsp70 expression in human cells: effects of heat shock, inhibition of protein synthesis, and adenovirus infection on translation and mRNA stability. Mol Cell Biol 7(12):4357–4368
Wang S, Diller KR, Aggarwal SJ (2003) Kinetics study of endogenous heat shock protein 70 expression. J Biomech Eng 125:794–797
Zhang X, Kluger Y, Nakayama Y, Poddar R, Whitney C, DeTora A, Weissman SM, Newburger PE (2004) Gene expression in mature neutrophils: early responses to inflammatory stimuli. J Leukoc Biol 75:358–372
Zhuravleva A, Clerico EM, Gierasch LM (2012) An interdomain energetic tug-of-war creates the allosterically active state in Hsp70 molecular chaperones. Cell 151:1296–1307

Part II
Oxidative Stress

Chapter 5
Relationship between Oxidative Stress-Induced Effects and Physical Exercise

Hyunseok Jee

Abstract This chapter focuses on various agents that induce intracellular oxidative stress within the cell environment, the roles of heat shock proteins (HSP) within such harsh cellular environments, and the relationship between HSP and the exercises used to alleviate the negative effects of oxidative stress in the living body. The keywords used in this study, such as the origin of oxidative stress, HSP, and exercise, will provide scientists insights into the optimized exercise-induced adaptation for minimizing the pathological effects of oxidative stress. This study can provide important clues to design optimized exercise programs, which can produce maximal effects against the cellular damages caused by oxidative stress.

Keywords Antioxidants · Exercise · Heat shock protein · Oxidative stress · Stress

Abbreviations

AMPK	AMP-activated protein kinase
ARE	Antioxidant response elements
CAT	Catalase
CSF	Cerebrospinal fluid
GSH	GSH-reductase
GSTr	Glutathione-s-transferase
HO1	Heme oxygenase −1
HSE	Heat shock element
HSF	Heat shock factor
HSP	Heat shock protein
LDH	Lactate dehydrogenase
mTORC 1	Mammalian target, the rapamycin complex 1
NFkB	Nuclear transcriptional factor kappa B

H. Jee (✉)
Frontier Research Institute of Convergence Sports Science (FRICSS), Yonsei University, Seoul, Republic of Korea

A. A. A. Asea, P. Kaur (eds.), *Heat Shock Proteins and Stress*,
Heat Shock Proteins 15, https://doi.org/10.1007/978-3-319-90725-3_5

NO	Nitric oxide
Nrf2	Nuclear factor-erythroid 2-related factor 2
PRXs	Peroxiredoxins
RNS	Reactive nitrogen species
ROS	Reactive oxygen species
Sesns	Sestrins
UV	Ultraviolet

5.1 Introduction

According to Sies H's study in 1985, oxidative stress is indicated by the transformation of the stable intracellular components into their oxidative forms by the intracellular redox system (Sies 1991). Many oxidative components showing the 'redox state', such as (high levels of glutathione converted to) glutathione disulfide, vitamin E, etc., in the cellular environment are considered markers of the level of oxidation. Oxidative stress is the result of numerous sequential events involving intra- and extra-cellular chemicals, with the subsequent metabolites derived from various toxic drugs, excessive energy consumption by excessive physical activity, lactate dehydrogenase (LDH) secreted by cells into the extracellular environment, etc. (Brune 1991).

Many studies attempted to find artificial ways of inducing oxidative stress from the aforementioned sequential events. These ways included irradiation with ultraviolet (UV) rays, and application of hydroperoxide, 8-hydroxydeoxyguanosine, hydrogen peroxide & reactive oxygen radicals, and mitochondria-derived superoxide. Antioxidants such as glutathione and glutathione transferase not only counter the effects of oxidative stress, but are also derived from agents that cause oxidative stress; antioxidants have also been used to study the effects of oxidative stress.

Heat shock proteins (HSP) have two release mechanisms: passive and active; these mechanisms seem to rely on the intensity of stimulation of the HSP. A related report by Alexzander AA Asea stated that trauma, or an infected state, such as necrosis or infection, is relevant to the passive release mechanism of HSP, and conditions such as inflammation, exercise, psychological stress, UV radiation, and the state of the brain and cerebrospinal fluid (CSF) are associated with the active release mechanism of HSP (Asea 2007). This study also indicates that free HSP are released into the exosome (the passive mechanism of HSP), and the heat shock factor (HSF) binds to the cyto-nucleus of the heat shock element (HSE) to synthesize and release HSP (as a danger signal messenger) via inflammatory mediators (e.g., IL-10, IL-12) and receptors (e.g., Toll-like receptors), as intermediating agents (Fig. 5.1).

There is a controversy about whether the intracellular expression of HSP should be considered as favorable or not (Bonorino and Souza 2007). Bonorino et al. claimed that there is a dilemma regarding the general role of HSP in tumors, owing to their role in increasing the survival rate of tumors, subsequently inducing their proliferation. Additionally, there are two possible outcomes of the oncogenic Ras pathway: one, the carcinocyte progresses towards cancer proliferation, and two, the carcinocyte progresses towards senescence and not towards cancer metastasis via

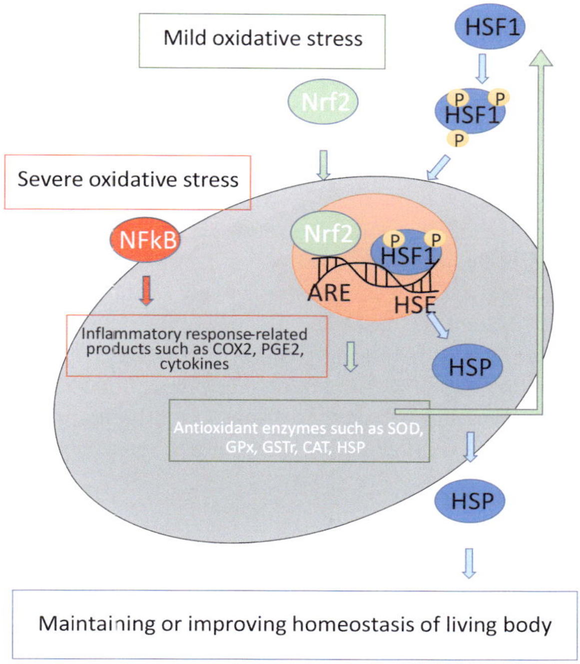

Fig. 5.1 HSP production modified from (Sagai and Bocci 2011; Jee 2016). HSF1 is activated by phosphorylation and is translocated into the nucleus to bind to HSEs. This process provides various types of HSP to maintain or improve homeostasis of the living body

p53 to p21 Chang et al. 1999a,b; Chang et al. 1999a,b; Sherman and Yaglom 2007). There is also an unexpected case wherein an elevated level of HSP is involved in the first case (promoting cancer metastasis), inhibiting the pathway to p53.

Studies regarding HSP in cancer make readers unsure about whether the role of HSP is negative or positive. The following study provides information on this query, and suggests that the expression of HSP increases due to various intracellular stressors. This increased expression is associated with a protective role, i.e. reinforcing the resistance against cellular damage, consequently acting as a regulating mechanism, and interfering with cellular apoptosis. Furthermore, there is a paucity of evidence about the fact that increased HSP accumulation by oxidative stress-inducing agents such as reactive oxygen species (ROS) leads to destined cell death (Samali and Orrenius 1998).

Based on the aforementioned studies about HSP, this review aims to provide updated information about the role of HSP levels in oxidative stress environments in cells and the relationship between these levels and physical activities.

5.1.1 8-Hydroxydeoxyguanosine, HSP and Exercise

ROS are controlled by scavenging systems such as superoxide dismutase, peroxiredoxins (PRXs), and glutathione-related antioxidant defenses. However, ROS attain an imbalanced redox state; they produce oxidative stress, which increases the damage to DNA, which in turn leads to the production of 8-hydroxydeoxyguanosine when the ROS exceed the capacity of the scavenging systems (Sova et al. 2010).

8-hydroxydeoxyguanosine is used as a biomarker for genetic instability or alteration (DNA damages) by oxidative stress. However, it also prevents colitis-associated carcinogenesis via inhibiting allergy-induced inflammation & TNF-alpha, regulating oxidative stress-induced gastritis, and boosting the anti-inflammatory action; thus, its role is controversial (Ock et al. 2012). Markedly higher 8-hydroxydeoxyguanosine levels were also shown in aged, sedentary mice and *ad libitum*-fed sedentary mice when compared to that in normal, young mice and mice from the exercise group, respectively. Another related study conducted in humans showed that DNA damages occurred not only in aged, but also in obese people. However, it was less damaged in lean runners whose running intensities were mild-to-moderate. In high-intensity exercises, induced oxidative stress provoked cellular protection and subsequently needed a protein turnover, which induced a high HSP expression (HSP25 in the liver), as a defense mechanism (Huffman et al. 2008).

5.1.2 Radicals, HSP and Exercise

Muscle contraction, a functional property of the living body, produces free radicals that can cause various diseases. Severe exercise causes hyperthermic conditions in molecular environments, which induces oxidative damages by the increased production of free radicals on a cellular level. A specific elevation in HSP72 expression is suggested to attenuate heat-induced protein degradations in these situations (Chang et al. 2007). Hamilton KL et al. also showed consistent results that exercise-induced oxidant production may stimulate the expression of HSP72, and that subsequent antioxidants attenuate exercise-induced HSP72 (Hamilton et al. 2003). This suggests that an increased expression of HSP may protect molecular substrate proteins that are degraded because of oxidative stress-derived radicals. Once this phenomenon continues, intracellular homeostasis would be sustainable, owing to increased HSP expression and the functions of antioxidants. Then, the expression of HSP would subsequently decrease. The expression of HSP seems to function according to the mechanism described above. Free radicals appear to act as activators of the adaptive response in contracting muscles, which are responsible for physical activities. This results in the increased production of antioxidants, such as heat shock proteins.

5.1.3 Hydrogen Peroxide (H_2O_2), HSP and Exercise

Physical activity-derived muscle contraction induces the production of free radicals, including reactive oxygen species (ROS) or reactive nitrogen species (RNS), which also generate superoxide and nitric oxide (NO). These primary radical species in the muscle fibers induce H_2O_2 production into the interstitial space (McArdle et al. 2001; Patwell et al. 2004; Vasilaki et al. 2006a,b; Jackson 2011). Hydrogen peroxide (which is also caused by muscle contraction) increases the pattern of oxidative damage, which causes cataract *in vivo* (Garner et al. 1982; Giblin et al. 1987; Spector 1991); it may also stimulate HSP synthesis (Ziemann et al. 2013). This H_2O_2 produced by exercise is also considered as a key regulator of muscle adaptation, involving antioxidants, and the chaperone defense systems of HSP70 and HSP27 (Powers et al. 2010; Ziemann et al. 2013).H_2O_2-derived changes due to physical activities on gene expression induce muscle contraction, and stimulate redox-related adaptive signaling factors such as NF-κB and Heat Shock Factor 1 (HSF1) (Cotto and Morimoto 1999), which are associated with the activities of cytoprotective proteins (e.g., possibly HSP) such as antioxidants (Vasilaki et al. 2006a,b). Activation of the redox signaling pathway by H_2O_2 is possibly a key regulator of muscle contraction-derived oxidative stress. The pathway is alleviated via the action of HSP (expressed by HSF1-related pathway stimulation by H_2O_2), as cellular homeostatic episodes.

5.1.4 UV Radiation-Derived Oxidative Stress, HSP, Exercise and Emerging Antioxidants

UV radiation is usually divided into the low-energy UVA radiation (320–400 nm) and the high-energy UVB radiation (290–320 nm). UVA causes oxidative stress, inducing the production of ROS, which damage the cell membrane, organelles, and DNA (Semenkov et al. 2015), consequently inducing photocarcinogenesis. UVB causes DNA-damaging sunburn that accelerates the aging process, finally causing skin cancer (Battie et al. 2014; Nishisgori 2015). AMP-activated protein kinase (AMPK)-regulating proteins called Sestrins (Sesns) have been recently discovered.; these proteins play a role in maintaining the cellular environment by reducing damages from free radical-induced oxidative stress via peroxiredoxins (PRXs). Physical activities behave like double-edged swords. On one hand, they have been shown to trigger oxidative stress, thereby inducing tissue damage, but on the other, they modulate a series of chemical reactions on a molecular level (e.g., Sesns are involved in the PRX-regenerating signal pathway), thereby reducing the damage from oxidative stress (Sanchis-Gomar 2013). Mlitz et al. reported that Sesn (type 2) is UVA or UVB specifically involved in its mammalian target, the rapamycin complex 1 (mTORC 1) (one of upstreaming activation cascade)-related pathway, and its expression is regulated via p53 suppression (Mlitz et al. 2016). The role of HSP70 in intracellular protection, such as neutralizing the negative effect of UV radiation-derived ROS from oxidative stress mechanisms (Njemini et al. 2007; Semenkov et al. 2015), has been known. However, only a few studies (as far as we know) have been performed about the relationship between HSP and the influence of Sesns on the oxidative stress caused by UV radiation. Considering their common role as antioxidants against UV radiation, an inter-modulating interaction between HSP and Sesns, both of which are involved as mTOR-Akt-regulated factors in the signaling pathway, can be suggested.

5.1.5 *HSP and Exercise as Antioxidants*

As mentioned previously, the extent of oxidative stress has an ambilateral dependence on the extent of exercise. Exercise seems to play completely opposite roles against oxidative stress: both an oxidant stress- causing and a protective effect-driving role. Fehrenbach and Northoff reported that regular exercise may have a protective effect on excessive exercise-induced DNA damage. HSP70, HSP27, or heme oxygenase −1 (HO1) are the repairing defense systems used in response to the excessive exercise-derived negative effect (Fehrenbach and Northoff 2001). Atalay et al. emphasized not only the exercise-induced upregulation of HSP as an antioxidant defense system during intensive exercise, which easily causes oxidative stress-induced micro-biological (endogenous) damages, but also the modality of exercise for ameliorating exercise induced-negative effects (endurance exercise induces HSP upregulation to offset the adverse effect of the diabetes-derived oxidative stress) (Atalay et al. 2004).

5.2 Conclusions

The HSP family has a common structural subset related to a chaperoning function that helps maintain the homeostasis of proteins under a stressful intracellular environment (Dimauro et al. 2016). We have thus provided the commonly overlapped amino acid sequences and 3D morphological figures of 3 randomly selected HSP (crystalline, HSP70, and HSP90) (Table 5.1, Fig. 5.2) because one of the commonly overlapped subsets can be a considerably indispensable chaperoning function-related part. This suggests that these sequences are possibly indispensable for the chaperoning function against the aforementioned oxidant stress-derived agents (e.g., 8-hydroxydeoxyguanosine). It also suggests that these common subset sequences are associated with the induction of activated HSP. For maintaining the optimal homeostatic state of the living body by inducing the activated HSP against oxidative stressors such as 8-hydroxydeoxyguanosine, it is necessary that physical activities are optimized for maximum stress-countering effects, by designing suitable exercise programs. Hence, to alleviate the negative effects of oxidative stress due to unfavorable, excessive physical activities, an appropriate exercise program with an optimized intensity, duration, and frequency should be designed on the basis of the knowledge about HSP.

Table 5.1 Sequential identity between human crystallin, HSP 70, and Hsp 90 proteins

HSP70 vs HSP90

```
40.7% identity in 27 residues overlap
HSP 70 Seq     503 SEENEEPMETDQNAKEEEKMQ-VDQEE
HSP 90 Seq     169 SDEEEEKKDGDKKKKKKIKEKYIDQEE
                   * * **    *   *   *    ****
```

Crystalline vs HSP70

```
45.5% identity in 11 residues overlap
Crystallin Seq  75 LDVKHFSPEDL
HSP 70     Seq 442 LEAYYSSPQDL
                   *     ** **

42.9% identity in 14 residues overlap
Crystallin Seq  64 VRSDRDKFVIFLDV
HSP 70     Seq 826 VPSDSDKKLPEMDI
                   * ** **     *

71.4% identity in 7 residues overlap
Crystallin Seq  40 LSSTISP
HSP 70     Seq 775 LTSTCSP
                   * ** **

40.0% identity in 10 residues overlap
Crystallin Seq  67 DRDKFVIFLD
HSP 70     Seq 644 DRNSFTLKLE
                   **  *   *

57.1% identity in 7 residues overlap
Crystallin Seq  25 QFFGEGL
HSP 70     Seq 356 KFFGKEL
                    ***  *

60.0% identity in 5 residues overlap
Crystallin Seq  94 VEIHG
HSP 70     Seq 484 VNVHG
                   *  **

57.1% identity in 7 residues overlap
Crystallin Seq  77 VKHFSPE
HSP 70     Seq 241 VNHFCEE
                   * **  *

42.9% identity in 7 residues overlap
Crystallin Seq  30 GLFEYDL
HSP 70     Seq 182 GIYKQDL
                   *    **

75.0% identity in 4 residues overlap
Crystallin Seq  94 VEIH
HSP 70     Seq 498 VEVH
                   ** *

50.0% identity in 6 residues overlap
Crystallin Seq  86 TVKVQD
HSP 70     Seq 649 TLKLED
                   * *  *

44.4% identity in 9 residues overlap
Crystallin Seq  18 YPSRLFDQF
HSP 70     Seq 454 YPDPAIAQF
                   **     **

80.0% identity in 5 residues overlap
Crystallin Seq  91 DDFVE
HSP 70     Seq  77 DPFVE
                   * ***
```

Table 5.1 (continued)

Crystalline vs HSP90

```
77.8% identity in 9 residues overlap
Crystallin Seq  75 LDVKHFSPE
HSP 90     Seq 230 LAVKHFSVE
                   * ***** *

40.0% identity in 15 residues overlap
Crystallin Seq  26 FFGEGLFEYDLLPFL
HSP 90     Seq 235 FSVEGQLEFRALLFV
                   *  **  *   * *

55.6% identity in 9 residues overlap
Crystallin Seq  64 VRSDRDKFV
HSP 90     Seq 128 VEKERDKEV
                   *   *** *

44.4% identity in 9 residues overlap
Crystallin Seq  94 VEIHGKHNE
HSP 90     Seq  54 VTVITKHND
                   *    ***

40.0% identity in 10 residues overlap
Crystallin Seq  61 ISEVRSDRDK
HSP 90     Seq 164 IEDVGSDEEE
                   *  * **

66.7% identity in 6 residues overlap
Crystallin Seq  18 YPSRLF
HSP 90     Seq 122 YPITLF
                   **  **

66.7% identity in 6 residues overlap
Crystallin Seq  15 GPFYPS
HSP 90     Seq 216 GEFYKS
                   * ** *

50.0% identity in 8 residues overlap
Crystallin Seq  95 EIHGKHNE
HSP 90     Seq 111 EIVKKHSQ
                   **  **

50.0% identity in 8 residues overlap
Crystallin Seq  63 EVRSDRDK
HSP 90     Seq 173 EEKKDGDK
                   *   * **

40.0% identity in 10 residues overlap
Crystallin Seq  28 GEGLFEYDLL
HSP 90     Seq   1 GIGMTKADLI
                   * *    **

62.5% identity in 8 residues overlap
Crystallin Seq  93 FVEIHGKH
HSP 90     Seq 413 FVERLRKH
                   ***   **
```

Evaluate the significance of the protein (crystalline vs HSP70 vs HSP90) sequence similarity score using PRSS (http://www.ch.embnet.org/software/PRSS_form.html). More than 40% homological identity is selected between compared HSPs.

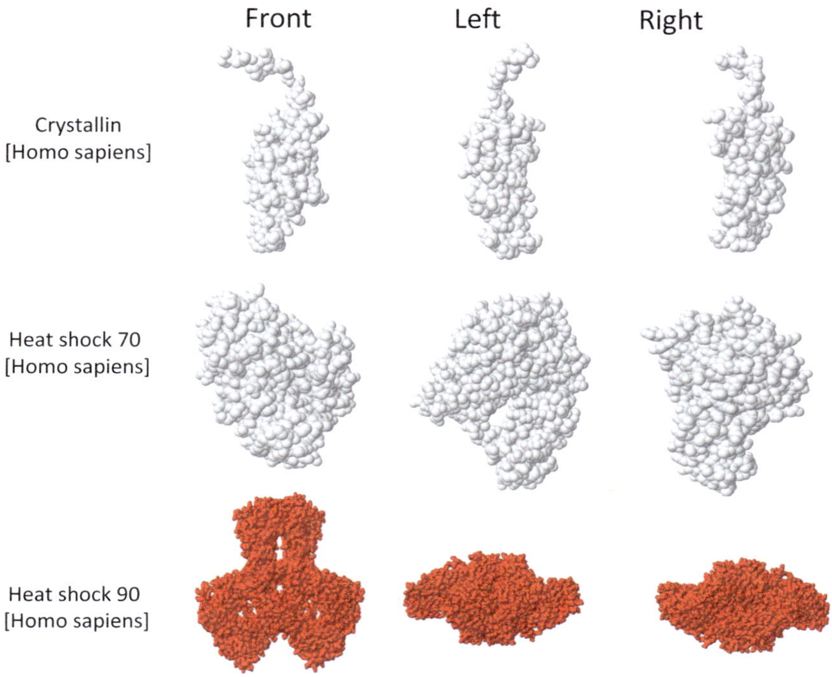

Fig. 5.2 The 3D morphologic view of crystallin, heat shock 70 and heat shock 90. The front, left, and right shape of human crystallin, heat shock 70, and heat shock 90 are shown. The source of human crystallin, heat shock 70, and heat shock 90 is from http://www.rcsb.org/pdb/explore/jmol.do?structureId=3L1G&opt=3&jmolMode=HTML5, http://www.rcsb.org/pdb/explore/jmol.do?structureId=4FSV&bionumber=1&view=symmetry and http://www.rcsb.org/pdb/explore/jmol.do?structureId=3Q6M&bionumber=1&view=symmetry respectively

Acknowledgements This study was supported by the Ministry of Education and the National Research Foundation of Korea (NRF-2015S1A5B8036349). I would like to thank God for helping me complete this article.

References

Asea AA (2007) Release of heat shock proteins: passive versus active release mechanisms. In: Heat shock proteins: potent mediators of inflammation and immunity Springer. Springer, Dordrecht, pp 3–20

Atlay M, Oksala NK, Laaksonen DE, Khanna S, Nakao C, Lappalainen J, Roy S, Hänninen O, Sen CK (2004) Exercise training modulates heat shock protein response in diabetic rats. J Appl Physiol 97:605–611

Battie C, Jutsukawa S, Bernerd F, Del Bino S, Marionnet C, Verschoore M (2014) New insights in photoaging, UVA induced damage and skin types. Exp Dermatol 23:7–12

Bonorino C, Souza AP (2007) Hsp70 in tumors: friend or foe? In: Heat shock proteins in cancer. Springer, pp 191–208

Brune B (1991) Oxidative stress in platelets. In: Oxidative stress, Oxidants and antioxidants, pp 421–443
Chang BD, Broude EV, Dokmanovic M, Zhu H, Ruth A, Xuan Y, Kandel ES, Lausch E, Christov K, Roninson IB (1999a) A senescence-like phenotype distinguishes tumor cells that undergo terminal proliferation arrest after exposure to anticancer agents. Cancer Res 59:3761–3767
Chang BD, Xuan Y, Broude EV, Zhu H, Schott B, Fang J, Roninson IB (1999b) Role of p53 and p21 waf1/cip1 in senescence-like terminal proliferation arrest induced in human tumor cells by chemotherapeutic drugs. Oncogene 18:4808
Chang CK, Chang CP, Liu SY, Lin MT (2007) Oxidative stress and ischemic injuries in heat stroke. Prog Brain Res 162:525–546
Cotto JJ, Morimoto RI (1999) Stress-induced activation of the heat-shock response: cell and molecular biology of heat-shock factors. *Biochem Society Symposia* JSTOR
Dimauro I, Mercatelli N, Caporossi D (2016) Exercise-induced ROS in heat shock proteins response. Free Radic Biol Med 98:46–55
Fehrenbach E, Northoff H (2001) Free radicals, exercise, apoptosis, and heat shock proteins. Exerc Immunol Rev 7:66–89
Garner M, Garner W, Spector A (1982) The effect of H2O2 on Na/K ATPase. Invest Ophthalmol Vis Sci 22:34
Giblin FJ, McCready JP, Schrimscher L, Reddy VN (1987) Peroxide-induced effects on lens cation transport following inhibition of glutathione reductase activity in vitro. Exp Eye Res 45:77–91
Hamilton KL, Staib JL, Philips T, Hess A, Lennon SL, Powers SK (2003) Exercise, antioxidants, and HSP72: protection against myocardial ischemia/reperfusion. Free Radic Biol Med 34:800–809
Huffman DM, Moellering DR, Grizzle WE, Stockard CR, Johnson MS, Nagy TR (2008) Effect of exercise and calorie restriction on biomarkers of aging in mice. Am J Physiol Regul Integr Comp Physiol 294:R1618–R1627
Jackson MJ (2011) Control of reactive oxygen, species production in contracting skeletal muscle. Antioxid Redox Signal 15:2477–2486
Jee H (2016) Size dependent classification of heat shock proteins: a mini-review. J Exerc Rehabil 12:255
McArdle A, Pattwell D, Vasklaki A, Griffiths R, Jackson M (2001) Contractile activity-induced oxidative stress: cellular origin and adaptive responses. Am J Physiol Cell Physiol 280:C621–C627
Mlitz V, Gendronneau G, Berlin I, Buchberger M, Eckhard L, Tschachler E (2016) The expression of the endogenous mTORC1 inhibitor sestrin 2 is induced by UVB and balanced with the expression level of sestrin 1. PLoS One 11:e0166832
Nishisgori C (2015) Current concept of photocarcinogenesis. Photochem Photobiol Sci 14:1713–1721
Njemini R, Bautmans I, Lambert M, Demanet C, Mets T (2007) Heat shock proteins and chemokine/cytokine secretion profile in ageing and inflammation. Mech Ageing Dev 128:450–454
Ock CY, Kim EH, Choi DJ, Lee HJ, Hahm KB, Chung MH (2012) 8-Hydroxydeoguanosine: not mere biomarker for oxidative stress, but remedy for oxidative stress-implicated gastrointestinal disease. World J Gastroenterol:WJG 18:302
Patwell DM, MrArdle A, Morgan JE, Patridge TA, Jackson MJ (2004) Release of reactive oxygen and nitrogen species from contracting skeletal muscle cells. Free Radic Biol Med 37:1064–1072
Powers SK, Duarte J, Kavazis AN, Talbert EE (2010) Reactive oxygen species are signalling molecules for skeletal muscle adaptation. Exp Physiol 95:1–9
Sagai M, Bocci V (2011) Mechanisms of action involved in ozone therapy: is healing induced via a mild oxidative stress? Med Gas Res 1:29
Samali A, Orrenius S (1998) Heat shock proteins: regulators of stress response and apoptosis. Cell Stress Chaperones 3:228
Sanchis-Gomar F (2013) Sestrins: novel antioxidant and AMPK-modulating functions regulated by exercise? J Cell Physiol 228:1647–16550

Semenkov VF, Michalski AI, Sapozhnikov AM (2015) Heating and ultraviolet light activate anti-stress gene functions in humans. Front Genet 6

Sherman M, Yaglom J (2007) Role of molecular chaperones in cell senescence. In: Heat shock proteins in cancer, pp 159–168

Sies H (1991) Oxidative stress: oxidants and antioxidants. *Academic Pr*

Sova H, Jukkola-Vuorinen A, Puistola U, Kauppila S, Karihtala P (2010) 8-Hydroxydeoxyguanosine: a new potential independent prognostic factor in breast cancer. Br J Cancer 102:1018

Spector A (1991) The lens and oxidative stress. In: Oxidative stress, oxidants and antioxidants, pp 529–558

Vasilaki A, Csete M, Pye D, Lee S, Palomero J, McArdle F, Van Remmen H, Richardson A, McArdle A, Faulkner J (2006a) Genetic modification of the manganese superoxide dismutase/glutathione peroxidase 1 pathway influences intracellular ROS generation in quiescent, but not contracting skeletal muscle cells. Free Radic Biol Med 41:1719–1725

Vasilaki A, McArdle F, Iwaneijko L, McArdle A (2006b) Adaptive responses of mouse skeletal muscle to contractile activity: the effect of age. Mech Ageing Dev 127:830–839

Ziemann E, Zembroñ-Lacny A, Kasperska A, Antosiewicz J, Grzywacz T, Garsztka T, Laskowski R (2013) Exercise training-induced changes in inflammatory mediators and heat shock proteins in young tennnis players. J Sports Sci Med 12:282

Chapter 6
Role of Heat Shock Proteins in Oxidative Stress and Stress Tolerance

Sumit Ghosh, Poulami Sarkar, Priyanka Basak, Sushweta Mahalanobish, and Parames C. Sil

Abstract Heat shock proteins (HSP) also referred to as stress proteins are a family of proteins produced by cells in response to exposure to stressful conditions like heat, cold, different kinds of environmental stress, such as infection, inflammation, exercise, exposure of the cell to toxins (ethanol, arsenic, trace metals, and UV light, among many others), starvation, hypoxia (oxygen deprivation), nitrogen deficiency (in plants), water deprivation and during wound healing or tissue remodeling. In this review, the authors have elucidated the role of heat shock proteins in stress tolerance both in eukaryotes and prokaryotes. Here, the role of heat shock proteins in survival during more extreme conditions and maintenance of normal cellular homeostasis have also been briefly discussed.

Keywords Heat shock element · Heat shock factor · Heat shock proteins · Oxidative stress · Stress tolerance

Abbreviations

ABA	Abscisic acid
AD	Alzheimer's disease
AHPND	Acute hepatopancreatic necrosis disease
ALS	Amyotrophic lateral sclerosis
aMCI	Amnestic mild cognitive impairment
AR	Amalaki Rasayana
Cd	Cadmium
Clp	Caseinolytic protease
COMT1	Caffeic acid O-methyltransferase1
Cpn	Chaperonin
DBMIB	2,5-dibromo-3-methyl-6-isopropyl-p-benzoquinone

S. Ghosh · P. Sarkar · P. Basak · S. Mahalanobish · P. C. Sil (✉)
Division of Molecular Medicine, Bose Institute, Kolkata, India
e-mail: parames@jcbose.ac.in

A. A. A. Asea, P. Kaur (eds.), *Heat Shock Proteins and Stress*,
Heat Shock Proteins 15, https://doi.org/10.1007/978-3-319-90725-3_6

ER	Endoplasmic reticulum
GolS	Galactinol synthase
GPx	Glutathione peroxidise
H_2O_2	Hydrogen peroxide
HD	Huntington's disease
HSC	Heat shock cognate
HSE	Heat shock elements
HSF	Heat shock factors
HSP	Heat shock proteins
HspA	Heat shock protein family A
HspB	Heat shock protein family B
MG	Methylglyoxal
miRNA	MicroRNA
MPK3	Mitogen-activated protein kinase
mRNA	Messenger RNA
NLHS	Non-lethal heat shock
NO	Nitric oxide
Nrf2	Nuclear factor (erythroid-derived 2)-like 2
PARK7	Parkinson disease protein 7
PD	Parkinson's disease
PEG	Polyethylene glycol
PSII	Photosystem II
RCS	Reactive carbonyl species
RNS	Reactive nitrogen species
ROS	Reactive oxygen species
Se	Selenium
sHsp	Small heat shock proteins
SOD	Superoxide dismutase
T2DM	Type 2 diabetes mellitus
UPR	Unfolded protein response

6.1 Introduction

Since the dawn of cellular life, all living organism in this world is constantly challenged by ever changing variables of the environment. The ability of the organism to successfully adapt to its changing environment is critical to its survival and likely represents an integral driving force in evolution. To cope up with different conditions of stress, organisms have been found to exhibit selective upregulation in the expression of a group of proteins called the heat shock proteins (Hsp). Hsp constitutes a large family of proteins that are usually classified based on their molecular weight: Hsp10, Hsp40, Hsp60, Hsp70, Hsp90, etc. Hsp have been reported to be involved in antigen presentation as chaperones in course of transfer of antigenic peptides to the class I and class II molecules of the major histocompatibility complexes (Arrigo 1994; Li et al. 2002). A link between heat shock proteins and oxidative stress has been established in a wide array of research (Sørensen 2010). Though the exact mechanisms are yet to be elucidated, it is suggested that oxidative stress can be a principal determinant of thermal tolerance (Abele et al. 2001; Pörtner 2001). The following chapter will inspect the role of heat shock proteins in oxidative stress and stress tolerance (especially thermal tolerance).

6.1.1 Heat Shock Proteins: An Overview

Heat shock proteins are ubiquitous proteins which render universal protective mechanisms from prokaryotes to eukaryotes to sustain cellular function and homeostasis (Table 6.1) (Lindquist and Craig 1988). These proteins were first discovered in *Drosophila* (Ritossa 1964) where it was found that increase in temperature of a few degrees above the physiological range induced the synthesis of these proteins from their salivary glands. The term "heat shock protein" is a misnomer. This family of proteins was originally named due to their expression after heat exposure. But now it is evident that Hsp synthesis is not only due to hyperthermia but also triggered by a wide range of environmental as well as metabolic stress conditions like heavy metal ions, ethanol, nicotine, anoxia, ischemia, glucose deprivation, surgical stress and viral agents (Feige et al. 2013; Santoro 2000). Consequently, the term

Table 6.1 Cellular location and function of mammalian heat shock proteins

HSP family	Molecular Size (kDa)	Cellular location	Proposed function
Ubiquitin	8	Cytosol/ nucleus	Facilitates targeting and removal of proteins denatured by stress
HSP10	10	Mitochondria	Cofactor for Hsp60
Small HSP	20 to 30	Cytosol/ nucleus	Some may be responsible for regulating the cellular cytoskeleton, and others regulate vascular tone and vesselwall remodeling
HSP56	56	Cytosol	Binds and stabilizes the steroid hormone receptor complex
HSP60	60	Mitochondria	Molecular chaperone
HSP70 family:			
HSP 72	72	Cytosol/ nucleus	Protein folding, cytoprotection (stress inducible)
HSP 73	73	Cytosol/ nucleus	Molecular chaperone (constitutively expressed)
HSP75(mHSP70)	75	Mitochondria	Molecular chaperone
HSP78(GRP78)	78	ER	Molecular chaperone, cytoprotection
HSP90	90	Cytosol/ER/ nucleus	Regulation of steroid hormone receptors, protein translocation
HSP110	110	Cytosol	Protein folding

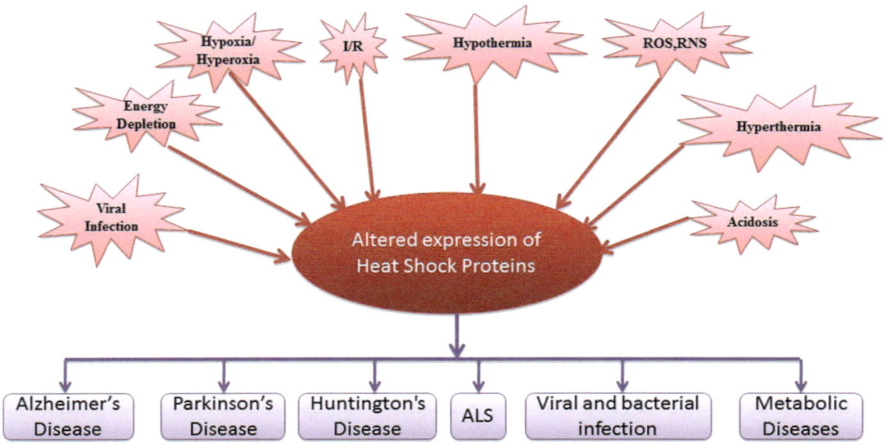

Fig. 6.1 Role of heat shock proteins in various conditions of stress

'stress protein' is preferably used (Whitley et al. 1999). Stress proteins belong to a multigene family and their nomenclature is given based on their molecular size (8-150 kDa). For example, Hsp 60, Hsp 70, Hsp 90 refers the family of heat shock proteins on the order of molecular size 60, 70 and 90 kDa respectively. Hsp are highly conserved and abundant in nearly all subcellular compartments. Many of these proteins show constitutive expression whereas expression level of others get increased under stressed condition (Fig. 6.1). In eukaryotic cells, heat shock factors (HSF) act as transcriptional activators for heat shock genes. There are different families of HSF that get activated under several stressed conditions. Among them, HSFs 1 and 3 function as stress responsive activators and show maximal heat shock responsiveness (Tanabe et al. 1998), whereas HSF 2 is activated during embryonic development and differentiation (Schuetz et al. 1991; Sistonen et al. 1992). HSF 4 seems to be preferentially expressed in the human heart, brain, skeletal muscle and pancreas (Nakai et al. 1997). HSF recognizes modular sequence elements referred to as HSE (a series of pentameric units arranged as inverted adjacent arrays of the sequence 5′-nGAAn-3′) located within the promoter region of heat shock gene (Amin et al. 1988; Perisic et al. 1989). In inactive stage, heat shock factors exist in monomeric form. But when activated, they trimerize into an active form that is capable of binding to the heat shock element (HSE) site of the stress protein gene and initiating transcription and translation of heat shock protein.

Stress proteins function as intracellular chaperones that permit the maturation and correct folding of most of the proteome. They play a crucial role in three dimensional folding of the newly translated polypeptide chain and assist to form proper protein conformation. In fact, they ensure that the newly synthesized polypeptide chain proceed correctly through successive folding and unfolding mechanism to achieve their functional conformation. They can recognize non-native conformation of the polypeptide and prevent their aggregation by binding to hydrophobic residues that have been exposed during translocation or stress induced damage (Pirkkala and Sistonen 2001). Upon exposure of extreme stress condition, the damaged protein loses its ability to be successfully refolded. Hsp promote ubiquitin mediated proteasomal degradation of these abnormal, unstable proteins. Furthermore, they mediate protein translocation across organellar membranes. It has been found that synthesis of these stress proteins under non-lethal temperature appears to protect the organism from upcoming high temperature or other stress that would be otherwise lethal. In the case of exposure to heat, this phenomenon has been called 'thermotolerance'. Clinical evidences suggest that stress protein synthesis under hyperthermic condition may provide potential safeguard against cardiac ischemia, arterial injury, endotoxic shock, renal and hepatic ischemia, ethanol-induced gastric ulcerations and skeletal muscle ischemia-reperfusion (Morimoto and Santoro 1998; Suzuki et al. 1997). Stress proteins play a pivotal role in the maintenance of cellular homeostasis by regulating immunity, cell cycle progression and apoptosis during development and differentiation. HSP have a significant role on adaptive immunity. During the time of antigen presentation, these molecular chaperons (HSP) associate with a wide range of antigenic peptides and facilitate the delivery of antigenic peptides to the class I and class II molecules of the major histocompatibility complexes. In addition, extracellular HSP can stimulate professional antigen presenting cells (macrophages, dendritic cells) to onset immune response (Li et al. 2002). Members of Hsp family especially Hsp 27 and Hsp 70 function as potent antiapoptotic molecules. Upon exposure to oxidative stress, Hsp 27 inhibits cytochrome c release where inducible Hsp70 prevents apoptosis by interfering with the signaling cascades and by impairing the activation of effector caspases (Pirkkala and Sistonen 2001).

6.1.2 Role of Heat Shock Proteins in Oxidative Stress

Heat stress activates HSFs and also lead to accumulation of reactive oxygen species (ROS), thereby inducing oxidative damage. The HSFs bind to the HSE of the promoter region of HSF, Hsp, ROS scavenging genes and miRNA 398. miRNA 398 downregulates SOD scavengers, thereby leading to rapid accumulation of ROS. This activates Hsp which in turn, activates the ROS scavengers to avoid cellular damage (Fig. 6.2) (Driedonks et al. 2015). A range of physiological stimuli (oxidative stress, hypothermia, hyperthermia, ischemia-reperfusion, hypoxia, acidosis etc.) activate HSFs, which then get separated from HSP. On phosphorylation by protein kinases, HSFs form trimers in the cytosol. The trimer complexes then enter the nucleus and bind to the HSE in the promoter region of the corresponding Hsp gene. Hsp mRNA gets transcribed and leaves the nucleus for the cytosol. Hsp usually function as molecular chaperones and assist in the assembly and translocation of newly synthesized proteins within the cell along with the repair and refolding of damaged proteins (Kregel 2002). Constitutively expressed housekeeping molecular chaperones maintain cellular redox status and integrity of mitochondria (cytosolic Hsp), protect

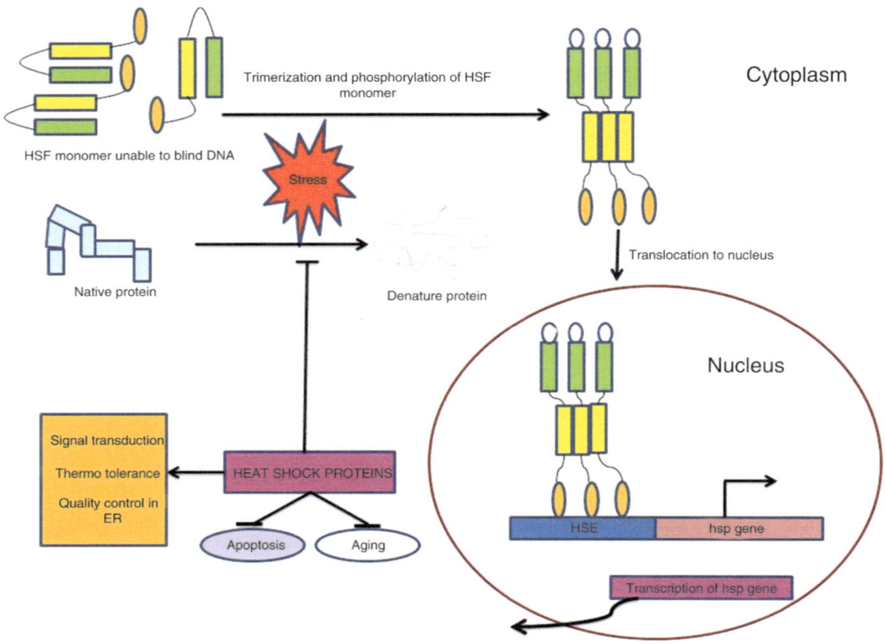

Fig. 6.2 Mechanism of action of heat sock proteins

and facilitate the assembly of respiratory complexes and facilitate their assembly (mitochondrial Hsp) (Kuzmin et al. 2004).

Selenium (Se), a dietary trace element exhibits antioxidant properties. Recent experiments on chicken liver have shown that Hsp play an important role in the oxidative stress induced damage during Se deficiency. It has been reported that expression of Hsp 27, 40, 60, 70, and 90 increase with increment in oxidative stress during Se deficiency. Hsp have been found to exert protective effects during the generation of oxidative stress due to Se deficiency (Liu et al. 2015). Studies have shown that accumulation of ROS in yeast (*Saccharomyces cerevisiae)* is associated with the induction of a heat shock reporter gene over a range of heat shock temperatures (33-44°C). ROS-mediated signaling ensures cooperation between heat shock and antioxidant responses (Moraitis and Curran 2004).

Hsp 70 protects cells from oxidative damage. They bind to stress exposed hydrophobic residues of denatured proteins, thereby preventing them from aggregating and promoting refolding. In the thermophile anaerobe (*Thermatoga maritima*), Hsp 70 exhibits redox sensitive binding to concerned peptides in response to oxidative stress. It interacts with an E3 ubiquitin ligase to transfer proteins towards proteolysis (Lüders et al. 2000; Rajkumar et al. 2015).

Most neurons contain high levels of Hsc70 and low levels of Hsp70 (Cruz et al. 1991). In motor neurons, both forms of Hsp are present, but no increase in endogenous expression is observed during heat shock. Extracellular Hsp70 inhibits apoptosis. Johnson *et al.* (Johnson et al. 1990; Johnson et al. 1995) have shown that Hsc/Hsp70 improves resistance to nutrient-deprivation stress when bound to arterial smooth muscle cells after being added to the culture medium *in vitro*. Similar observations have been found in case of cultured monocytes (Guzhova et al. 1998). Tidwell *et al.* have shown that *in vivo* administration of a mixture of Hsc/Hsp70 inhibits motor and sensory neuronal degeneration after sciatic nerve axotomy (Tidwell et al. 2004; Yu et al. 2001). Oxidative stress is characteristic of different protein misfolding associated diseases such as Alzheimer's disease (AD), Parkinson's disease (PD), Huntington's disease (HD) and Amyotrophic lateral sclerosis (ALS) (Andersen 2004; Rauchova et al. 2012; Yu et al. 2001). Oxidative stress disturbs calcium homeostasis, mitochondrial and cytoskeletal functions and can induce neuronal cell death. Thus, apoptosis have been reported in different neurodegenerative diseases (Friedlander 2003; Mattson 2000; Wyttenbach and Arrigo 2009). rhHsp70 protects motor neurons subjected to oxidative stress, common in neurodegenerative diseases such as AD, PD, ALS etc. (Robinson et al. 2005). The levels of different Hsp have been reported to increase in the brain of neurodegenerative disease patients (Brownell et al. 2012; Sun and MacRae 2005; Tóth et al. 2015; Wilhelmus et al. 2006). Several members of the HSPB family have been observed in the senile plaques and cerebral amyloid angiopathy of AD patients. HSPA1 have been found to colocalize with Aβ peptides and α-synuclein. HSPA8 has been found in intracellular inclusions in ALS (Muchowski and Wacker 2005). Overexpression of members of DNAJ protein family suppress the nuclear aggregation of mutant ataxin-1 and ataxin-3 thereby decreasing their toxicity in vitro model (Chai et al. 1999; Cummings et al. 2001). Overexpression of HSPA has been found to decrease polyglutamine induced neurodegeneration in *Drosophila* model of polyglutamine disease (Warrick et al. 1999).

ROS generation in the brain of AD patients lead to the increased expression of HSC 71, Hsp 27, HDJ 1 etc. (Najarzadegan et al. 2016). Poly(Q) mutation can generate ROS and is therefore prevented by Hsp 27 (Wyttenbach et al. 2002). HDJ-1 preserves mitochondrial function and proteasomal activity following oxidative injury (Ding and Keller 2001). Increased level of oxidation in AD brain is induced by HSC 71 (Castegna et al. 2002). Recent observations of the mechanism of induction of HSP and the decrease in the levels of Thioredoxin 1 in amnestic mild cognitive impairment (aMCI) brain suggest that alteration in the chaperone systems may contribute to the pathogenesis and progression of AD (Di Domenico et al. 2010).

Glial cells have been speculated to supply neighboring neurons with Hsp 70 by means of motor neurons during conditions of stress (Tytell et al. 1986). On the other hand, the HSPB protect cells from apoptosis and oxidative insult by binding to the cytoskeletal proteins (Tóth et al. 2015). HSPB decrease the level of ROS and regulate intracellular redox homeostasis, thereby protecting cytoskeleton, an important target for oxidative stress (Golenhofen et al. 2006; Mymrikov et al. 2011). HSPB 5/HSPB 2 deficiency affects cardiac function pointing towards their possible role in maintaining normal cardiac functions (Golenhofen et al. 2006; Morrison et al. 2004; Ray et al. 2001).

Ethanol consumption results in increased generation of ROS (Wu and Cederbaum 2003). The protective effects of HSP against ethanol induced toxicity have been reported in case of brain and hepatic cells (Russo et al. 2001). Vogt and coworkers (2001) have reported that metabolic stress caused by the high intensity activity, rather than stimulated hypoxia, is the primary cause of activation of Hsp 70. The accumulation of lactate and alterations of pH during intensive exercise act as the primary stimuli for Hsp expression (Mc Naughton et al. 2006). The heat shock protein 70–2 (*HSPA1B*) and 70-hom (*HSPA1L*) genes and their functional polymorphisms have been reported to play a role in the pathogenesis of type 2 diabetes mellitus (T2DM) (Umapathy et al. 2012). Toxic metals are associated with induction of oxidative stress. Accumulation of Cu^{2+} is associated with genetic disorders such as Wilson's and Menke's disease. Small Hsp (sHsp) such as Hsp27 and αB-crystallin bind to Cu^{2+} and inhibit Cu^{2+} mediated generation of ROS (Bakthisaran et al. 2015).

6.1.3 Role of Heat Shock Proteins in Stress Tolerance in Prokaryotes

Study of Sakthivel et al. have shown that HSPA and small heat shock proteins in general, play an important role in oxidative stress tolerance and help to stabilize membrane proteins like the photosystems and phycobilisomes in the cyanobacterium. In this study, both wild type and an *hspA* deleted strain from *Synechocystis* sp. PCC 6803 were being compared in the presence of methyl viologen or peroxide stress. The results showed that the cell growth, viability and pigment contents in thylakoid membranes were markedly reduced in the mutant even after enhancing the expression of other molecular chaperones. But *Synechococcus elongatus* strain ECT16–1, which constitutively expresses HSPA showed a much better cell growth and viability. The reduction of pigment contents was also arrested after 24 h in this strain (Sakthivel et al. 2009). It has been reported that *Salmonella typhimurium* gets adapted to oxidative stress when pre-treated with non-lethal dose of hydrogen peroxide by virtue of the induction of 30 proteins. Of these, 9 are constitutively overexpressed in hydrogen peroxide resistant *oxyR* mutants. Mutant *oxyR1* is resistant to oxidizing agents and overexpresses 3 Hsp (D 64a, E 89, F 52a). The *oxyR* regulatory network is a previously uncharacterized global regulatory network in enteric bacteria (Christman et al. 1985).

6.1.4 Role of Heat Shock Proteins in Stress Tolerance in Plants

Heat stress induces protein denaturation which in turn, phosphorylates and activates Hsp 60, 70, 90, 100. Hsp 70 can be obtained from tomato rice etc., Hsp 90 from maize, wheat etc., Hsp 70 from Soyabean, Hsp 60 from barley, rye and sHsp from millet, sorghum etc. 7Hsp 70, Hsp 60 (Chaperonin), Hsp 90, Hsp 100/Clp and small Hsp act as molecular chaperones and impart plant tolerance to various modes of stress. The various subfamilies (DnaK, Hsp 110) of Hsp 70 are distributed in the cytosol, chloroplast, mitochondria and endoplasmic reticulum. They are associated with the prevention of protein aggregation and assistance in refolding, translocation, transcriptional activation and signal transduction. The various subfamilies (Cpn 60 etc.) of Hsp 60 are distributed in the cytosol and mitochondria and assist in protein folding. The various subfamilies (AtHsp 90–1, 5, 6, 7) of Hsp 90 are found in the cytosol, chloroplast, mitochondria and endoplasmic reticulum. They facilitate genetic buffering and signal transduction. The various subfamilies (Clp B, Clp D etc.) of Hsp 100 are found in the cytosol, chloroplast and mitochondria and assist in protein unfolding. The various subfamilies (Hsp 17.6, 17.9 etc.) of sHsp are found in the cytosol, chloroplast, mitochondria and endoplasmic reticulum. They prevent protein aggregation and stabilize non-native proteins (Usman et al. 2014).

Hsp104, Hsp70, and Hsp40 are predominant during replication. On the other hand, Hsp90 is associated with transcriptional and post-transcriptional processes. Hsp70 and Hsp90, in combination, plays a major role in morphogenesis. Heat stress in fungi lead to increased expression of Hsp10, Hsp30, Hsp60, Hsp90, and Hsp104 proteins, while expression of Hsp12 increases in response to cold stress. Osmotic stress is controlled by small heat shock proteins and Hsp60. Expression of Hsp30, Hsp70, and Hsp90 increases in response to pH stress. Hsp104 is also associated with conditions of high pressure. Recent experimentations suggest that Hsp90 can be a potential antifungal target due to its role in morphogenesis (Tiwari et al. 2015).

In another study, the expression level of a small heat shock protein *EsHsp16.9,* isolated from Siberian wild rye (*Elymus sibiricus*) increases when exposed to stressors like heat, drought, arsenic, methyl viologen and H2O2 treatment. *In vitro* study, this protein has reduced thermal aggregation of malate dehydrogenase. When this *EsHsp16.9* is expressed in E. coli, it shows better cell growth under salt, arsenic, polyethylene glycol and heat (Lee et al. 2015). The yeast metallothionein gene CUP1 is transcriptionally activated during oxidative stress by means of a functional HSE and HSF. CUP1 imparts metal detoxification and protects against oxidative insult (Liu and Thiele 1996).

Heat stress is a primary factor of deterioration in summer creeping bentgrass (*Agrostis stolonifera*) (Heat Shock Proteins in Relation to Heat Stress Tolerance of Creeping Bentgrass at different nitrogen levels). Nitrogen medium helped the grass to better survive during long term heat stress. Fv/Fm, a correlation between chloroplast-sHsp production and PSII efficiency indicates that this chloroplast-sHsp plays a direct role in the protection of PSII (Heckathorn et al. 1998). 38It is also reported that heterologous expression of MsHsp 23in creeping bentgrass protects against thermal stress by chaperone activity, associated with the induction of ascorbate peroxidase.

Now-a-days, cross-stress tolerance possesses an important role in producing stress-resistant crops. Heat or cold priming-induced cross-tolerance is very common in plants. Multiple stress signaling pathways involving ROS, RNS, reactive carbonyl species (RCS), most particularly H_2O_2, nitric oxide (NO) and methylglyoxal (MG) provide resistance to abiotic stresses via modulating the expression of genes as well as modifying proteins post-translationally (Hossain et al. 2017). ROS influences the stress tolerance in the yeast. H_2O_2 promotes accumulation of Hsp 70 and Hsp 90 mRNA (Moraitis and Curran 2004). ROS, H_2O_2, and 2, 5-dibromo-3-methyl-6-isopropyl-p-benzoquinone (DBMIB), an endogenous H_2O_2 propagator induces HsfA3 transcription in seedlings leading to oxidative stress tolerance. The overexpression of several genes encoding galactinol synthase (GolS) increase galactinol levels (Song et al. 2016).

Studies in Arabidopsis (*Arabidopsis thaliana*) have shown that induction of HSFA4A enhances tolerance to salt and oxidative agents via altering transcription of a large set of genes. Its inactivation results in hypersensitivity to salt stress. hsfa4a mutants elevate H_2O_2 accumulation and lipid peroxidation (Pérez-Salamó et al. 2014). Chaperone production and improvement of stress resistance in arabidopsis can be elevated by overexpressing AtHSFA1b, AtHSFA2, AtHSFA3, AtHsfA4a and AtHSFA6a (Jacob et al. 2016) but this may lead to moderate to severe dwarfism (Yoshida et al. 2011). Alfalfa, (*Medicago sativa*) overexpressing Hsp gene (MsHSP70) is less sensitive to polyethylene glycol (PEG) than control group (Li et al. 2017). Arabidopsis Hsp 21 is a crucial component of thermomemory. It is a plastidial small heat shock protein that rapidly accumulates after heat stress (Sedaghatmehr et al. 2016). Lack of functional FtsH6, a plastid-localized metalloprotease promotes Hsp 21 accumulation during the later stages of thermomemory and increases thermomemory capacity.

In yeast (*Saccharomyces cerevisiae*) two-hybrid and bimolecular fluorescence complementation assays, homomeric interaction of HSFA4A is reduced by alanine replacement of three conserved cysteine residues. MPK3 and MPK6 phosphorylate HSFA4A in vitro on three distinct sites, among them serine-309 is the major phosphorylation site. They can form in vivo complexes with ZAT10/salt tolerance zinc finger and thus regulate plant defense responses (Mittler et al. 2006; Nguyen et al. 2012).

CsHSP17.2, a molecular chaperone present in the cytosol and the nucleus is essential for thermotolerance by maintaining maximum photochemical efficiency and protein synthesis, enhancing the scavenging ROS and inducing the expression of HS-responsive genes in tea plants (*Camellia sinensis*) (Wang et al. 2017). Its expression is significantly upregulated by heat stress. Transgenic E. coli and Pichia pastoris expressing CsHSP17.2 exhibit enhanced thermotolerance.

DcHsp17.7, another cytosolic Class I heat shock protein accumulates in carrot (*Daucus carota*) leaf tissue under oxidative and osmotic (PEG) stress conditions (Ahn and Song 2012). Heterologously expressed DcHsp17.7 enhanced E. coli cell viability under various abiotic stress conditions such as salinity (Song and Ahn 2011), heat (Kim and Ahn 2009), cold (Song and Ahn 2010), oxidation and osmotic stress. Transgenic carrot plants overexpressing DcHsp17.7 showed higher thermotolerance (Malik et al. 1999) and transgenic potato plants showed enhanced cellular membrane stability and tuberization under heat stress (Ahn et al. 2004).

Genetic engineering develops crop plants possessing enhanced stress tolerance (Bhatnagar-Mathur et al. 2008). During abiotic stress, methylglyoxal accumulates to toxic levels in affected cells. In stress affected cells it is routinely detoxified through the action of DJ1/PARK7/Hsp31 proteins via targeting mitochondria and inducing expression of key stress-related genes. Hsp31 confers dual biotic and abiotic stress tolerance in*Nicotiana tabacum* as well as stress tolerance against diverse biotic stress inducers such as viruses, bacteria and fungi in tobacco plants (Melvin et al. 2017). 39On the other hand, the Hsp 90 and isocitrate lyase genes of *Nicotiana tobacum* plants have been reported to be associated with the development and reproduction of a root knot nematode *Meloidogyne incognita*.

HSFA6b plays a pivotal role in the response to ABA and in thermotolerance (Huang et al. 2016). ABA-responsive element-binding protein1 activates the HSFA6B promoter, as a result increased HSFA6B directly binds to the promoter of dehydration-responsive element-binding protein2A and enhances its expression. Under salt stress conditions, ER-sHsp transgenic plants have more vigorous roots, maintain a higher relative water content, absorb less sodium, accumulate more osmolytes, and sustain less damage to the photosystem, compared to wild-type non-transgenic plants. This study proves that UPR signaling pathway affects plant abiotic tolerance (Fu et al. 2016).

The tomato (*Lycopersicon esculentum*) chloroplast small heat shock protein (sHsp), Hsp 21 is induced by thermal stress in leaves and in developing fruits in course of transition of chloroplasts to chromoplasts. Hsp 21 protects PSII from temperature dependent oxidative stress. Analysis of leaf phenotype, chlorophyll content, and photosynthetic efficiency reveals that silencing of heat-shock factor A1a (HsfA1a) gene decreases melatonin levels and cadmium (Cd) tolerance, whereas its overexpression enhances the expression of the melatonin biosynthetic gene caffeic acid O-methyltransferase1 (COMT1) and melatonin accumulation leading to Cd tolerance in tomato plants (Cai et al. 2017; Neta-Sharir et al. 2005).

6.1.5 Role of Heat Shock Proteins in Stress Tolerance in Animals

A study by Dwivedi and Lakhotia have shown that Amalaki Rasayana (AR), an ayurvedic herbal formulation, improves tolerance in *Drosophila melanogaster* against crowding, thermal (Dwivedi et al. 2012; Wang et al. 2004) or oxidative stress. The expression of Hsp 27 which is important in stress tolerance like starvation, oxidative stress and in life span determination (Hao et al. 2007; Pandey et al. 2016; Shi et al. 2011; Wang et al. 2004) is elevated in both AR-fed wild type control and AR-treated heat shocked larvae (Dwivedi and Lakhotia 2016).

Antarctic sea urchin (*Sterechinus neumayeri)* are reported to be capable of overexpressing stress proteins (Hsp 90, Hsc 70, Hyou 1etc.) as a result of thermal stress (González et al. 2016).

It is reported that Hsp 70 family of proteins are essential for the cellular survival from heat stress and other types of physiological challenges in fish. Although programs of gene therapy have made impressive advances in recent years, the over expression of Hsp70 has proven to be problematic. Thus, its therapeutic role needs further investigation (Basu et al. 2002).

However, it has been shown that oxidative stress mediates upregulation of Hsp 90α in the hepatocytes of *Mugil cephalus* living in a polluted estuary. The stress resistance imparted by Hsp 90α involves proteins such as apoptosis signal-regulating kinase 1 (ASK 1), c-Jun NH2-terminal protein kinase 1/2, signal transducers and activators of transcription (STAT), extracellular signal-regulated kinase 1/2 (ERK 1/2), protein kinase B, nuclear factor-kappa B (NF kB), Ets-like protein 1, and B cell lymphoma-2 (Padmini and Rani 2011).

According to one study, both acute and chronic non-lethal heat shock (NLHS) treatment enhances the survival rate of shrimps *Peneaus vannamei* against acute hepatopancreatic necrosis disease (AHPND) caused by *Vibrio parahaemolyticus* than non-treated shrimps (Junprung et al. 2017). The authors have shown that the expression levels of heat shock proteins like LvHSP70, LvHSP90 and immune-related genes like LvproPO and LvCrustin are upregulated in chronic NLHS treated shrimps. But in shrimps with LvHSP70 and LvHSP90 gene knockdown, tolerance against chronic NLHS was absent, suggesting a vital role of heat shock proteins like Hsp70 and Hsp90 in bacterial defence in shrimp. Not only that, knockdown of Hsp 70 also influences the heat and bacterial infection tolerance in *Artemia franciscana* nauplii (Iryani et al. 2017). Fishes of hot spring run-off areas (*Puntius sophore*) have shown high expression level of Hsp 90 and Hsp 47 to tolerate heat shock (Mahanty et al. 2017).

Nrf2 (Nuclear factor (erythroid-derived 2)-like 2) helps in stress tolerance. A study by Mohanty et al. have shown that dietary curcumin supplementation in *Puntius sophore* enhances the expression levels of Nrf-2, SOD, catalase, gpx, Hsp 60, Hsp 70, Hsp 90 and Hsp 110 in gill and liver compared to control group. The result suggests that curcumin supplemented group exerts better stress tolerance via Nrf2 induced upregulation of various antioxidative enzymes and heat shock proteins (Mahanty et al. 2017). According to another study, in thermal tolerance, the expression mode of Hsp 70 is much more crucial than Hsp 70 itself. When *P. olivaceus* and the hybrids (*P.olivaceus* ♀ × *P. dentatus* ♂) are subjected to both acute and chronic heat stress, the hybrids shows better thermotolerence via increased temperature thresholds, shorter durations, stronger magnitudes and increased delays of Hsp 70expression (Liu et al. 2017).

Kurino et al. have shown in their study that *Caenorhabditis elegans* exertsmore longevity and thermotolerance when treated with isoamyl alcohol and acetic acid as odor stimuli. In presence of isomethyl alcohol, the expression levels of SOD 3 and Hsp 12.6 are increased and heat stress tolerance is mediated through DAF-16 activation whereas acetic acid odor stimulus increases the expression level of Hsp 16.2 and the stress tolerance is mediated through HSF-1 activation (Kurino et al. 2017).

6.2 Conclusions

Owing to their evolutionary significance, heat shock proteins have been extensively studied. Their link with oxidative stress has been reported in a wide range of studies. However, their detailed molecular mechanisms are yet to be investigated. Oxidative stress has also been reported to be a principal component of thermal tolerance. This chapter has ventured into the ocean of information put forward by a wide range of research articles related to the role of heat shock proteins in oxidative stress. The related studies have imparted valuable knowledge of the interplay of heat shock proteins and oxidative insult in a wide range of diseases like Alzheimer's, Parkinson, diabetes etc. The various mechanisms of stress tolerance associated with heat shock proteins have also been provided with respect to prokaryotes, plants and animals.

Acknowledgements The authors thank Bose Institute, Kolkata for providing the facility needed to gather information in relation to this chapter.

References

Abele D, Tesch C, Wencke P, Pörtner HO (2001) How does oxidative stress relate to thermal tolerance in the Antarctic bivalve Yoldia eightsi? Antarct Sci 13:111–118

Ahn Y-J, Song N-H (2012) A cytosolic heat shock protein expressed in carrot (Daucus carota L.) enhances cell viability under oxidative and osmotic stress conditions. Hortscience 47:143–148

Ahn Y-J, Claussen K, Zimmerman JL (2004) Genotypic differences in the heat-shock response and thermotolerance in four potato cultivars. Plant Sci 166:901–911

Amin J, Ananthan J, Voellmy R (1988) Key features of heat shock regulatory elements. Mol Cell Biol 8:3761–3769

Andersen JK (2004) Oxidative stress in neurodegeneration: cause or consequence? Nat Med 10

Arrigo A-P (1994) Expression and function of the low-molecular-weight heat shock proteins. The biology of heat shock proteins and molecular chaperones, 335–373

Bakthisaran R, Tangirala R, Rao CM (2015) Small heat shock proteins: role in cellular functions and pathology. Biochim Biophys Acta Proteins Proteomics 1854:291–319

Basu N, Todgham A, Ackerman P, Bibeau M, Nakano K, Schulte P, Iwama GK (2002) Heat shock protein genes and their functional significance in fish. Gene 295:173–183

Bhatnagar-Mathur P, Vadez V, Sharma KK (2008) Transgenic approaches for abiotic stress tolerance in plants: retrospect and prospects. Plant Cell Rep 27:411–424

Brownell SE, Becker RA, Steinman L (2012) The protective and therapeutic function of small heat shock proteins in neurological diseases. Front Immunol 3:74

Cai SY, Zhang Y, Xu YP, Qi ZY, Li MQ, Ahammed GJ, Xia XJ, Shi K, Zhou YH, Reiter RJ (2017) HsfA1a upregulates melatonin biosynthesis to confer cadmium tolerance in tomato plants. J Pineal Res 62

Castegna A, Aksenov M, Thongboonkerd V, Klein JB, Pierce WM, Booze R, Markesbery WR, Butterfield DA (2002) Proteomic identification of oxidatively modified proteins in Alzheimer's disease brain. Part II: dihydropyrimidinase-related protein 2, α-enolase and heat shock cognate 71. J Neurochem 82:1524–1532

Chai Y, Koppenhafer SL, Bonini NM, Paulson HL (1999) Analysis of the role of heat shock protein (Hsp) molecular chaperones in polyglutamine disease. J Neurosci 19:10338–10347

Christman MF, Morgan RW, Jacobson FS, Ames BN (1985) Positive control of a regulon for defenses against oxidative stress and some heat-shock proteins in salmonella typhimurium. Cell 41:753–762

Cruz T, Kandel R, Brown I (1991) Interleukin 1 induces the expression of a heat-shock gene in chondrocytes. Biochem J 277:327–330

Cummings CJ, Sun Y, Opal P, Antalffy B, Mestril R, Orr HT, Dillmann WH, Zoghbi HY (2001) Over-expression of inducible HSP70 chaperone suppresses neuropathology and improves motor function in SCA1 mice. Hum Mol Genet 10:1511–1518

Di Domenico F, Sultana R, Tiu GF, Scheff NN, Perluigi M, Cini C, Butterfield DA (2010) Protein levels of heat shock proteins 27, 32, 60, 70, 90 and thioredoxin-1 in amnestic mild cognitive impairment: an investigation on the role of cellular stress response in the progression of Alzheimer disease. Brain Res 1333:72–81

Ding Q, Keller JN (2001) Proteasome inhibition in oxidative stress neurotoxicity: implications for heat shock proteins. J Neurochem 77:1010–1017

Driedonks N, Xu J, Peters JL, Park S, Rieu I (2015) Multi-level interactions between heat shock factors, heat shock proteins, and the redox system regulate acclimation to heat. Front Plant Sci 6

Dwivedi V, Lakhotia SC (2016) Ayurvedic Amalaki Rasayana promotes improved stress tolerance and thus has anti-aging effects in Drosophila melanogaster. J Biosci 41:697–711

Dwivedi V, Anandan E, Mony RS, Muraleedharan T, Valiathan M, Mutsuddi M, Lakhotia SC (2012) In vivo effects of traditional Ayurvedic formulations in Drosophila melanogaster model relate with therapeutic applications. PLoS One 7:e37113

Feige, U., Morimoto, R.I., Polla, B. (2013) Stress-inducible cellular responses. *Birkhäuser*

Friedlander RM (2003) Apoptosis and caspases in neurodegenerative diseases. N Engl J Med 348:1365–1375

Fu C, Liu X, Yang W, Zhao C, Liu J (2016) Enhanced salt tolerance in tomato plants constitutively expressing heat-shock protein in the endoplasmic reticulum. Genet Mol Res 15

Golenhofen N, Redel A, Wawrousek E, Drenckhahn D (2006) Ischemia-induced increase of stiffness of αB-crystallin/HSPB2-deficient myocardium. Pflugers Arch 451:518–525

González K, Gaitán-Espitia J, Font A, Cárdenas CA, González-Aravena M (2016) Expression pattern of heat shock proteins during acute thermal stress in the Antarctic Sea urchin, Sterechinus neumayeri. Rev Chil Hist Nat 89:2

Guzhova IV, Arnholdt AC, Darieva ZA, Kinev AV, Lasunskaia EB, Nilsson K, Bozhkov VM, Voronin AP, Margulis BA (1998) Effects of exogenous stress protein 70 on the functional properties of human promonocytes through binding to cell surface and internationalization. Cell Stress Chaperones 3:67

Hao X, Zhang S, Timakov B, Zhang P (2007) The Hsp27 gene is not required for drosophila development but its activity is associated with starvation resistance. Cell Stress Chaperones 12:364–372

Heckathorn SA, Downs CA, Sharkey TD, Coleman JS (1998) The small, methionine-rich chloroplast heat-shock protein protects photosystem II electron transport during heat stress. Plant Physiol 116:439–444

Hossain MA, Li Z-G, Hoque TS, Burritt DJ, Fujita M, Munné-Bosch S (2017) Heat or cold priming-induced cross-tolerance to abiotic stresses in plants: key regulators and possible mechanisms. *Protoplasma*:1–14

Huang Y-C, Niu C-Y, Yang C-R, Jinn T-L (2016) The heat stress factor HSFA6b connects ABA signaling and ABA-mediated heat responses. Plant Physiol 172:1182–1199

Iryani MTM, MacRae TH, Panchakshari S, Tan J, Bossier P, Wahid MEA, Sung YY (2017) Knockdown of heat shock protein 70 (Hsp70) by RNAi reduces the tolerance of Artemia franciscana nauplii to heat and bacterial infection. J Exp Mar Biol Ecol 487:106–112

Jacob P, Hirt H, Bendahmane A (2016) The heat shock protein/chaperone network and multiple stress resistance. Plant Biotechnol J

Johnson AD, Berberian PA, Bond MG (1990) Effect of heat shock proteins on survival of isolated aortic cells from normal and atherosclerotic cynomolgus macaques. Atherosclerosis 84:111–119

Johnson AD, Berberian PA, Tytell M, Bond MG (1995) Differential distribution of 70-kD heat shock protein in atherosclerosis. Arterioscler Thromb Vasc Biol 15:27–36

Junprung W, Supungul P, Tassanakajon A (2017) HSP70 and HSP90 are involved in shrimp Penaeus vannamei tolerance to AHPND-causing strain of Vibrio parahaemolyticus after non-lethal heat shock. Fish Shellfish Immunol 60:237–246

Kim H, Ahn Y-J (2009) Expression of a gene encoding the carrot HSP17. 7 in Escherichia coli enhances cell viability and protein solubility under heat stress. Hortscience 44:866–869

Kregel KC (2002) Invited review: heat shock proteins: modifying factors in physiological stress responses and acquired thermotolerance. J Appl Physiol 92:2177–2186

Kurino C, Furuhashi T, Sudoh K, Sakamoto K (2017) Isoamyl alcohol odor promotes longevity and stress tolerance via DAF-16 in Caenorhabditis elegans. Biochem Biophys Res Commun 485:395–399

Kuzmin EV, Karpova OV, Elthon TE, Newton KJ (2004) Mitochondrial respiratory deficiencies signal up-regulation of genes for heat shock proteins. J Biol Chem 279:20672–20677

Lee S-H, Lee K-W, Lee D-G, Son D, Park SJ, Kim K-Y, Park HS, Cha J-Y (2015) Identification and functional characterization of Siberian wild rye (Elymus sibiricus L.) small heat shock protein 16.9 gene (EsHsp16. 9) conferring diverse stress tolerance in prokaryotic cells. Biotechnol Lett 37:881–890

Li Z, Menoret A, Srivastava P (2002) Roles of heat-shock proteins in antigen presentation and cross-presentation. Curr Opin Immunol 14:45–51

Li Z, Long R, Zhang T, Wang Z, Zhang F, Yang Q, Kang J, Sun Y (2017) Molecular cloning and functional analysis of the drought tolerance gene MsHSP70 from alfalfa (Medicago sativa L.). J Plant Res 130:387–396

Lindquist S, Craig E (1988) The heat-shock proteins. Annu Rev Genet 22:631–677

Liu X-d, Thiele DJ (1996) Oxidative stress induced heat shock factor phosphorylation and HSF-dependent activation of yeast metallothionein gene transcription. Genes Dev 10:592–603

Liu C, Fu J, Xu F, Wang X, Li S (2015) The role of heat shock proteins in oxidative stress damage induced by se deficiency in chicken livers. Biometals 28:163–173

Liu Y, Ma D, Zhao C, Xiao Z, Xu S, Xiao Y, Wang Y, Liu Q, Li J (2017) The expression pattern of hsp70 plays a critical role in thermal tolerance of marine demersal fish: multilevel responses of Paralichthys olivaceus and its hybrids (P. Olivaceus♀× P. Dentatus♂) to chronic and acute heat stress. Mar Environ Res 129:386–395

Lüders J, Demand J, Höhfeld J (2000) The ubiquitin-related BAG-1 provides a link between the molecular chaperones Hsc70/Hsp70 and the proteasome. J Biol Chem 275:4613–4617

Mahanty A, Purohit GK, Yadav RP, Mohanty S, Mohanty BP (2017) hsp90 and hsp47 appear to play an important role in minnow Puntiussophore for surviving in the hot spring run-off aquatic ecosystem. Fish Physiol Biochem 43:89–102

Malik MK, Slovin JP, Hwang CH, Zimmerman JL (1999) Modified expression of a carrot small heat shock protein gene, Hsp17. 7, results in increased or decreased thermotolerance. Plant J 20:89–99

Mattson MP (2000) Apoptosis in neurodegenerative disorders. Nat Rev Mol Cell Biol 1:120

Mc Naughton L, Lovell R, Madden L (2006) Heat shock proteins in exercise: a review. J Exerc Sci Physiother 2:13

Melvin P, Bankapalli K, D'Silva P, Shivaprasad P (2017) Methylglyoxal detoxification by a DJ-1 family protein provides dual abiotic and biotic stress tolerance in transgenic plants. Plant Mol Biol:1–17

Mittler R, Kim Y, Song L, Coutu J, Coutu A, Ciftci-Yilmaz S, Lee H, Stevenson B, Zhu J-K (2006) Gain-and loss-of-function mutations in Zat10 enhance the tolerance of plants to abiotic stress. FEBS Lett 580:6537–6542

Moraitis C, Curran BP (2004) Reactive oxygen species may influence the heat shock response and stress tolerance in the yeast Saccharomyces cerevisiae. Yeast 21:313–323

Morimoto RI, Santoro MG (1998) Stress–inducible responses and heat shock proteins: new pharmacologic targets for cytoprotection. Nat Biotechnol 16:833–838

Morrison LE, Whittaker RJ, Klepper RE, Wawrousek EF, Glembotski CC (2004) Roles for αB-crystallin and HSPB2 in protecting the myocardium from ischemia-reperfusion-induced damage in a KO mouse model. Am J Phys Heart Circ Phys 286:H847–H855

Muchowski PJ, Wacker JL (2005) Modulation of neurodegeneration by molecular chaperones. Nat Rev Neurosci 6:11

Mymrikov EV, Seit-Nebi AS, Gusev NB (2011) Large potentials of small heat shock proteins. Physiol Rev 91:1123–1159

Najarzadegan M, Ataei E, Akbarzadeh F (2016) The role of heat shock proteins in Alzheimer disease: a systematic review. J Syndr 3:6

Nakai A, Tanabe M, Kawazoe Y, Inazawa J, Morimoto RI, Nagata K (1997) HSF4, a new member of the human heat shock factor family which lacks properties of a transcriptional activator. Mol Cell Biol 17:469–481

Neta-Sharir I, Isaacson T, Lurie S, Weiss D (2005) Dual role for tomato heat shock protein 21: protecting photosystem II from oxidative stress and promoting color changes during fruit maturation. Plant Cell 17:1829–1838

Nguyen XC, Hoang MHT, Kim HS, Lee K, Liu X-M, Kim SH, Bahk S, Park HC, Chung WS (2012) Phosphorylation of the transcriptional regulator MYB44 by mitogen activated protein kinase regulates Arabidopsis seed germination. Biochem Biophys Res Commun 423:703–708

Padmini E, Rani MU (2011) Heat-shock protein 90 alpha (HSP90α) modulates signaling pathways towards tolerance of oxidative stress and enhanced survival of hepatocytes of Mugil cephalus. Cell Stress Chaperones 16:411–425

Pandey A, Saini S, Khatoon R, Sharma D, Narayan G, Chowdhuri DK (2016) Overexpression of hsp27 rescued neuronal cell death and reduction in life-and health-span in drosophila melanogaster against prolonged exposure to dichlorvos. Mol Neurobiol 53:3179–3193

Pérez-Salamó I, Papdi C, Rigó G, Zsigmond L, Vilela B, Lumbreras V, Nagy I, Horváth B, Domoki M, Darula Z (2014) The heat shock factor A4A confers salt tolerance and is regulated by oxidative stress and the mitogen-activated protein kinases MPK3 and MPK6. Plant Physiol 165:319–334

Perisic O, Xiao H, Lis JT (1989) Stable binding of drosophila heat shock factor to head-to-head and tail-to-tail repeats of a conserved 5 bp recognition unit. Cell 59:797–806

Pirkkala L, Sistonen L (2001) Heat shock proteins (HSPs): structure, function and genetics. *eLS*

Pörtner H (2001) Climate change and temperature-dependent biogeography: oxygen limitation of thermal tolerance in animals. Naturwissenschaften 88:137–146

Rajkumar U, Vinoth A, Rajaravindra K, Shanmugham M, Rao S (2015) Effect of in ovo inoculation of vitamin E on expression of Hsp-70 m RNA and juvenile growth in coloured broiler chicken. Indian J Poult Sci 50:104–108

Rauchova H, Vokurkova M, Koudelova J (2012) Hypoxia-induced lipid peroxidation in the brain during postnatal ontogenesis. Physiol Res 61:S89

Ray PS, Martin JL, Swanson EA, Otani H, Dillmann WH, Das DK (2001) Transgene overexpression of αB crystallin confers simultaneous protection against cardiomyocyte apoptosis and necrosis during myocardial ischemia and reperfusion. FASEB J 15:393–402

Ritossa F (1964) Experimental activation of specific loci in polytene chromosomes of drosophila. Exp Cell Res 35:601–607

Robinson MB, Tidwell JL, Gould T, Taylor AR, Newbern JM, Graves J, Tytell M, Milligan CE (2005) Extracellular heat shock protein 70: a critical component for motoneuron survival. J Neurosci 25:9735–9745

Russo A, Palumbo M, Scifo C, Cardile V, Barcellona M, Renis M (2001) Ethanol-induced oxidative stress in rat astrocytes: role of HSP70. Cell Biol Toxicol 17:153–168

Sakthivel K, Watanabe T, Nakamoto H (2009) A small heat-shock protein confers stress tolerance and stabilizes thylakoid membrane proteins in cyanobacteria under oxidative stress. Arch Microbiol 191:319–328

Santoro MG (2000) Heat shock factors and the control of the stress response. Biochem Pharmacol 59:55–63

Schuetz TJ, Gallo GJ, Sheldon L, Tempst P, Kingston RE (1991) Isolation of a cDNA for HSF2: evidence for two heat shock factor genes in humans. Proc Natl Acad Sci 88:6911–6915

Sedaghatmehr M, Mueller-Roeber B, Balazadeh S (2016) The plastid metalloprotease FtsH6 and small heat shock protein HSP21 jointly regulate thermomemory in Arabidopsis. Nat Commun 7:12439

Shi Y, Nishida K, Di Giammartino DC, Manley JL (2011) Heat shock-induced SRSF10 dephosphorylation displays thermotolerance mediated by Hsp27. Mol Cell Biol 31:458–465

Sistonen L, Sarge K, Phillips B, Abravaya K, Morimoto R (1992) Activation of heat shock factor 2 during hemin-induced differentiation of human erythroleukemia cells. Mol Cell Biol 12:4104–4111

Song N-H, Ahn Y-J (2010) DcHsp17. 7, a small heat shock protein from carrot, is upregulated under cold stress and enhances cold tolerance by functioning as a molecular chaperone. Hortscience 45:469–474

Song N-H, Ahn Y-J (2011) DcHsp17. 7, a small heat shock protein in carrot, is tissue-specifically expressed under salt stress and confers tolerance to salinity. New Biotechnol 28:698–704

Song C, Chung WS, Lim CO (2016) Overexpression of heat shock factor gene HsfA3 increases Galactinol levels and oxidative stress tolerance in Arabidopsis. Mol Cells 39:477

Sørensen JG (2010) Application of heat shock protein expression for detecting natural adaptation and exposure to stress in natural populations. Curr Zool 56:703–713

Sun Y, MacRae TH (2005) The small heat shock proteins and their role in human disease. FEBS J 272:2613–2627

Suzuki K, Sawa Y, Kaneda Y, Ichikawa H, Shirakura R, Matsuda H (1997) In vivo gene transfection with heat shock protein 70 enhances myocardial tolerance to ischemia-reperfusion injury in rat. J Clin Investig 99:1645

Tanabe M, Kawazoe Y, Takeda S, Morimoto RI, Nagata K, Nakai A (1998) Disruption of the HSF3 gene results in the severe reduction of heat shock gene expression and loss of thermotolerance. EMBO J 17:1750–1758

Tidwell JL, Houenou LJ, Tytell M (2004) Administration of Hsp70 in vivo inhibits motor and sensory neuron degeneration. Cell Stress Chaperones 9:88–98

Tiwari S, Thakur R, Shankar J (2015) Role of heat-shock proteins in cellular function and in the biology of fungi. Biotechnol Res Int 2015:1

Tóth ME, Gombos I, Sántha M (2015) Heat shock proteins and their role in human diseases. Acta Biol Szeged 59:121–141

Tytell M, Greenberg S, Lasek R (1986) Heat shock-like protein is transferred from glia to axon. Brain Res 363:161–164

Umapathy D, Krishnamoorthy E, Muthukumaran P, Rajaram R, Padmalayam I, Viswanathan V (2012) Association of A1538G and C2437T single nucleotide polymorphisms in heat shock protein 70 genes with type 2 diabetes. Lab Med 43:250–255

Usman MG, Rafii M, Ismail M, Malek M, Latif MA, Oladosu Y (2014) Heat shock proteins: functions and response against heat stress in plants. Int J Sci Technol Res 3:204–218

Wang H-D, Kazemi-Esfarjani P, Benzer S (2004) Multiple-stress analysis for isolation of drosophila longevity genes. Proc Natl Acad Sci U S A 101:12610–12615

Wang M, Zou Z, Li Q, Sun K, Chen X, Li X (2017) The CsHSP17. 2 molecular chaperone is essential for thermotolerance in Camellia sinensis. Sci Rep 7:1237

Warrick JM, Chan HE, Gray-Board GL, Chai Y, Paulson HL, Bonini NM (1999) Suppression of polyglutamine-mediated neurodegeneration in drosophila by the molecular chaperone HSP70. Nat Genet 23:425–428

Whitley D, Goldberg SP, Jordan WD (1999) Heat shock proteins: a review of the molecular chaperones. J Vasc Surg 29:748–751
Wilhelmus M, Otte-Höller I, Wesseling P, De Waal R, Boelens W, Verbeek M (2006) Specific association of small heat shock proteins with the pathological hallmarks of Alzheimer's disease brains. Neuropathol Appl Neurobiol 32:119–130
Wu D, Cederbaum AI (2003) Alcohol, oxidative stress, and free radical damage. Alcohol Res Health 27:277–284
Wyttenbach A, Arrigo AP (2009) The role of heat shock proteins during neurodegeneration in Alzheimer's, Parkinson's and Huntington's disease. In: Heat shock proteins in neural cells, pp 81–99
Wyttenbach A, Sauvageot O, Carmichael J, Diaz-Latoud C, Arrigo A-P, Rubinsztein DC (2002) Heat shock protein 27 prevents cellular polyglutamine toxicity and suppresses the increase of reactive oxygen species caused by huntingtin. Hum Mol Genet 11:1137–1151
Yoshida T, Ohama N, Nakajima J, Kidokoro S, Mizoi J, Nakashima K, Maruyama K, Kim J-M, Seki M, Todaka D (2011) Arabidopsis HsfA1 transcription factors function as the main positive regulators in heat shock-responsive gene expression. Mol Gen Genomics 286:321–332
Yu Q, Kent CR, Tytell M (2001) Retinal uptake of intravitreally injected Hsc/Hsp70 and its effect on susceptibility to light damage. Mol Vis 7:48–56

Chapter 7
Heat Shock Proteins in Oxidative and Nitrosative Stress

Asmaa A. A. A. Kattaia, Samia A. Abd El-Baset, and Eman M. Mohamed

Abstract Multiple human pathologies have been attributed to oxidative/nitrosative stress. Stress occurs when the production of reactive oxygen species (ROS) and/or reactive nitrogen species (RNS) exceeds the antioxidant capacity. Heat shock proteins (HSP) are closely related families of constitutively expressed and stress-induced cellular proteins that protect cellular homeostasis against heat and other stress stimuli. Oxidative stress is considered a key mediator of HSP induction. This chapter reviews the role of Hsp70 in different disorders associated with oxidative/nitrosative stress with reference to factors affecting Hsp70 gene expression, the role of proteasomes, the anti-apoptotic and anti-inflammatory effects and the relation between heat stress and oxidative stress. Moreover, we address the overexpression of other HSP such as Hsp32, 27, 40 and 60 following oxidative/nitrosative stress.

Keywords Heat shock proteins · Hsp70 · Nitrosative stress · NO · Oxidative stress

Abbreviations

Ag NP	Silver nanoparticles
ARE	Antioxidant responsive element
HBA	Heat shock protein 90-binding agent
HO-1	Heme oxygenase-1
HSE	Heat shock elements
HSF	Heat shock factor
HSP	Heat shock proteins
I/R	Ischemia/reperfusion
iNOS or NOS II	Inducible nitric-oxide synthase
NF-κB	Nuclear factor kappa beta
NO	Nitric oxide

A. A. A. A. Kattaia (✉) · S. A. Abd El-Baset · E. M. Mohamed
Department of Histology and Cell Biology, Faculty of Medicine, Zagazig University, Zagazig, Egypt

A. A. A. Asea, P. Kaur (eds.), *Heat Shock Proteins and Stress*,
Heat Shock Proteins 15, https://doi.org/10.1007/978-3-319-90725-3_7

NOS I/PKG	Nitric-oxide synthase I/ protein kinase G
ONHs	Optic nerve heads
RNS	Reactive nitrogen species
ROS	Reactive oxygen species
RVLM	Rostral ventrolateral medulla
TNF	Tumor necrosis factor

7.1 Introduction

Stress occurs when the net flux of reactive oxygen species (ROS) and/or reactive nitrogen species (RNS) production exceeds the capacity of antioxidant defences (Wellen and Thompson 2010). Under oxidative stress, excessive ROS/RNS compromise the intracellular environment and can lead to damage of DNA and degradation of lipid and protein (Guo et al. 2013). Moreover, changes in the cellular redox state can also be a signal for the activation of the inflammatory and apoptotic pathways (Mattson 2006; Varga et al. 2015).

Oxidative stress is considered as a risk factor for the development and progression of human pathologies such as cardiovascular diseases, aging, neurodegeneration, and cancer (Acharya et al. 2010; Salmon et al. 2010).

Nitrosative stress is the reaction of body tissues to nitric oxide (NO) (Eu et al. 2000). The direct effect or the primary mode of NO action results from reactions between NO itself and other biological molecules and results in most of the physiological effects of NO (Pacher et al. 2007). Whereas, the indirect or the secondary mode of NO action results from reactions of NO-derived RNS with other molecules producing marked cellular injury. It has been shown that the direct effects are mediated at low concentrations of NO for short periods; meanwhile the indirect effects are sustained by high local NO concentrations over prolonged periods (Kiang et al. 2004). The reaction of NO with superoxide produces the highly reactive peroxynitrite which generates massive ROS and massive S-nitrosylation causing cellular damage (Calabrese et al. 2000a; Wen et al. 2015)

Heat shock proteins (HSP) are closely related families of constitutively expressed and stress-induced cellular proteins. They function as molecular chaperones, facilitating protein synthesis, protein folding and transport processes (Beckmann et al. 1990). They protect cellular homeostasis against heat and other stress stimuli including reactive oxygen species (ROS), ischemia and reperfusion, sepsis, and acute inflammation (Plumier et al. 1995). Oxidative stress is considered a key mediator of HSP induction. Cells exposed to oxidative/nitrosative stress at normal temperatures respond by increasing the expression of HSP (Omar and Pappolla 1993; Wen et al. 2015).

7.1.1 Heat Shock Protein 70 (Hsp70)

Hsp70 represents the major protective stress protein. Hsp70 family includes the cytosolic constitutive form Hsc70 (cognate Hsp70; 73 kDa), the stress inducible form (Hsp72; 72 kDa), GRP78 (glucose-regulated protein in the endoplasmic reticulum; 78 kDa) and the mitochondrial Hsp70 (Hsp75; 75 kDa) (Calabrese et al. 2003). Hsp70-mediated chaperoning may be exerted at the mitochondrial membrane and/or within the mitochondria as well as within the cytoplasm (Stuart et al. 1994). Polla et al. (1996) suggested that HSP protect cells from oxidative damage by targeting mitochondria. The protection by Hsp70 may be stronger than the mitochondria-specific Hsp65. Hsp70 acts as a sensor of cellular redox changes and play a role in cellular processes that occur during and after exposure to oxidative stress (Kalmar and Greensmith 2009). These chaperones may suppress the toxicity of oxidative stress (Liu et al. 1997). They enable the cell to cope with excess misfolded or denatured proteins (Asea 2005). Hsp70 correct misfolded proteins by refolding or directing damaged proteins to the degenerative pathways (Mayer and Bukau 2005). Hsp70 also has anti-apoptotic and anti-inflammatory actions and thus can actively prevent protein damage following oxidative stress (Kalmar and Greensmith 2009).

7.1.1.1 Activation of Heat Shock Factor 1 (HSF1)

Two HSFs can be recognised in humans, HSF1 and HSF2. HSF1 is activated by heat shock and other stresses. HSF2 activation is dependent on non-stressful conditions, including cell-cycle events during embryonic development, growth factors, or oncogene expression (Morimoto et al. 1994). Following stress, denatured proteins serve as a stimulus for induction of HSF1 and cause dissociation of HSF1 and Hsp90 (Zou et al. 1998). Freed HSF1 is phosphorylated and forms trimers. The trimers enter the nucleus and bind to heat shock elements within the promoter of the Hsp70 gene to stimulate transcription of heat shock proteins (Morimoto et al. 1994; Wu et al. 1994). It has been demonstrated that NO may lead to trimerization and activation of HSF-1 by interacting with its SH-groups to form the intermediate S-nitrosothiol that degrades to highly reactive thiol radicals which in turn accelerate the formation of disulphide links between HSF-1 molecules (Malyshev et al. 1996). Indirect activation of (HSF-1) occurs when ROS/RNS oxidize critical amino acids, such as cysteine thiols, resulting in activation of Keap1/Nrf2 and Hsp90/HSF-1 transcriptional pathways and production of Hsp70 (Um et al. 2011; Zhang et al. 2011). Many inducers of Nrf 2-dependent genes have been shown to increase the protein levels of Hsp70 (Zhang et al. 2004).

7.1.1.2 Hsp70 Gene Expression

Nature of the Oxidative Challenge

External oxidants such as hydrogen peroxide and menadione exhibited activation of HSF-1 and accumulation of Hsp70 mRNA and protein in most systems tested (Jornot et al. 1991; McDuffee et al. 1997), with a few exceptions (Bruce et al. 1993) who evidenced that external oxidants activate the HSF but fail to cause transcription of the heat shock genes or synthesis of Hsp70 protein in NIH-3 T3 cells. On the other hand, when NIH-3 T3 cells were exposed to hypoxia/reperfusion, Hsp70 protein increased (Donati et al. 1990). A possible explanation is that oxidants cause only the initial oligomerization step during the activation of the HSF but not the subsequent phosphorylation which is required before transcription (Bruce et al. 1993). In Bovine coronary venular endothelial cells incubated with xanthine oxidase and hypoxanthine, oxidative injury increased Hsp70 mRNA synthesis, but no changes in Hsp70 protein levels were detected (Aucoin et al. 1995). This was attributed to inhibition of protein synthesis by oxidative stress at the translational level, specifically at the level of chain initiation (Jornot et al. 1991).

Interaction between NO and Hsp70 Genes

Several studies have shown that NO and Hsp70 affect the gene expression of each other (Adrie et al. 2000). Hsp70 themselves can inhibit inducible nitric-oxide synthase (iNOS or NOS II) gene expression. For example, in rostral ventrolateral medulla (RVLM), Hsp70 protects macrophages against peroxynitrite cytotoxicity by enhancing the nitric-oxide synthase I/ protein kinase G (NOS I/PKG) signaling pathway and inhibiting the NOS II/peroxynitrite cascade (Li et al. 2005).

7.1.1.3 Proteasomes and Hsp70

Hsp70 is important in sorting out proteins for both the proteasome and lysosome-mediated protein degradation systems (Kalmar and Greensmith 2009). When the proteasome becomes overloaded following exposure to prolonged oxidative stress, the lysosome may be utilized to free the cell from damaged proteins in a process called "chaperone mediated autophagy" (Kaushik and Cuervo 2008).

Following oxidative injury, proteasome activity is inhibited which may result in accumulation of damaged proteins (Okada et al. 1999). Application of proteasome inhibitors was found to induce Hsp70 expression (Lee and Goldberg 1998). Other HSP also attenuate oxidative injury by preserving proteasome activity (Ciechanover et al. 2000; Ding and Keller 2001).

7.1.1.4 Hsp70 Inhibits Inflammation and Apoptosis Following Oxidative/Nitrosative Stress

Oxidative/nitrosative stress is followed by activation of inflammatory and apoptotic cell death pathways (Varga et al. 2015; Wen et al. 2015). The anti-inflammatory actions of Hsp70 are mediated by binding to inflammatory mediators such cyclooxygenase- 2 and the nuclear factor kappa beta (NF-κB) and the subsequent inhibition of them (Ialenti et al. 2005; Jo et al. 2006). Hsp70 plays an important role as anti-apoptotic by binding proteins activated via the mitochondrial pathway of apoptosis or preventing caspase independent cell death (Li et al. 2000; Matsumori et al. 2006).

In hematopoietic cells, Hsp70 is able to protect against TNF-induced apoptosis (Wen et al. 2015). It is possible that the complex formed between Hsp70 and iNOS might decrease the enzymatic activity of iNOS and subsequently decrease NO production resulting in turn in decreasing TNF-α level and consequently inhibit apoptosis (Kiang et al. 2004). On the other hand, Nitration of Hsp90 over a certain threshold is sufficient to convert this pro-survival chaperone into a pro-apoptotic protein and mediate the activation of death pathways to remove damaged and subfunctional cells. Nitrated Hsp90 induces apoptosis by activation of Fas pathway of cell death (Franco et al. 2013).

7.1.1.5 Relation between Heat Stress and Oxidative Stress

Heat Stress Induces Oxidative Stress

Heat stress is a major source of oxidative stress via generation of oxygen free radicals and lipid peroxidation, creating a redox imbalance (Altan et al. 2003). The use of free radical scavengers inhibited lipid peroxidation and HSP induction by heat (Burdon et al. 1987). Hsp70 significantly elevated antioxidant enzyme activities and inhibited lipid peroxidation to protect the intestinal mucosa from heat-stress injury in broilers (Gu et al. 2012).

Oxidative Stress Affects Heat Tolerance

Several studies suggest a synergistic relationship between heat stress and oxidative stress that enhances cell death (McAnulty et al. 2005). After heat exposure, the refolding activity of proteins was inhibited in H_2O_2-pretreated cells (human malignant glioma T98G and A172 cells) and almost recovered by pre-treatment with L-N-acetylcystein. Under heat stress, oxidative stress attenuated the HSR and induced ER stress. The multiple stress signals that develop in cells augment cell death. (Adachi et al. 2009)

7.1.1.6 Role of Hsp70 in Disorders Associated with Oxidative/Nitrosative Stress

Calabrese et al. (2000b) reported that Hsp70 has been involved in the protection against oxidative/nitrosative stress. Treatment with the heat shock protein 90-binding agent (HBA) geldanamycin and its analogues (17-AAG and 17-DMAG) mediated up-regulation of Hsp70 and Hsp27 expressions in a mouse model of kidney ischemia/reperfusion (I/R) injury and in renal adenocarcinoma (ACHN) cells exposed to H_2O_2 in vitro. Harrison et al. (2008) suggested that renal cells were protected from oxidative injury through a stress protein-mediated mechanism independent of heme oxygenase-1 (HO-1). During oxidative stress induced by varying selenium concentrations, the testicular cells synthesized high levels of Hsp70, which protected against the oxidative injury in mice (Kaushal and Bansal 2009).

Selenium protected liver cells from oxidative damage caused by chronic arsenic poisoning in rats by increasing the expression of Hsp70 to improve the activities of antioxidant enzymes (Xu et al. 2013). In another study, the hepatic cells counteracted the depletion of glutathione induced by selenium deficiency by the up-regulation of Hsp70, 40, 60, 27, and 90 genes in chicken (Liu et al. 2015). Increased expression of Hsp70 protected cultured human respiratory epithelium against oxidant injury secondary to hyperoxia by decreasing the production of 8-iso-PGF2a, a sensitive marker of cellular oxidant stress (Wong et al. 1998). In contrast, Strand et al. (1994) found that HSP induction did not protect whole rat lungs or cultured rat fibroblasts against hyperoxia. This was explained by the methodological differences between the two studies (Wong et al. 1998). Also induced Hsp70 level in the brain tissue of rats in response to decrease in the glutathione/glutathione disulfide ratio and oxidative stress was reported (Calabrese et al. 2000b).

NO could initiate the synthesis of the stress protein, Hsp70. Injection of NO donors such as dinitrosyl iron complex resulted in accumulation of Hsp70 in the rat heart (Malyshev et al. 1996). Pretreatment of hepatocytes with NO altered the redox state and was accompanied by oxidation of glutathione and formation of S-nitrosoglutathione which induced Hsp70 mRNA (Kim et al. 1997). Calabrese et al. (2000a) added that NO-induced activation of the heat shock response is thought to be mediated by the cellular oxidant/antioxidant balance, primarily by the glutathione status and the antioxidant enzymes. Further, stress caused by peroxynitrite exposure up-regulated Hsp70 expression that participated in elimination and transportation of misfolded proteins to afford neuroprotection (Wen et al. 2015).

Our data revealed significant enhancement of immune-histochemical expression of Hsp70 following oxidative/nitrosative stress induced by nitrate exposure in rat liver. The co-administration of the anti-oxidant s-allylcysteine decreased Hsp70 level to normal. We attributed that to the ability of s-allylcysteine to improve oxidative stress resistance of tissues by inhibiting iNOS gene expression and restoring hepatic glutathione level (Kattaia et al. 2017). Chan et al. (2004) indicated that Hsp70 produce protection against sepsis via inhibition of iNOS gene expression in the RVLM.

7.1.2 *Other HSP Linked to Oxidative/Nitrosative Stress*

Heme oxygenase-1 (HO-1; Hsp32; 32 kD) is a redox sensor protein with enzymatic activity involved in the degradation of pro-oxidant heme into potent antioxidant CO and bilirubin. HO-1 is inducible and synthesized in response to oxidative stress, whereas HO-2 is constitutively expressed (Gozzelino et al. 2010). Hsp32 expression is also induced by oxidative/nitrosative stress and regulated by activation of the common transcription factor for HSP, HSF1. The Hsp32 gene contains the antioxidant responsive element (ARE) in its promoter region, similar to other antioxidant enzymes (Calabrese et al. 2003; Kalmar and Greensmith 2009). In human lung carcinoma (A549) cells and HepG2 cells, Hsp70 and HO-1 played an important role in the protection against silver nanoparticles (Ag NP)-induced oxidative injury (Xin et al. 2015).

The small HSP such as Hsp27 have been also reported to act as antioxidants and protect against protein misfolding following oxidative stress (Arrigo 2007). This protective effect of Hsp27 is mediated by the ability to keep glutathione in its reduced form under oxidative stress (Arrigo et al. 2005). In cultured human optic nerve heads (ONHs) astrocytes, H_2O_2 induced Hsp27 mRNA expression (Alice et al. 2008). Hsp27 is synthesized mainly in astrocytes in response to ischemic situations (Calabrese et al. 2005). Hsp27 has been implicated as a therapeutic target in ischemic stroke and neurodegenerative disorders (Shimura et al. 2017). Hsp27 was also induced by oxidative stress in different cellular systems such as vascular endothelial cells (Nguyen et al. 2004), ventricular myocytes (Blunt et al. 2007) and lymphomonocytes from patients with liver disease (Federico et al. 2005). The up-regulation of Hsp27 expression correlated strongly with protection of retinal cells and corneal epithelial cells from stress injury (Li et al. 2003; Shi et al. 2006).

Neural cells transfected with human HDJ-1, 1, a member of the Hsp40 family, were more resistant to oxidative damage. Cells expressing increased levels of HDJ-1 showed a greater preservation of proteasomal activity and mitochondrial function following exposure to oxidative stressors (Ding and Keller 2001). Further, cold stress resulted in oxidative damage in the chicken heart and increased Hsp40 and Hsp70 levels (Zhao et al. 2013). Oxidative stress and hyperglycaemia can also lead to up-regulation of Hsp60 and Hsp70 in HeLa cells (Hall and Martinus 2013). Hsp60 may act as a barrier to mitochondrial oxidative stress mediated apoptosis (Sarangi et al. 2013).

7.2 Conclusions

Hsp70 acts as a sensor of cellular redox changes and has been implicated in the protection against oxidative/nitrosative stress. Indirect activation of HSF-1 occurs when ROS/RNS oxidize critical amino acids, such as cysteine thiols, resulting in activation of transcriptional pathways and production of Hsp70. The nature of the oxidative challenge affects Hsp70 gene expression. Moreover, nitric oxide (NO) and Hsp70 affect the gene expression of each other. Hsp70 is important in sorting out proteins for both the proteasome and lysosome-mediated protein degradation systems. Application of proteasome inhibitors was found to induce Hsp70 expression. Hsp70 also has anti-apoptotic and anti-inflammatory actions and thus can actively prevent protein damage following oxidative/nitrosative stress. Heat stress is a major source of oxidative stress. Several studies suggest a synergistic relationship between heat stress and oxidative stress that enhances cell death. We highlighted the overexpression of Hsp70 in different organs and the underlying mechanisms following oxidative/nitrosative stress. Other HSP such as Hsp32, 27, 40 and 60 have also been linked to oxidative/nitrosative stress.

Acknowledgements The authors would like to thank the support from Histology and Cell Biology Department, Faculty of Medicine, Zagazig University. We also thank many lab members and colleagues for critical and productive collaborations.

References

Acharya A, Das I, Chandhok D, Saha T (2010) Redox regulation in cancer: a double-edged sword with therapeutic potential. Oxidative Med Cell Longev 3:23–34

Adachi M, Liu Y, Fujii K, Calderwood SK, Nakai A, Imai K, Shinomura Y (2009) Oxidative stress impairs the heat stress response and delays unfolded protein recovery. PLoS One 4:e7719

Adrie C, Richter C, Bachelet M, Banzet N, François D, Dinh-Xuan AT, Dhainaut JF, Polla BS, Richard MJ (2000) Contrasting effects of NO and peroxynitrites on HSP70 expression and apoptosis in human monocytes. Am J Phys Cell Physiol 279:C452–C460

Alice LY, Fuchshofer R, Birke M, Kampik A, Bloemendal H, Welge-Lüssen U (2008) Oxidative stress and TGF-β2 increase heat shock protein 27 expression in human optic nerve head astrocytes. Invest Ophthalmol Vis Sci 49:5403–5411

Altan Ö, Pabuçcuoğlu A, Altan A, Konyalioğlu S, Bayraktar H (2003) Effect of heat stress on oxidative stress, lipid peroxidation and some stress parameters in broilers. Br Poult Sci 44:545–550

Arrigo AP (2007) The cellular "networking" of mammalian Hsp27 and its functions in the control of protein folding, redox state and apoptosis. In: Molecular aspects of the stress response: chaperones, membranes and networks. Springer, New York, pp 14–26

Arrigo AP, Virot S, Chaufour S, Firdaus W, Kretz-Remy C, Diaz-Latoud C (2005) Hsp27 consolidates intracellular redox homeostasis by upholding glutathione in its reduced form and by decreasing iron intracellular levels. Antioxid Redox Signal 7:414–422

Asea A (2005) Stress proteins and initiation of immune response: chaperokine activity of hsp72. Exerc Immunol Rev 11:34

Aucoin MM, Barhoumi ROLA, Kochevar DT, Granger HJ, Burghardt RC (1995) Oxidative injury of coronary venular endothelial cells depletes intracellular glutathione and induces HSP 70 mRNA. Am J Physiol Heart Circ Physiol 268:H1651–H1658

Beckmann RP, Mizzen LA, Welch WJ (1990) Interaction of Hsp 70 with newly synthesized proteins: implications for protein folding and assembly. Science 248:850–854

Blunt BC, Creek AT, Henderson DC, Hofmann PA (2007) H_2O_2 activation of HSP 25/27 protects desmin from calpain proteolysis in rat ventricular myocytes. Am J Physiol Heart Circ Physiol 293:H1518–H1525

Bruce JL, Price BD, Coleman CN, Calderwood SK (1993) Oxidative injury rapidly activates the heat shock transcription factor but fails to increase levels of heat shock proteins. Cancer Res 53:12–15

Burdon RH, Gill VM, Rice-Evans C (1987) Oxidative stress and heat shock protein induction in human cells. Free Radic Res Commun 3:129–139

Calabrese V, Copani A, Testa D, Ravagna A, Spadaro F, Tendi E, Nicoletti VG, Giuffrida Stella AM (2000a) Nitric oxide synthase induction in astroglial cell cultures: effect on heat shock protein 70 synthesis and oxidant/antioxidant balance. J Neurosci Res 60:613–622

Calabrese V, Testa G, Ravagna A, Bates TE, Stella AMG (2000b) HSP70 induction in the brain following ethanol administration in the rat: regulation by glutathione redox state. Biochem Biophys Res Commun 269:397–400

Calabrese V, Scapagnini G, Colombrita C, Ravagna A, Pennisi G, Stella AG, Galli F, Butterfield DA (2003) Redox regulation of heat shock protein expression in aging and neurodegenerative disorders associated with oxidative stress: a nutritional approach. Amino Acids 25:437–444

Calabrese V, Lodi R, Tonon C, D'Agata V, Sapienza M, Scapagnini G, Mangiameli A, Pennisi G, Stella AG, Butterfield DA (2005) Oxidative stress, mitochondrial dysfunction and cellular stress response in Friedreich's ataxia. J Neurol Sci 233:145–162

Chan JY, Ou CC, Wang LL, Chan SH (2004) Heat shock protein 70 confers cardiovascular protection during endotoxemia via inhibition of nuclear factor-κB activation and inducible nitric oxide synthase expression in the rostral ventrolateral medulla. Circulation 110:3560–3566

Ciechanover A, Orian A, Schwartz AL (2000) The ubiquitin-mediated proteolytic pathway: mode of action and clinical implications. J Cell Biochem 77:40–51

Ding Q, Keller JN (2001) Proteasome inhibition in oxidative stress neurotoxicity: implications for heat shock proteins. J Neurochem 77:1010–1017

Donati YR, Slosman DO, Polla BS (1990) Oxidative injury and the heat shock response. Biochem Pharmacol 40:2571–2577

Eu JP, Liu L, Zeng M, Stamler JS (2000) An apoptotic model for nitrosative stress. Biochemistry 39:1040–1047

Federico A, Tuccillo C, Terracciano F, D'Alessio C, Galdiero M, Finamore E, D'Isanto M, Peluso L, Blanco CDV, Loguercio C (2005) Heat shock protein 27 expression in patients with chronic liver damage. Immunobiology 209:729–735

Franco MC, Ye Y, Refakis CA, Feldman JL, Stokes AL, Basso M, de Mera RM, Sparrow NA, Calingasan NY, Kiaei M, Rhoads TW (2013) Nitration of Hsp90 induces cell death. Proc Natl Acad Sci 110:E1102–E1111

Gozzelino R, Jeney V, Soares MP (2010) Mechanisms of cell protection by heme oxygenase-1. Annu Rev Pharmacol Toxicol 50:323–354

Gu XH, Hao Y, Wang XL (2012) Overexpression of heat shock protein 70 and its relationship to intestine under acute heat stress in broilers: 2. Intestinal oxidative stress. Poult Sci 91:790–799

Guo C, Sun L, Chen X, Zhang D (2013) Oxidative stress, mitochondrial damage and neurodegenerative diseases. Neural Regen Res 8:2003

Hall L, Martinus RD (2013) Hyperglycaemia and oxidative stress upregulate HSP60 & HSP70 expression in HeLa cells. Springerplus 2(1):431

Harrison EM, Sharpe E, Bellamy CO, McNally SJ, Devey L, Garden OJ, Wigmore SJ (2008) Heat shock protein 90-binding agents protect renal cells from oxidative stress and reduce kidney ischemia-reperfusion injury. Am J Phys Renal Phys 295:F397–F405

Ialenti A, Di Meglio P, D'acquisto F, Pisano B, Maffia P, Grassia G, Ianaro A (2005) Inhibition of cyclooxygenase-2 gene expression by the heat shock response in J774 murine macrophages. Eur J Pharmacol 509:89–96

Jo SK, Ko GJ, Boo CS, Cho WY, Kim HK (2006) Heat preconditioning attenuates renal injury in ischemic ARF in rats: role of heat-shock protein 70 on NF-κB–mediated inflammation and on tubular cell injury. J Am Soc Nephrol 17:3082–3092

Jornot L, Mirault ME, Junod AF (1991) Differential expression of hsp70 stress proteins in human endothelial cells exposed to heat shock and hydrogen peroxide. Am J Respir Cell Mol Biol 5:265–275

Kalmar B, Greensmith L (2009) Induction of heat shock proteins for protection against oxidative stress. Adv Drug Deliv Rev 61:310–318

Kattaia AA, Abd El-Baset SA, Mohamed EM, Abdul-Maksou RS, Elfakharany YM (2017) Molecular mechanisms underlying histological and biochemical changes induced by nitrate in rat liver and the efficacy of S-Allylcysteine. Ultrastruct Pathol 41:10–22

Kaushal N, Bansal MP (2009) Diminished reproductive potential of male mice in response to selenium-induced oxidative stress: involvement of HSP70, HSP70-2, and MSJ-1. J Biochem Mol Toxicol 23:125–136

Kaushik S, Cuervo AM (2008) Chaperone-mediated autophagy. Autophagosome and Phagosome 445:227–244

Kiang JG, Bowman PD, Zhao B, Atkins JL Tsokos GC (2004) Heat shock Protein-70 inducers and iNOS inhibitors as therapeutics to ameliorate hemorrhagic shock. WALTER REED ARMY INST OF RESEARCH SILVER SPRING MD

Kim YM, de Vera ME, Watkins SC, Billiar TR (1997) Nitric oxide protects cultured rat hepatocytes from tumor necrosis factor-α-induced apoptosis by inducing heat shock protein 70 expression. J Biol Chem 272:1402–1411

Lee DH, Goldberg AL (1998) Proteasome inhibitors cause induction of heat shock proteins and trehalose, which together confer thermotolerance inSaccharomyces cerevisiae. Mol Cell Biol 18:30–38

Li CY, Lee JS, Ko YG, Kim JI, Seo JS (2000) Heat shock protein 70 inhibits apoptosis downstream of cytochrome c release and upstream of caspase-3 activation. J Biol Chem 275:25665–25671

Li Y, Roth S, Laser M, Ma JX, Crosson CE (2003) Retinal preconditioning and the induction of heat-shock protein 27. Invest Ophthalmol Vis Sci 44:1299–1304

Li FC, Chan JY, Chan SH, Chang AY (2005) In the rostral ventrolateral medulla, the 70-kDa heat shock protein (HSP70), but not HSP90, confers neuroprotection against fatal endotoxemia via augmentation of nitric-oxide synthase I (NOS I)/protein kinase G signaling pathway and inhibition of NOS II/peroxynitrite cascade. Mol Pharmacol 68:179–192

Liu H, Bowes RC, van de Water B, Sillence C, Nagelkerke JF, Stevens JL (1997) Endoplasmic reticulum chaperones GRP78 and calreticulin prevent oxidative stress, Ca2+ disturbances, and cell death in renal epithelial cells. J Biol Chem 272:21751–21759

Liu CP, Fu J, Xu FP, Wang XS, Li S (2015) The role of heat shock proteins in oxidative stress damage induced by se deficiency in chicken livers. Biometals 28:163–173

Malyshev IY, Malugin AV, Golubeva LY et al (1996) Nitric oxide donor induces HSP70 accumulation in the heart and in cultured cells. FEBS Lett 391:21–23

Matsumori Y, Northington FJ, Hong SM, Kayama T, Sheldon RA, Vexler ZS, Liu J (2006) Reduction of caspase-8 and-9 cleavage is associated with increased c-FLIP and increased binding of Apaf-1 and Hsp70 after neonatal hypoxic/ischemic injury in mice overexpressing Hsp70. Stroke 37:507–512

Mattson MP (2006) Neuronal life-and-death signaling, apoptosis, and neurodegenerative disorders. Antioxid Redox Signal 8:1997–2006

Mayer MP, Bukau B (2005) Hsp70 chaperones: cellular functions and molecular mechanism. Cell Mol Life Sci 62:670

McAnulty SR, McAnulty L, Pascoe DD, Gropper SS, Keith RE, Morrow JD, Gladden LB (2005) Hyperthermia increases exercise-induced oxidative stress. Int J Sports Med 26:188–192

McDuffee AT, Senisterra G, Huntley S, Lepock JR, Sekhar KR, Meredith MJ, Borrelli MJ, Morrow JD, Freeman ML (1997) Proteins containing non-native disulfide bonds generated by oxidative stress can act as signals for the induction of the heat shock response. J Cell Physiol 171(2):143–151

Morimoto RI, Jurivich DA, Kroeger PE (1994) Regulation of heat shock gene transcription by a family of heat shock factors. In: Morimoto RI, Tissieres A, Georgopoulos C (eds) The biology of heat shock proteins and molecular chaperones. *Cold Spring Harbor Laboratory Press*, Cold Spring Harbor, pp 417–455

Nguyen A, Chen P, Cai H (2004) Role of CaMKII in hydrogen peroxide activation of ERK1/2, p38 MAPK, HSP27 and actin reorganization in endothelial cells. FEBS Lett 572:307–313

Okada K, Wangpoengtrakul C, Osawa T, Toyokuni S, Tanaka K, Uchida K (1999) 4-Hydroxy-2-nonenal-mediated impairment of intracellular proteolysis during oxidative stress Identification of Proteasomes As Target Molecules. J Biol Chem 274:23787–23793

Omar R, Pappolla M (1993) Oxygen free radicals as inducers of heat shock protein synthesis in cultured human neuroblastoma cells: relevance to neurodegenerative disease. Eur Arch Psychiatry Clin Neurosci 242(5):262–267

Pacher P, Beckman JS, Liaudet L (2007) Nitric oxide and peroxynitrite in health and disease. Physiol Rev 87:315–424

Plumier JC, Ross BM, Currie RW et al (1995) Transgenic mice expressing the human heat shock protein 70 have improved post-ischemic myocardial recovery. J Clin Invest 95:1854

Polla BS, Kantengwa S, Francois D, Salvioli S, Franceschi C, Marsac C, Cossarizza A (1996) Mitochondria are selective targets for the protective effects of heat shock against oxidative injury. Proc Natl Acad Sci U S A 93:6458–6463

Salmon AB, Richardson A, Pérez VI (2010) Update on the oxidative stress theory of aging: does oxidative stress play a role in aging or healthy aging? Free Radic Biol Med 48:642–655

Sarangi U, Singh MK, Abhijnya KVV, Reddy LPA, Prasad BS, Pitke VV, Sreedhar AS (2013) Hsp60 chaperonin acts as barrier to pharmacologically induced oxidative stress mediated apoptosis in tumor cells with differential stress response. Drug target insights 7:35

Shi B, Han B, Schwab IR, Isseroff RR (2006) UVB irradiation-induced changes in the 27-kd heat shock protein (HSP27) in human corneal epithelial cells. Cornea 25:948–955

Shimura H, Tanaka R, Urabe T, Hattori N (2017) Heat shock protein 27 (HSP27) as a therapeutic target in ischemic stroke and neurodegenerative disorders. Juntendo Med J 63:17–21

Strand C, Warshaw JB, Snow K, Jacobs HC (1994) Heat shock does not induce tolerance to hyperoxia. Lung 172:79–89

Stuart RA, Cyr DM, Neupert W (1994) Hsp70 in mitochondrial biogenesis: from chaperoning nascent polypeptide chains to facilitation of protein degradation. Cell Mol Life Sci 50:1002–1011

Um HC, Jang JH, Kim DH, Lee C, Surh YJ (2011) Nitric oxide activates Nrf2 through S-nitrosylation of Keap1 in PC12 cells. Nitric Oxide 25:161–168

Varga ZV, Giricz Z, Liaudet L, Haskó G, Ferdinandy P, Pacher P (2015) Interplay of oxidative, nitrosative/nitrative stress, inflammation, cell death and autophagy in diabetic cardiomyopathy. Biochim Biophys Acta 1852:232–242

Wellen KE, Thompson CB (2010) Cellular metabolic stress: considering how cells respond to nutrient excess. Mol Cell 40:323–332

Wen J, Li H, Zhang Y, Li X, Liu F (2015) Modification of HSP proteins and Ca2+ are responsible for the NO-derived peroxynitrite mediated neurological damage in PC12 cell. Int J Clin Exp Pathol 8:4492

Wong HR, Menendez IY, Ryan MA, Denenberg AG, Wispé JR (1998) Increased expression of heat shock protein-70 protects A549 cells against hyperoxia. Am J Phys Lung Cell Mol Phys 275:L836–L841

Wu C, Clos J, Giorgi G, Haroun RI (1994) Structure and regulation of heat shock transcription factor. In: Morimoto RI, Tissieres A, Georgopoulos C (eds) The biology of heat shock proteins and molecular chaperones. Cold Spring Harbor Laboratory Press, Cold Spring Harbor, pp 395–416

Xin L, Wang J, Wu Y, Guo S, Tong J (2015) Increased oxidative stress and activated heat shock proteins in human cell lines by silver nanoparticles. Hum Exp Toxicol 34:315–323

Xu Z, Wang Z, Li JJ, Chen C, Zhang PC, Dong L, Wang ZL (2013) Protective effects of selenium on oxidative damage and oxidative stress related gene expression in rat liver under chronic poisoning of arsenic. Food Chem Toxicol 58:1–7

Zhang X, Lu L, Dixon C, Wilmer W, Song H, Chen X, Rovin BH (2004) Stress protein activation by the cyclopentenone prostaglandin 15-deoxy-delta12, 14-prostaglandin J2 in human mesangial cells. Kidney Int 65:798–810

Zhang Y, Ahn YH, Benjamin IJ, Honda T, Hicks RJ, Calabrese V, Dinkova-Kostova AT (2011) HSF1-dependent upregulation of Hsp70 by sulfhydryl-reactive inducers of the KEAP1/NRF2/ARE pathway. Chem Biol 18:1355–1361

Zhao FQ, Zhang ZW, Wang C, Zhang B, Yao HD, Li S, Xu SW (2013) The role of heat shock proteins in inflammatory injury induced by cold stress in chicken hearts. Cell Stress Chaperones 18:773–783

Zou J, Guo Y, Guettouche T, Smith DF, Voellmy R (1998) Repression of heat shock transcription factor HSF1 activation by HSP90 (HSP90 complex) that forms a stress-sensitive complex with HSF1. Cell 94:471–480

Part III
Stress Response Pathway

Chapter 8
High Content Screening of Small Molecule Modulators Targeting Heat Shock Response Pathway

Daniel Zhang and Bin Zhang

Abstract Phenotypic high content screening (HCS) has resurged in recent years as an important platform in the drug discovery paradigm, particularly to address the serious challenges in the current target based approaches. Using the highly conserved heat shock response (HSR) pathway as a therapeutic intervention point, we established a cell based, high throughput, multiplexing, and disease relevant phenotypic screening platform for small molecule HSF1 modulators in cancer and neurodegenerative diseases. In this chapter, the authors reviewed their systematic design of methodology and workflow in detail, including the characterization of cellular phenotypic changes, image analysis and quantification, assay development and automation, screening operation and quality control, counter-screen and lead optimization, target identification and mechanism of action study. Selected compounds from the phenotypic screening, including novel HSF1 activators and HSF1 inhibitors, were discussed in regard to their chemical structures, therapeutic effects, and cytotoxicity for potential drug development. The authors also addressed the uncertainties and risks in target deconvolution, which still remains to be the most difficult hurdle in phenotypic screening. Recent HCS advances in tissues, organs and whole organisms, such as 3D tissue imaging, organ-on-a chip, IPS derived cell models, relevant animal models, etc., may bring an effective solution for delivering target-specific, first-in-class drugs in the future.

Keywords Drug discovery · HCS · HSF1 · HSP · HSR · HTS · Small molecule · Target ID

D. Zhang
Department of Biology, Alpine Therapeutics, Inc., San Diego, CA 92129, USA

Massachusetts Institute of Technology, Cambridge, MA 02139, USA

B. Zhang (✉)
Department of Biology, Alpine Therapeutics, Inc., San Diego, CA 92129, USA

A. A. A. Asea, P. Kaur (eds.), *Heat Shock Proteins and Stress*,
Heat Shock Proteins 15, https://doi.org/10.1007/978-3-319-90725-3_8

Abbreviations

CV	Coefficient of variation
EC50	Half maximal effective concentration
FDA	Food and drug administration
GFP	Green fluorescent protein
HCS	High content screening
HSF1	Heat shock factor 1
HSP	Heat shock protein
HSR	Heat shock response
HTS	High throughput screening
IC50	Half maximal inhibitory concentration
iPS cell	Induced pluripotent stem cell
MOA	Mechanism of action
PD	Pharmacodynamics
PK	Pharmacokinetics
SAR	Structure activity relationship
shRNA	Small hairpin RNA
siRNA	Small interfering RNA
target ID	Target identification

8.1 Introduction

Target based drug discovery has been very successful in the past three decades, with delivery of several highly effective molecular therapeutics in the clinical applications (Workman 2005; Sawyers 2005), including small molecule drugs (Gleevec, Iressa, and Tarceva) and therapeutic antibodies (Herceptin, Erbutix, and Avastin). However, in recent years, "low-hanging fruit" in terms of validated and druggable targets have become depleted, particularly in the less understood disease areas such as neurodegeneration, metabolic syndromes and certain types of cancer. The discovery of first-in-class compounds or biologicals has become a significant challenge, with high attrition rate and low production observed in the process. As a result, only 40% of FDA approved first-in-class drugs (small molecule and biologicals combined) from 1999–2008 are through target based approaches (Swinney and Anthony 2011; Eder et al. 2014). Interestingly, the majority of the approved drugs with novel mechanisms of action (MOA) are the result of phenotypic screening. Thus, the phenotypic based approach has regained broad interests both in the academic institutions and pharmaceutical companies, with the hope of addressing the poor productivity of new drugs in the current target centric era (Scannell et al. 2016; Wagner 2016; Wagner and Schreiber 2016; Moffat et al. 2017).

Phenotypic screening features the disease relevant phenotype at the cellular level or more complex biological systems, in which a small molecule or antibody can be

Table 8.1 Phenotypic screening vs target based screening in drug discovery

	Phenotypic Screen	Target-based Screen
Prior knowledge of target or pathway	Unknown or limited information	Essential or well studied
	Good for novel target exploration	Conventional, relatively low risk
Methodology	Unbiased image based assays with cell lines, primary cells, iPS cells, 3D tissues or whole organisms	Target focused biochemical assays with recombinant proteins
Primary screen assay	Disease relevant phenotypes are essential	Relatively less expensive *in vitro* assays
	Complex setups with advanced instrumentations	High or ultra-high throughput
		Data analysis can be straightforward
	Medium or high throughput with large data set	Hit ID and ranking for single target
	Potential hits for different targets or MOAs	
Counter screen	Usually an extensive multi-assay screen	*In vivo* cellular data can be very different
	Removing hits with irrelevant phenotypes/MOAs	Removing false positive or cytotoxic hits
	Target ID and hit validation can be challenging	Cell permeability can be another issue
Lead selection/ optimization	Compound profiling & subcategory are essential for maintaining the same MOA	Relatively standard and predicable SAR and lead triage efforts
	Lead selection based on potency with minimum cytotoxicity and off-target effects.	Different pharmacophores can be combined for lead optimization
	Targets/MOAs are not essential for Go/ No-Go decision	Target ID is essential with defined MOA and minimal off-target effects

screened to induce or reverse the phenotypic changes to identify and validate new lead compounds or new drug targets (Boutros et al. 2015; Fetz et al. 2016; Pegoraro and Misteli 2017). Comparing to the target based drug discovery, phenotypic screening offers many advantages (Table 8.1): First, a particular phenotypic change is usually tightly linked to the relevant disease, providing an excellent starting point for small molecule/biologicals screening and intervention. With ever improving imaging tools, complicate phenotypic changes can be accurately captured and quantitatively measured for possible assay development (Mattiazzi et al. 2016). Second, knowledge of molecular targets is unnecessary at the start of the screen, making it possible to identify novel molecular targets or novel MOAs in an unbiased fashion. This feature is extremely applicable to neurodegenerative diseases, where distinguished cellular aggregations in the neuronal cells are well known whereas little is understood in regard to the underlying molecular targets or MOAs (Khurana et al. 2015). Third, a phenotypic screen is performed with intact cells, enabling

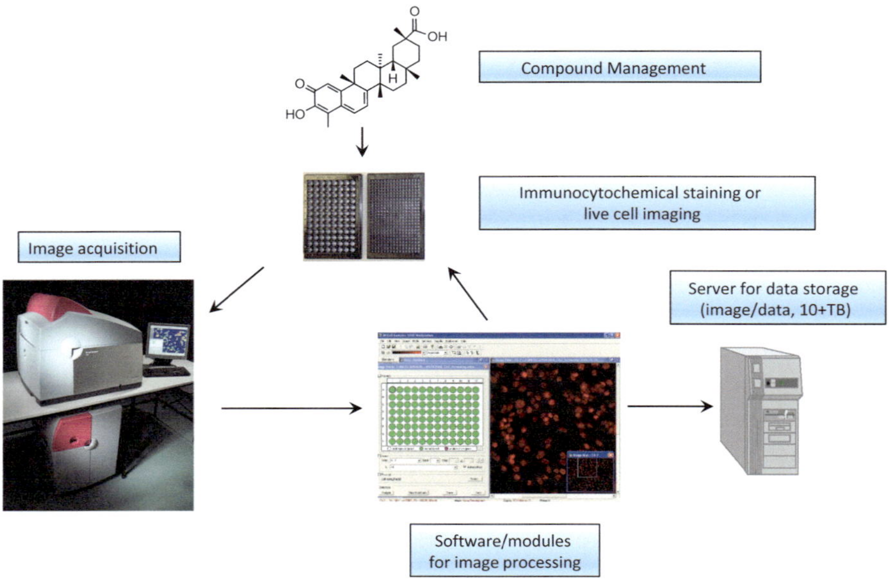

Fig. 8.1 Illustration of a typical HCS instrumentation setup and compound screening operation

direct evaluation of screening hits on cell permeability and efficacy. In comparison, target based drug discovery relies on a purified recombinant protein outside its relevant cellular context, which may cause undesirable downstream consequences. However, phenotypic screening has considerable challenges, particularly with target identification and hit validation (Nijman 2015). Subcategory of screening hits with same phenotypes is the key for maintaining the same MOA, thus producing predictable responses after compound treatment. In addition to difficult target deconvolution, expensive imaging instruments and complicated software setups create a burden in the costly drug discovery programs (Moffat et al. 2017).

High throughput phenotypic screening has become a reality due to recent developments in high content image based screening (HCS) (Thomas 2010). As shown in Fig. 8.1, this platform utilizes high density plates (384 wells or 1536 wells), robotic plate delivery, automatic image acquisition instruments and advanced image data processing software, thus substantially removing the bottle necks in the large scale phenotypic screening. As a result, large compound libraries can be screened and analyzed in a short period of time, sometimes closely matching the throughput level of a biochemical assay (Chessel 2017). Several disease relevant phenotypic models have been successfully investigated for small molecule screening with HCS format, such as anti-infectious disease models (Gao et al. 2010), PCSK9 reduction in lowering LDL cholesterol level (Petersen et al. 2016) and TGFβ signaling in rare monogenic disease RDEB (Nyström et al. 2015), etc.

8.1.1 Heat Shock Response (HSR) and Small Molecule Intervention

Heat shock factor 1 (HSF1) is a master transcriptional regulator that activates gene transcription of a family of heterogeneous molecular chaperones called heat shock proteins (HSP), in response to a wide range of intracellular proteotoxic stress (Gomez-Pastor et al. 2017). HSP can recognize hydrophobic regions from the interior of proteins and protect or rescue these client proteins from misfolding and denaturation. As a result, HSP can facilitate the majority of intracellular proteins toward functional conformations and restore homeostasis (Lindquist and Kelly 2011; Saibil 2013). This highly conserved pathway is well controlled at multiple levels. Upon cellular stress, HSF1 undergoes a stepwise activation process including trimerization, nuclear translocation and inducible posttranslational modifications (phosphorylation, sumoylation, and acetylation). The active HSF1 trimer recognizes the promoter binding site, a consensus sequence block called heat shock element for induction of HSP (Dayalan and Dinkova-Kostova 2017). Beyond the regulation of HSP genes, HSF1 also targets a large group of other genes for various cytoprotective functionalities (Anckar and Sistonen 2011). The important role of HSF1 in cellular adaptation and survival has led to the hypothesis that some diseases, particularly cancer and neurodegenerative diseases, are the accumulated results from abnormal regulation of HSF1 and HSR pathway (Calderwood and Murshid 2017).

The impairment of HSF1 activity is a common defect in protein misfolding based neurodegenerative diseases, as in the case of Alzheimer's disease, Parkinson's disease, Amyotrophic lateral sclerosis, etc. Decreased expression of chaperone proteins can result in protein aggregation, proteotoxicity, neuronal dysfunction and cell death (Neef et al. 2011). Several therapeutic active compounds were reported through the activation of HSF1/HSSP stress response, including celastrol, 17-AAG, arimoclomol, etc. (Dayalan and Dinkova-Kostova 2017). Positive results from animal and clinical studies suggest this approach is an attractive strategy to reverse the degenerative process (Phukan 2010; Xu et al. 2011).

In contrast, overexpression of HSF1 with increase of HSP is observed in a broad range of tumors and tumor cell lines, enabling uncontrolled malignant growth, metastasis and angiogenesis (Vydra et al. 2014). Several studies were performed with HSF1 deficiency transgenic mouse models and human cancer cell lines with shRNA knockdown of HSF1. The results confirmed that tumor growth is significantly inhibited in HSF1−/− mice, and tumor cells have a much greater dependence on HSF1 function compared to normal cells (Dai et al. 2007). This "HSF1 addiction" phenomenon suggests an effective anticancer strategy by inhibiting HSF1. A number of small molecule inhibitors of HSF1 were identified with positive results in the preclinical setting (De Thonel et al. 2011). In summary, the reciprocal dysregulation of HSR in cancer and neurodegenerative diseases enhances our view that HSF1 plays a central role in proteostasis and pathological conditions, with great therapeutic potentials for small molecule intervention in these two important disease areas (Dayalan and Dinkova-Kostova 2017).

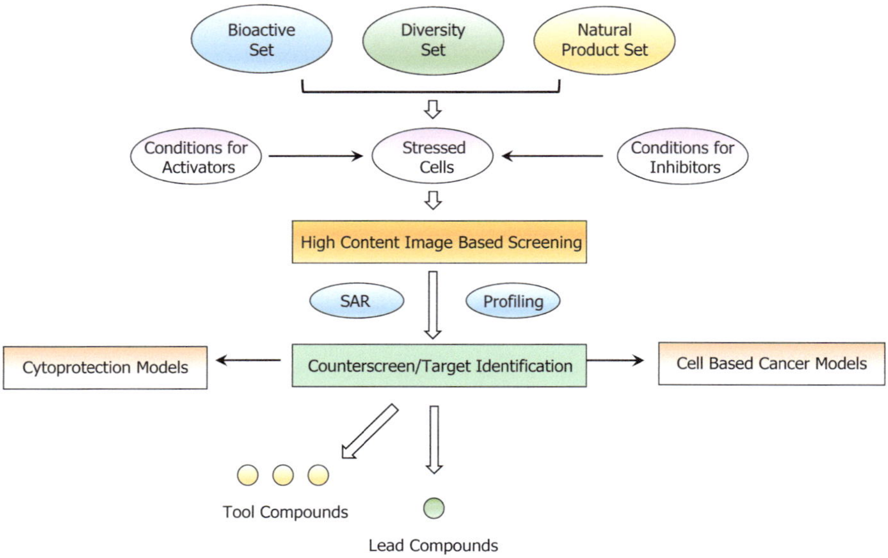

Fig. 8.2 Design of a HCS platform for small molecule modulators of HSR pathway. Bold arrows indicate the workflow starting from a combined set of libraries (bioactive, diversified or natural derived compounds) to lead identification and validation

Although efforts have been made to screen for pharmaceutical HSF1 modulators, trimeric nature of HSF1 active molecule, along with its complicated posttranslational modification state, make it difficult to develop an *in vitro* biochemical assay. Instead, many studies were performed using engineered cell lines, containing luciferase or GFP (green fluorescent protein) reporter system under the control of a selected HSP promoter (Yoon et al. 2011; Santagata et al. 2012; Santagata et al. 2013). Several small molecule inhibitors were discovered, which exhibit potent effects on tumor cell lines and mouse tumor models (Yoon et al. 2011; Santagata et al. 2012). However, some intrinsic limitations may lead to biased results. For example, the engineered cell line may behave differently compared to the native cells, which can generate artificial effects that are unrelated to the disease's phenotype. Furthermore, the choice of heat shock element (HSE) in the promoter region can be biased and makes it impossible to monitor different HSP simultaneously.

We decided to use a cell based HCS platform as a primary screening tool for HSF1 modulators. The goals of our research were two-fold. On one hand, we would like to discover novel HSF1 modulators with high potency and unique chemotypes, for molecular intervention of cancer and neurodegenerative diseases. On the other hand, we would also like to develop a set of tool compounds that can functionally dissect HSR, allowing thorough understanding of MOA with defined targets in this important pathway. Figure 8.2 demonstrates our overall design for a cell based HCS of small molecules targeting HSR pathway, with a focus on the therapeutic intervention of neurodegenerative diseases and cancer. For HCS applications in more complex models, such as 3D tissues, organs, whole organisms or iPS derived cell models (induced pluripotent stem cell), the readers can refer to several recent reviews (Bhatia and Ingber 2014; Bilgin et al. 2016; Li et al. 2016; Shi et al. 2017).

8.1.2 HSF1 Granule: Stress Specific Cellular Phenotype

HSF1 forms distinguished nuclear granules when transcriptionally activated in stress-stimulated cells (Cotto et al. 1997; Sandqvist and Sistonen 2004). When stained with HSF1 immunofluorescent antibodies, these nuclear granules are large and bright, but are not colocalized with other known subnuclear components. Jolly et al. showed these stress granules are primarily located at the 9q12 region through direct DNA-protein interaction with satellite III repeats (Jolly et al. 2002; Jolly et al. 2004). In addition, Denegri et al. reported the same target region and also suggested that the centromeric regions of chromosomes 12 and 15 could be targets for the HSF1 granules (Denegri et al. 2002). Although there is no evidence that HSF1 stress granules colocalize with the transcription sites of the known HSP, the kinetics of HSF1 granule formation correlates well with the transcriptional activity of HSF1. For example, HSF1 can be detected in granules within 30 s of heat shock (Jolly et al. 1999). During attenuation and recovery process from heat shock, the HSF1 granules disappear gradually (Cotto et al. 1997). The similar kinetics of HSF1 granule formation were observed in a number of human primary and cancer cell lines, including *Hela*, A431, HOS, primary fibroblasts and epithelial cells, etc. (Cotto et al. 1997). In summary, all these evidences strongly suggest that HSF1 granule is a relevant phenotype in the stressed cells.

HSP70 protein is one of the most important and abundant molecular chaperones (Saibil 2013). The major function of HSP70 is to provide cytoprotection against stress-induced protein misfolding or denaturation. In addition, constitutively expressed HSP70 protein also plays important housekeeping roles in nonstressed cells (Daugaard et al. 2007). Following heat shock, there is a significant increase of HSF1 induced HSP 70 expression, and the majority of newly synthesized HSP70 rapidly migrate into the nucleus. Zeng et al. reported that GFP-HSP70 exhibits a significantly increased signal and becomes highly concentrated in the nucleoli and forms HSP70 granule (Zeng et al. 2004). The formation of HSP70 granule strictly coincides with the presence of HSF1 granules and is significantly downregulated in the thermotolerant cells (Alastalo et al. 2003). Therefore, both HSF1 and HSP70 granules are well documented phenotypic changes in the heat shocked cells, which can be used as highly relevant cellular markers of HSR for quantitative image based analysis.

8.1.3 HCS Assay Development

The following assay conditions were carefully designed and validated with experimental details published in several previously published articles (Au et al. 2008; Au et al. 2009; Zhang et al. 2008; Zhang et al. 2009a; Zhang et al. 2016a).

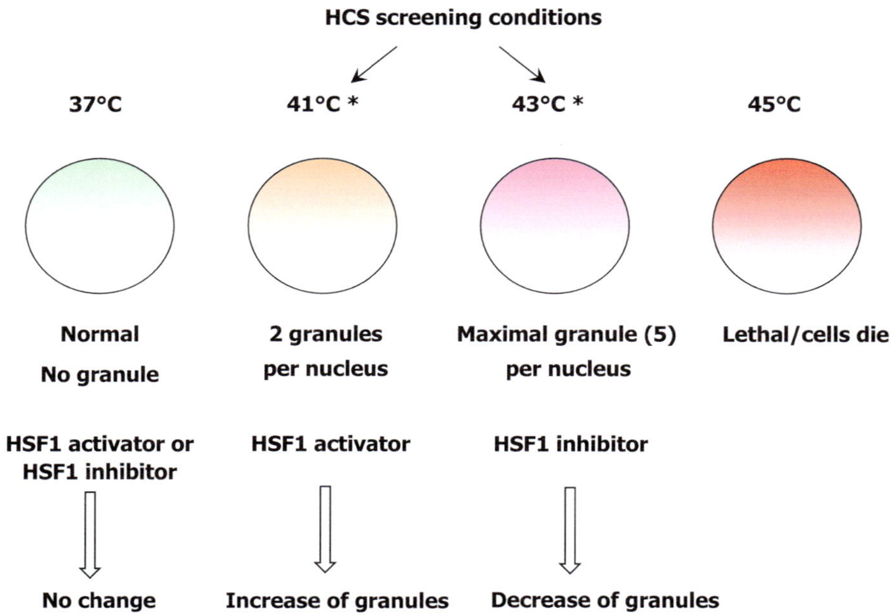

Fig. 8.3 Establishment of heat shock conditions for a robust HCS screening. To eliminate edge effects, customized aluminum plates were used for even heat transduction during heat shock process. For activator screening, average granule number per nucleus in *HeLa* cells is 2.4 ± 0.22 at 41°C for 2 h (Au et al. 2008). For inhibitor screening, average granule number nucleus in *HeLa* cells is 5 ± 0.14 at 43°C for 2 h (Au et al. 2009). The design allows for the screening of either HSF1 activators or inhibitors with quick change of heat shock temperature

1) Heat shock conditions for large scale compound screening:

Since we need to perform image acquisition of the stressed cells, constant and even heat shock conditions are critical. When regular air heat (from an incubator) was used for the heat shock experiments, we observed a large intraplate variability (CV >40%, coefficient of variation) of the HSF1/HSP70 granule count throughout the 96-well or 384-well plates (data not shown). Our solution was to use custom-made aluminum plates for even heat transduction. The aluminum plates can be preheated in an incubator at the designated temperature the day before experiment. As a result, significantly reduced CV (<7.94%) was documented in the 96-well or 384-well plates with no edge effects.

We then investigated the heat shock temperature and duration time with different cell lines. HeLa cells demonstrated the most constant phenotypic changes, i.e. HSF1 and HSP70 granule formation corresponding to the heat shock conditions. As illustrated in Fig. 8.3, HeLa cells showed an average of 2 granules per nucleus at 41°C for 2 h, and exhibited maximum granule (average of 5) granules per nucleus at 43°C at 2 h. At the lethal temperature of 45°C, most of stressed cells died. Our results were in line with previous studies in other laboratories.

For HSP70 and other HSPs (such as HSP90), we performed comprehensive kinetic studies to determine the induction window. Recovery conditions were set to incubate at 37 °C for 0 h recovery (R0, no recovery time), 2 h (R2), 4 h (R4), 6 h (R6), and 24 h recovery (R24), respectively. In our experience, HSP70 expression/granule formation peaked at R2, which was the time point chosen for HCS screening.

2) Assay controls

Celastrol was selected as a positive control for screening of small molecule activators (Westerheide et al. 2004). Figure 8.4A shows the granule formation in HeLa cells of HSF1 (R0, b) and HSP70 (R2, d) induced by 2 μM of celastrol at 41°C for 2 h, as compared to DMSO (negative control, a and c). These conditions were used for HSF1 activator screening (Au et al. 2008).

For inhibitors of HSR pathway, we performed kinetic studies with different concentrations of triptolide, a known inhibitor of HSR (Westerheide et al. 2006). As shown in Fig. 8.4B, our results confirmed this compound can decrease of HSF1/HSP granule formation at 43°C from 1 to 4 h, with 2 h as an optimal condition for HSF1 inhibitor screening (Au et al. 2009).

3) Multiplexing immunocytochemical assay

Immunocytochemical staining assays with HeLa cells were performed in 384 well plates following the previous publications (Zhang et al. 2008; Au et al. 2008; Au et al. 2009). Multiplexing DAPI/HSF1 (rhodamine)/HSP70 (FITC) tricolor staining was performed and validated, with automation and liquid handling instruments for large scale operation. To further understand the dynamics of HSP70 and HSP90 after HSF1 activation or inhibition, we also performed DAPI/HSP70 (rhodamine)/HSP90 (FITC) multiplexing experiments after heat shock.

4) High throughput image acquisition, processing and quality control

Image acquisition was performed using an INcell 1000 (GE Healthcare, Piscataway, NJ) system integrated with a Twister II (Caliper Life Sciences, Hopkinton, MA) robot for automated plate delivery. This real-time, walk-away assay has a throughput up to 50–100384 -well plates per day. An online image analysis was carried out using Multi Target Analysis (MTA) module from Workstation 3.6 (GE Healthcare, Piscataway, NJ). Algorithms for HSF1/HSP70 total granule count, granule area, and nuclear intensity coefficient of variation (CV) were established according to assay conditions and manufacture instructions. It should be noted that multi-channel high content images need large data storage space (up to tens to hundreds of terabytes), particularly when initiating a large-scale HCS campaign.

As for data quality control, Z' factor was adopted for monitoring robustness and reproducibility of HCS assay over the entire activator or inhibitor screening campaign (Zhang et al. 1999). For example, an average Z′ of 0.65 (for an ideal assay, Z' > 0.5) was obtained for the positive control with a satisfactory signal/noise ratio of 11.25 and a relatively low CV of 12%, proving that the adaptability of our HCS assay for automatic operation of HSF1 inhibitor screening (Au et al. 2009).

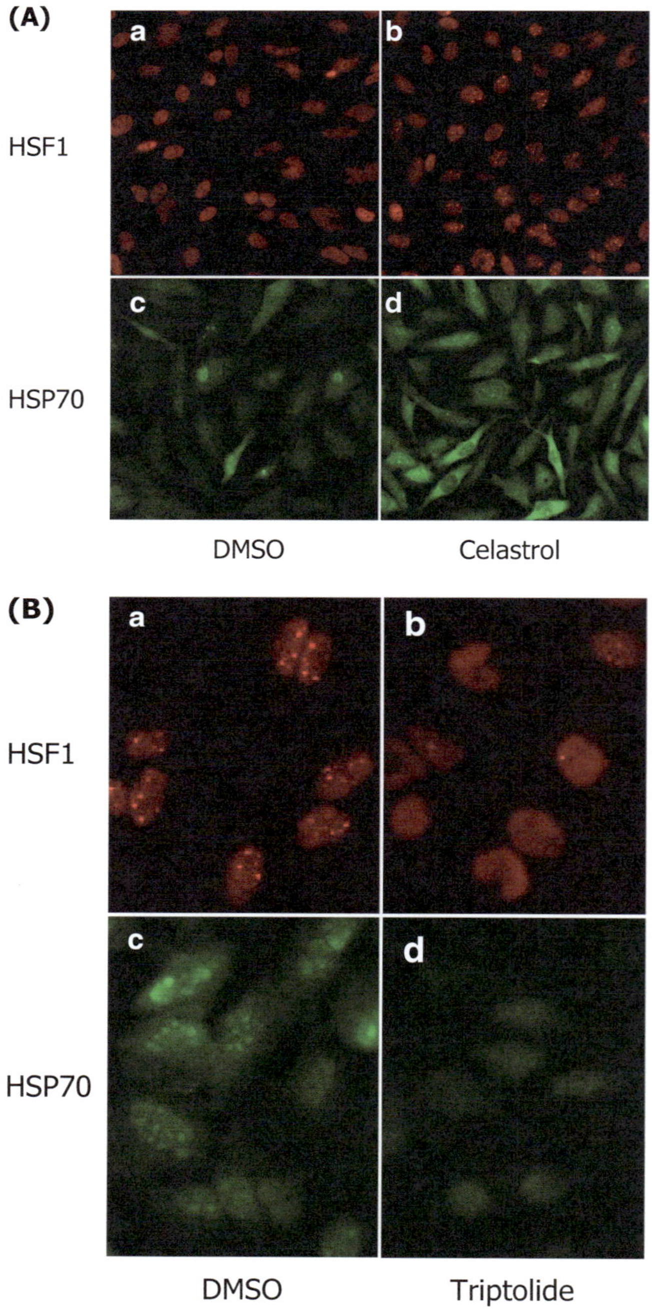

Fig. 8.4 Small molecule induced nuclear granule number changes of in the heat shocked *Hela* cells. (**A**) Increase of nuclear granule formation of HSF1 (b) and HSP70 (d) by 2 μM celastrol, a positive control for HSF1 activator screening. DMSO treated cells were shown in (a) and (c) (**B**) Inhibition of nuclear granule formation of HSF1 (b) and HSP70 (d) of by 10 μM triptolide, a positive control for HSF1 functional inhibitor screening. DMSO treated cells were shown in (a) and (c)

5) Compound ranking and subcategory of primary screening hits.

For the primary screen, we screened compounds at 10 μM in duplicate with multiplexing HCS assay on the heat shocked cells. As a control, the same compounds were applied to the unstressed cells (at 37°C instead of at the heat shock temperature). Compounds that activated the HSR pathway at 37°C were considered unspecific chemical stressors, and were removed from further studies. Selected hits were tested in a dose-dependent format (5 nM to 100 μM concentration range), with a hit list ranked by EC50 (half maximal effective concentration) or IC50 (half maximal inhibitory concentration) of HSF1 granule formation.

Next, screening hits were subcategorized based on their phenotypic behaviors, such as nuclear HSF1 granules, cellular HSP70 staining intensity, or nuclear fragmentation, etc. Several commercially available software tools were particularly useful for 3D visualization of multi-parameter image data. We integrated a customized Pipeline Pilot (Accelrys, San Diego, CA) dataflow platform with Spotfire software package (TIBCO, Somerville, MA), for real-time image data analysis and visualization.

HCS can be also used as a secondary screen, which follows the non-targeted primary screening with functional assays. The combined set of data, visualized and subcategorized by Spotfire software, can lead to an informative process during SAR, lead optimization and MOA study. For example, a set of 4000 bioactive compounds was screened by MG132 neuronal assay (ER stress cell based model) and MTS cytotoxicity assay, followed by HSF1/HSP70 testing (Fig. 8.5A). A set of 99 cytoprotective compounds was selected based on the preset cutoff values of MG-132 and MTS assays, followed by sorting these compounds into four different phenotypic groups (Fig. 8.5B). For the subgroup of HSF1+/HSP+, we discovered 11 compounds which are potential HSF1 activators through the desirable MOA. Other chemical series may exhibit different or unexpected phenotypes, such as compounds in HSF1-/HSP+ or HSF1-/HSP- subcategory, indicating a cytoprotection mechanism independent of HSF1. In addition, we also discovered several different chemical series with different kinetics of HSP induction, or inducing different types of HSPs. Efforts have been made to combine these data sets with the corresponding gene expression profiling results, thus facilitating pathway analysis and target ID studies (data not shown).

8.1.4 Small Molecule Library Construction

When we started building our in-house compound library, we decided to select three different categories of commercially available compounds to cover a broad chemical space (Table 8.2). The first category is called "chemogenomics library," containing compounds with known biological activities or known drug targets. Many published papers are available for these compounds, in addition to the public database "The Chemical Probes Portal" (http://www.chemicalprobes.org). Several small compound collections including LOPAC (Sigma-Aldrich, San Louis, MO), Pharmakon Collection (Microsource Discovery Systems, Gaylordsville, CT), Prestwick Chemical Library (Prestwick, San Diego, CA), and NIH compound

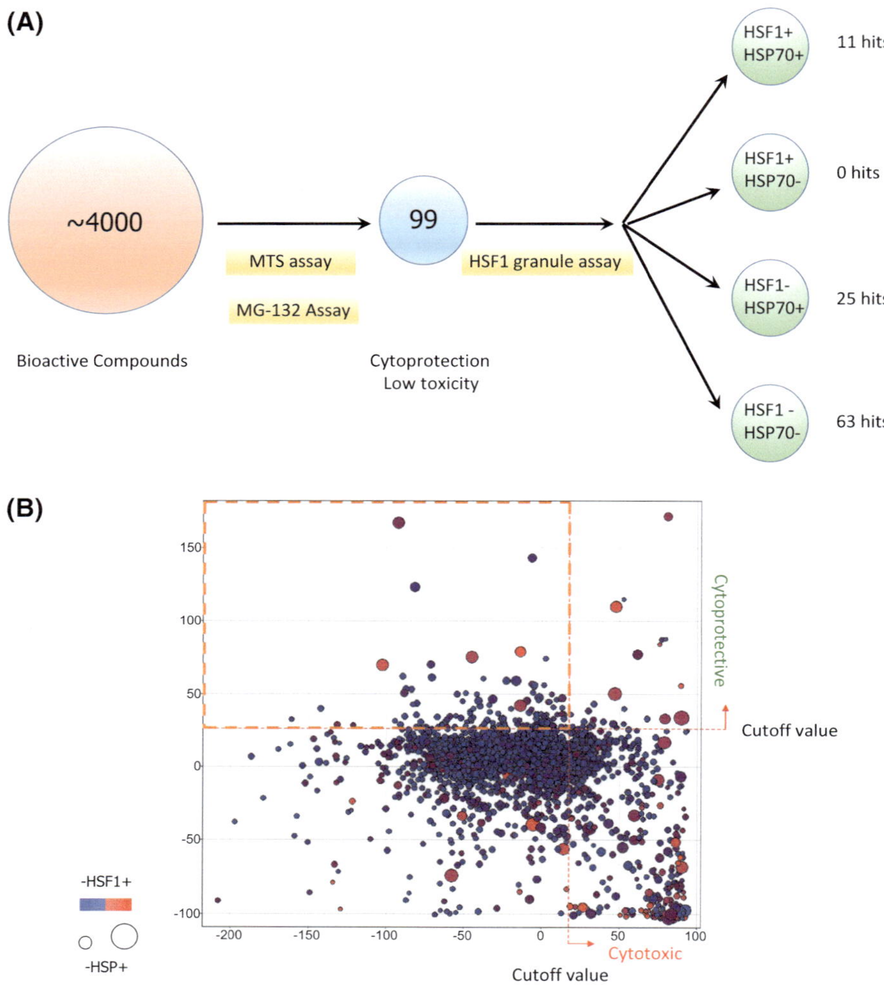

Fig. 8.5 (**A**) Subcategory of screening hits from a library of ~4,000 bioactive compounds. MG-132 cytotoxicity assay and MTS cytotoxicity assay were both performed as primary screening methods. Selected hits were applied to the HSF1/HSP70 granule assay (secondary screening method). (**B**) 2D visualization of four parameters for all 4,000 compounds, which includes Y axis, percent increase in neuronal cells (SK-N-SH); X axis, percent decrease in total live normal cells (HEK293). Quantification of HSF1 and HSP 70 granule formation (HeLa) were indicated with a color scale bar for HSF1 (from blue or low to red or high in HSF1 granule formation) and a size scale bar for HSP70 (from small or low to large or high in HSP70 granule formation). Cut off values (percentage of compound value compared to DMSO): MG132 assay, >30%; MTS assay, >80%; HSF1 assay, >20%; HSP assay, >30%. The selected subcategory of screening hits is inside the orange box with desirable phenotypes

Table 8.2 High content screening identified small molecule modulators in heat shock pathways

Targeted disease	Compound library	HCS Assay	Counterscreen Assay (Cell line/protein)	Lead compounds/Possible targets	References
Neurodegenerative diseases	Bioactive compounds:	Cell line: HeLa (adenocarcinoma)	MTS cytotoxicity assay (multiple lines)	A) Pyrimido[5,4e] [1,2,4] triazine-5,7(1H,6H)-dione derivatives	Au et al. (2008) Cher et al. (2010)
	LOPAC	Phenotypic change:	Colony formation assay (CD34+)		
	Prestwick	HSF1/HSP70 granule formation	HSP expression profiling (HeLa)	B) Gedunin derivatives	Zhang et al. (2009a)
	Microsource	Heat shock inducer:	MG-132 toxicity assay (SK-N-SH)	C) Chloro-oxime derivatives	Zhou et al. (2009a)
	In-house library	41°C for 2 h (HSF1 activator)	Western blot analysis (multiple lines)	Targets to be determined	Zhou et al. (2009b)
	(Arimoclomol-like compounds)	43°C for 2 h (HSF1 inhibitor) Recovery time 2 h (R2)	Transient transfected HTT model (HeLa)		Zhou et al. (2009c)
	NIH clinical collection	Control: 37°C for 2 h	Doxcycline induced HTT model (PC12)		Yoon et al. (2010)
	Chemically diverse compounds	Positive control compounds Celastrol (activator screening)	Rotenone stress assay (SK-N-SH)		
		Triptolide (inhibitor screening) Immunocytochemistry staining	Oxygen & glucose deprivation (SHSY5Y)		

(continued)

Table 8.2 (continued)

Targeted disease	Compound library	HCS Assay	Counterscreen Assay (Cell line/protein)	Lead compounds/Possible targets	References
Cancer	Chembridge	Multiplexing imaging	MTS cytotoxicity assay (multiple lines)	D) PW3405	Au et al. (2009)
	Life Chemicals	DAPI/HSF1/HSP70	Colony formation assay (CD34+)	E) AC-93253	Zhang et al. (2009a)
	ApexScreen	Image analysis and processing	HSP expression profiling (HeLa)	Primary targets to be determined	Zhang et al. (2016)
	Natural products and derivatives		Anchorage independent growth		Unpublished data
			Phosphatase assay (Phosphatases)	Possible targets involved AC-93253 inhibits Sirt 2	
	TimTec		Sirtuin biochemical assay (Sirtuins)		
	Analyticon		Kinase biochemical assay (kinases)		
	Microsource				
	Combined set of 100,000 compounds				

collections are extremely valuable, which may provide clues for a particular MOA. In addition, some of these compounds are FDA approved drugs so we can explore drug-repurposing opportunities.

The second category of compounds are from structurally diversified libraries with different chemotypes. Although most of these compound libraries have been heavily screened with the target based approaches, the inactive compounds, or "dark chemical matter," may contain hits with novel cellular targets (Wassermann et al. 2015). In addition, it has been suggested that the molecular targets of known drugs or tool compounds only account for 3 and 6% of the whole human proteome, respectively (Santos et al. 2017). Thus, many targets, some of them belonging to underprivileged target families, beg for innovative chemogenomic approaches (Jones and Bunnage 2017). In our design, we selected diversified compound libraries from Chembridge (San Diego, CA), Maybridge (Altrincham, United Kingdom), Life Chemicals (Burlington, Ontario, Canada), etc. We excluded promiscuous compounds with low cellular permeability, such as charged compounds or large molecular weight molecule (>500 Da).

Lastly, the third category of compounds are from natural products and derivatives. Natural compounds make up more than 25% of the new chemical entities that FDA has approved since 2010 (Harvey et al. 2015). Particularly, some of our control compounds are natural products, such as celastrol and triptolide, making this category critical for the success of our HCS campaign. Commercially available compounds included libraries from Microsource Discovery Systems, TimTec (Newark, DE) and Analyticon Discovery (Rockville, MD). Customized natural product libraries, both the diverse set and the screening hit focused set, were also synthesized for HSF1 inhibitor screening. The major concern for this category is in hit triaging and lead optimization phase, when a natural product series is structurally too complicated for structure-activity relation (SAR) study.

To facilitate target ID and deconvolution, chemogenomics libraries with compounds of known targets and MOAs, were usually used as reference libraries for the pilot screen. The large-scale screening of chemically diverse set or natural product set was always performed after data were available from reference libraries.

8.1.5 Counter-Screen and Lead Confirmation

As phenotypic screening opens up a significantly larger target space for investigation, lead series identification and validation becomes much more challenging than that of target based screening. In addition, the confirmed hits may bind to a target with significant safety liability. A rigorous counter-screen is essential for the success of our HCS campaign. A set of *in vitro* biochemical and cell based assays were incorporated to remove false positive hits or hits with undesirable safety profiles in early stage, followed by *in vivo* testing on disease relevant animal models. As shown in Table 8.2, our counter-screen assays evaluated selected hits with several

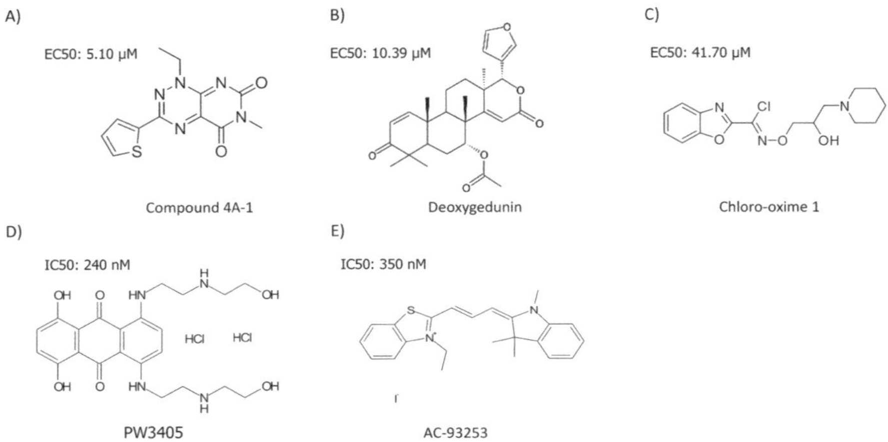

Fig. 8.6 Representative lead compounds from HCS screening with the chemical structures of HSF1 activators shown in (**A–C**) with their respective EC50 values in the HSF1 granule assay. The chemical structure of HSF1 inhibitors are shown in (**D**) and (**E**) with their IC50 values

follow-up assays on cytotoxicity, HSP expression profiling, HSF1 dependency, disease relevant cell based models, and MOA studies. Representative lead compounds and their structure are shown in Fig. 8.6.

1) Cytotoxicity evaluation

As the hit rate of our HCS was over 0.5%, much higher than conventional target based screening, the cytotoxicity profiling was prioritized in all discovered hits. A high throughput MTS viability assay was adopted to test compounds in several normal cell lines including HuVec (human umbilical vein endothelial cells), HMEC (human mammary epithelial cells), PrEC (human primary prostate epithelial cells) and WI-38 (human fibroblast cells). In our experience, hits should be removed from further consideration if they don't meet the stringent safety profiles, with a calculated therapeutic index over 100. If a bioactive compound has a known cellular target, all public available safety data, including cardiotoxicity and hepatoxicity, were analyzed to enable predictive toxicology (Persson and Hornberg 2016; Donato et al. 2017). To differentiate compound structure based vs MOA based cytotoxicity, active hits along with their structure-related inactive counterparts were often tested in parallel. Although not in the scope of this review, cytotoxicity evaluation in the relevant 3D tissues or mouse models should be introduced earlier in the lead selection process (Li et al. 2016). As a result, we can focus on a much shorten hit list for lead selection and validation. Special attention was also given to the safety profiling data during lead optimization phase, since there was a high probability that compound derivatives from SAR study bind to a different target, with different safety issues.

2) HSP expression profiling

HSP expression profiling is the key parameter for hit confirmation, which was achieved through a high throughput image based assay and a low throughput western blot assay. In our HCS approach, we selected 2 recovery time points after heat shock to measure HSP70 and HSP90 expression, including R2 and R4 (recovery at 2 or 4 h at 37°C), allowing response time for transcription and translation of HSPs. This design also captured compounds with different MOAs, i.e. immediate induction vs delayed induction. HSP expression was evaluated by granule formation (HSP70), cellular intensity or cellular intensity CV, following MTA modules of Workstations (GE Healthcare). Interestingly, screening hits demonstrated different expression profiles with HSP70 and HSP90. For example, some HSF1 inhibitors completely turn off HSF1 transcription activity with minimal HSP70 and HSP90 expression observed, while other hits only inhibit HSP70 or HSP90 (unpublished data). The different HSP expression profiles of selected hits were further confirmed by western blot analysis. In summary, we believe proper expression of HSP is a shared mechanism between HCS assay readout and relevant disease models. Accurate category of screening hits by individual HSP expression levels and kinetics provided us with a crucial tool to track underline MOA or potential target(s).

3) HSF1 dependency

It is crucial to validate that HSF1 is the relevant target responsible for compound-induced cellular response such as increased HSP expression (by HSF1 activator) or decreased HSP expression (by HSF1 inhibitor). Both siRNA (for transient transfection) and shRNA (for stable transfection) were used for HSF1 knock down. Particularly, we constructed a matching pair of HeLa cell lines stably transduced with a retroviral vector that contains either shRNA to specifically knock down HSF1 expression or control vector with normal HSF1 levels. Western blot confirmed that more than 90% of HSF1 was depleted in HeLa-HSF1 shRNA line. All selected hits were tested with this pair of cell lines for their HSF1 dependency. For example, in our HSF1 inhibitor screening, HSF1 functional inhibitor PW3405 (Fig. 8.6(D)) caused significantly more cancer cell killing compared to DMSO controls in Hela-vector shRNA control cells than that of HeLa-HSF1 shRNA cells lacking HSF1. The fact that PW3405 increased heat shock–induced cell death only in the presence of HSF1 further confirmed an HSF1-dependent MOA.

Our lead selection was based on the stringent criteria that the compounds should meet all phenotypic profiles (HSF1 and HSP profiles with no cytotoxicity) before lead optimization phase. Moreover, each chemotype from our phenotypic screen may represent a distinct mode of action with a unique target, which made our hit follow-up study extremely complicated. Careful selection of lead candidates can allow us to effectively use our resources at the early stage of the HCS campaign, thus facilitating a focused SAR study for lead optimization. In addition to lead candidates, some of the screening hits exhibited distinguished phenotypes with highly reproducible testing results, but failed to meet the lead selection criteria due to chemical liability. These tool compounds were documented for the purposes of assay development and subcategory reference. It proved to be a considerably valuable collection of compounds with their cellular signatures in HSR pathway. Similar strategy was also recently reported with NIH LINCS (The Library of Integrated Network-Based Cellular Signatures) L1000 big data approach, for compound profiling on the panel of more than 3000 potential drug targets (Liu et al. 2015).

4) Disease relevant cell based systems

A number of cell based assays relevant to cancer and neurodegenerative diseases were developed for confirming cytotoxicity of tumor cells (for HSF1 inhibitor screening) or cytoprotection of neuronal cells (for HSF1 activator screening). The details and related publications are listed in Table 8.2.

5) MOA study of HSF1 activators

In our practice, MOA study was performed in parallel with SAR of lead candidates. The goal was to obtain a clinical candidate with understanding of its MOA. As stated earlier, the complex mechanisms of HSF1 regulation at the posttranslational level make it difficult for target deconvolution. For the case of HSF1 activators, our understanding of compound induced HSF1 activation process is extremely limited. Figure 8.6(A)~8.6(D) demonstrate the chemical structures of three selected lead compounds. These compounds have 5~40 μM potency in HSF1 granule assay (Zhang et al. 2009a; Zhou et al. 2009b; Zhou et al. 2009c). After treatment with these compounds, prolonged activation of HSF1, along with amplified HSP expression, was confirmed in the heat shocked cells. Importantly, no effects were observed in cells without stress. As a result, potent cytoprotective activities were recorded in various disease relevant cell based models (Table 8.2), such as transiently or stably transfected poly Q-Htt models (relevant to Huntington's disease), rotenone neuronal cytotoxicity model (relevant to Parkinson's disease), MG-132 cytotoxicity model (relevant to ER stress), oxygen glucose deprivation model (relevant to ischemia and stroke), etc. Multiple HSF1 knockdown studies supported the notion of HSF1 dependency for compound induced cytoprotection in different cell based systems, although no individual cellular target was successfully identified (Zhang et al. 2009a).

SAR and lead optimization led to the discovery of a pyrimido[5,4-e] [1,2,4] triazine-5,7(1H,6H)-dione derivative (Compound 4A-1, Fig. 8.6(A)), which has a sub-micromolar potency (EC50 value of 0.41 μM) in the rotenone neuronal cytotoxicity model (Zhou et al. 2009c). The results from rat liver microsome study demonstrated that this compound has favorable metabolic stability with medium-high aqueous solubility. The good drug-like property makes this compound a potential clinical candidate in the treatment of Parkinson's disease (Zhou et al. 2009a).

6) MOA study of HSF1 inhibitors

Extensive target ID studies were also pursued on HSF1 functional inhibitors, with two representative lead compounds and their IC50 values shown in Fig. 8.6(D) and 8.6(E). These two compounds potently inhibit HSF1 granule formation at

200–400 nM range, with selective killing of tumor cells in multiple cancer cell lines (Zhang et al. 2009a; Zhang et al. 2016a). Given the major posttranslational modifications of HSF1 are phosphorylation and acetylation during heat shock process (Gomez-Pastor et al. 2017), significant efforts were focused on the two types of enzyme, kinase and deacetylase. Recombinant enzymes and *in vitro* biochemical assays were commercially available or developed in house, allowing us to establish a streamlined matrix system for measuring the binding affinity and specificity of selected compounds against the specific enzyme panels.

A total of 20 Ser or Thr phosphorylation sites were reported in the posttranslational modification of HSF1, making kinase the most prioritized category for target ID search (Gomez-Pastor et al. 2017; Dayalan and Dinkova-Kostova 2017). The activating sites are Ser320, Ser326, and Ser419 while the repressive sites are Ser121, Ser172, Ser303, Ser307 and Ser363 (Gomez-Pastor et al. 2017). Phosphorylation of Ser326 proved to be the most reliable indicator when HSF1 is activated, as confirmed in our HCS image assay and western blot study (Au et al. 2009; Zhang and Zhang 2016). For example, PW3405 (Fig. 8.6D) from our inhibitor screening significantly reduced the phosphorylation level of Ser326, followed by reduced expression of HSP proteins. shRNA knockdown of HSF1 confirmed the HSF1 dependency for selective killing of cancer cells (Zhang and Zhang 2016). The potency of PW3405 in HSF1 granule assay suggests that it would bind to its cellular targets with high affinity in the sub-micromolar range. *In vitro* kinase panel screen indicated that no high affinity kinase targets were found, including several well-known kinase targets (such as ERK1 and ERK2) in the HSF1 regulation. There is a possibility that PW3405 can activate ERK signal pathway through binding to an upstream target, thus enhancing the phosphorylation of Ser326 of HSF1. It is also possible that PW3405 exhibits polypharmacology, that is it exerts the therapeutic effects on the synergistic mechanisms of two or more targets (Zhang et al. 2016).

Another interesting category is acetylase/deacetylase, with nine lysine residues known for regulation of stress-activation of HSF1 (Gomez-Pastor et al. 2017). The most well studied site is Lys80 in the DNA binding domain of HSF1 molecule (Westerheide et al. 2009). It has been reported that SIRT1, a NAD-dependent protein deacetylase sirtuin 1, can deacetylate HSF1 and prolong its DNA binding capacity on the promoter of HSP70 (Westerheide et al. 2009). An *in vitro* counter-screening of sirtuins revealed that AC-92353 (Fig. 8.6E) can potently inhibit Sirt 2 with IC50 values of 6 μM, as well as two closely related members Sirt 1 and Sirt 3 with IC50 values of 45.3 μM or 24.6 μM (Zhang et al. 2009b), respectively. AC-93253 also exhibited sub-micromolar selective cytotoxicity towards several tumor cell lines tested with a therapeutic window of up to 200-fold, as compared to the three normal cell types. High content analysis further confirmed that AC-93253 significantly triggered apoptosis in these cancer cells (Zhang et al. 2009b). However, there is a more than 10-fold difference of IC50 values in the HSF1 granule assay compared to the *in vitro* sirtuin assay, suggesting that Sirt 2 may not be the primary cellular target. Efforts have been made to use AC-93253 as a novel chemical scaffold for SAR and lead development (data not shown). Although the roles of Sirt2 and related deacetylases are less understood in tumorigenesis, small molecule inhibitors targeting this new type of targets can potentially be first-in-class drugs with novel MOAs (Haigis and Sinclair 2010).

Although intensive studies were performed to identify the potential targets of our lead compounds, no molecular target could be confidently assigned to the lead compounds in our HCS drug discovery program. However, recent studies on HSF1 inhibitor screening have made some promising progresses. Using HSP72 expression in U2OS human osteosarcoma cell line as readout in a different HCS approach, Rye and colleagues identified a hit compound based on a 4, 6-disubstituted pyrimidine scaffold as a potent inhibitor of HSF1. One compound derivative possessed potent inhibitory activity against CDK9, although more work is needed to confirm CDK9 as a key molecular target in HSR pathway (Rye et al. 2016). Another HSF1 inhibitor, bisamide (CCT251236), was also discovered by the same group. This lead compound exhibited a high affinity to pirin as confirmed by surface plasmon resonance and crystallography. However, cellular SAR data were more complex for determining the role of purin in the HSF1 pathway (Cheeseman et al. 2017). In summary, the potential risk of a lead candidate without a defined target is still very high in HCS. On the other hand, target knowledge is not essential for moving a lead candidate to preclinical and clinical testing (Moffat et al. 2017). Although not in the scope of this paper, several of our lead candidates show excellent results in the preclinical testing with relevant animal models (data not shown).

8.2 Conclusions

HCS has made considerable progresses in recent years, credit to the advancements in multiplexing imaging technology, data processing software, and robotic instrumentations. More importantly, quantitative proteomics, gene expression analysis and next-generation sequencing have enabled the discovery of novel targets and MOAs through the integration of these system biology data with deep phenotypic information (Moffat et al. 2017; Schirle and Jenkins 2016; Schenone et al. 2013). These drug discovery efforts are starting to receive payoffs, particularly in many disease areas where standard target based approaches fail to deliver (Moffat et al. 2014; Moffat et al. 2017). As HCS starts to take an increasingly important role in the early drug discovery process, many critical issues are expected to be resolved through innovative approaches, such as 3D tissue imaging (Bilgin et al. 2016), iPS derived cell models (Avior et al. 2016; Horvath et al. 2016), CRISPR/Cas9 gene edited models (Moore 2015; Fellmann et al. 2017), organ-on-a-chip (Bhatia and Ingber 2014), HCS toxicity testing (Persson and Hornberg 2016), advanced data analysis and visualization technique (Moutsatsos and Parker 2016), etc. For the drug discovery of HSR pathway, these issues include: how to identify a small molecule that can locally inhibit HSF1 for cancer therapeutics, with no side effects (such as reduced expression of HSP) in the central nervous system? On the flip side, problems need to be addressed for possible tumor promotion and metastasis by small molecule HSP activators. Given that the 3D cocrystal structure of human

HSF1 DNA binding domain with DNA is available, can we discover novel HSF1 modulators by structure-based design and virtual screening (Neudegger et al. 2016; Vilaboa et al. 2017)? In the future, creative solutions will lead us to the discovery of first-in-class medications for the unmet medical needs in cancer and neurodegenerative diseases.

Acknowledgements The authors would like to thank all the supports received from everyone at Alpine Therapeutics, Inc. Special acknowledgements are given to Dr. Dorothy Wang for her critical review of the manuscript.

References

Alastalo TP, Hellesuo M, Sandqvist A, Hietakangas V, Kallio M, Sistonen L (2003) Formation of nuclear stress granules involves HSF2 and coincides with the nucleolar localization of Hsp70. J Cell Sci 116(Pt 17):3557–3570

Anckar J, Sistonen L (2011) Regulation of HSF1 function in the heat stress response: implications in aging and disease. Annu Rev Biochem 80:1089–1115

Au Q, Kanchanastit P, Barber JR, Ng SC, Zhang B (2008) High-content image-based screening for small-molecule chaperone amplifiers in heat shock. J Biomol Screen 13(10):953–959

Au Q, Zhang Y, Barber JR, Ng SC, Zhang B (2009) Identification of inhibitors of HSF1 functional activity by high-content target-based screening. J Biomol Screen 14(10):1165–1175

Avior Y, Sagi I, Benvenisty N (2016) Pluripotent stem cells in disease modelling and drug discovery. Nat Rev Mol Cell Biol 17(3):170–182

Bhatia SN, Ingber DE (2014) Microfluidic organs-on-chips. Nat Biotechnol 32(8):760–772

Bilgin CC, Fontenay G, Cheng Q, Chang H, Han J, Parvin B (2016) BioSig3D: high content screening of three-dimensional cell culture models. PLoS One 11(3):e0148379

Boutros M, Heigwer F, Laufer C (2015) Microscopy-based high-content screening. Cell 163(6):1314–1325

Calderwood SK, Murshid A (2017) Molecular chaperone accumulation in cancer and decrease in Alzheimer's disease: the potential roles of HSF1. Front Neurosci 11:192

Cheeseman MD, Chessum NE, Rye CS, Pasqua AE, Tucker MJ, Wilding B, Evans LE, Lepri S, Richards M, Sharp SY, Ali S, Rowlands M, O'Fee L, Miah A, Hayes A, Henley AT, Powers M, Te Poele R, De Billy E, Pellegrino L, Raynaud F, Burke R, van Montfort RL, Eccles SA, Workman P, Jones K (2017) Discovery of a chemical probe Bisamide (CCT251236): an orally bioavailable efficacious Pirin ligand from a heat shock transcription factor 1 (HSF1) phenotypic screen. J Med Chem 60(1):180–201

Cher C, Tremblay MH, Barber JR, Ng SC, Zhang B (2010) Identification of chaulmoogric acid as a small molecule activator of protein phosphatase 5. Appl Biochem Biotechnol 160(5):1450–1459

Chessel A (2017) An overview of data science uses in bioimage informatics. Methods 115:110–118

Cotto J, Fox S, Morimoto R (1997) HSF1 granules: a novel stress-induced nuclear compartment of human cells. J Cell Sci 110(Pt 23):2925–2934

Dai C, Whitesell L, Rogers AB, Lindquist S (2007) Heat shock factor 1 is a powerful multifaceted modifier of carcinogenesis. Cell 130(6):1005–1018

Daugaard M, Rohde M, Jäättelä M (2007) The heat shock protein 70 family: highly homologous proteins with overlapping and distinct functions. FEBS Lett 581(19):3702–3710

Dayalan NS, Dinkova-Kostova AT (2017) Regulation of the mammalian heat shock factor 1. FEBS J 284(11):1606–1627

De Thonel A, Mezger V, Garrido C (2011) Implication of heat shock factors in tumorigenesis: therapeutical potential. Cancers (Basel) 3(1):1158–1181

Denegri M, Moralli D, Rocchi M, Biggiogera M, Raimondi E, Cobianchi F, De Carli L, Riva S, Biamonti G (2002) Human chromosomes 9, 12, and 15 contain the nucleation sites of stress-induced nuclear bodies. Mol Biol Cell 13(6):2069–2079

Donato MT, Gómez-Lechón MJ, Tolosa L (2017) Using high-content screening technology for studying drug-induced hepatotoxicity in preclinical studies. Expert Opin Drug Discovery 12(2):201–211

Eder J, Sedrani R, Wiesmann C (2014) The discovery of first-in-class drugs: origins and evolution. Nat Rev Drug Discov **13**:577–587

Fellmann C, Gowen BG, Lin PC, Doudna JA, Corn JE (2017) Cornerstones of CRISPR-Cas in drug discovery and therapy. Nat Rev Drug Discov 16:89–100

Fetz V, Prochnow H, Bronstrup M, Sasse F (2016) Target identification by image analysis. Nat Prod Rep 33:655–667

Gao M, Nettles RE, Belema M, Snyder LB, Nguyen VN, Fridell RA, Serrano-Wu MH, Langley DR, Sun JH, O'Boyle DR 2nd, Lemm JA, Wang C, Knipe JO, Chien C, Colonno RJ, Grasela DM, Meanwell NA, Hamann LG (2010) Chemical genetics strategy identifies an HCV NS5A inhibitor with a potent clinical effect. Nature 465(7294):96–100

Gomez-Pastor R, Burchfiel ET, Thiele DJ (2017) Regulation of heat shock transcription factors and their roles in physiology and disease. Nat Rev Mol Cell Biol Aug 30. https://doi.org/10.1038/nrm.2017.73 [Epub ahead of print]

Haigis MC, Sinclair DA (2010) Mammalian sirtuins: biological insights and disease relevance. Annu Rev Pathol 5:253–295

Harvey AL, Edrada-Ebel R, Quinn RJ (2015) The re-emergence of natural products for drug discovery in the genomics era. Nat Rev Drug Discov 14:111–129

Horvath P, Aulner N, Bickle M, Davies AM, Nery ED, Ebner D, Montoya MC, Östling P, Pietiäinen V, Price LS, Shorte SL, Turcatti G, von Schantz C, Carragher NO (2016) Screening out irrelevant cell-based models of disease. Nat Rev Drug Discov 15(11):751–769

Jolly C, Usson Y, Morimoto RI (1999) Rapid and reversible relocalization of heat shock factor 1 within seconds to nuclear stress granules. Proc Natl Acad Sci U S A 96(12):6769–6774

Jolly C, Konecny L, Grady DL, Kutskova YA, Cotto JJ, Morimoto RI, Vourc'h C (2002) In vivo binding of active heat shock transcription factor 1 to human chromosome 9 heterochromatin during stress. J Cell Biol 156(5):775–781

Jolly C, Metz A, Govin J, Vigneron M, Turner BM, Khochbin S, Vourc'h C (2004) Stress-induced transcription of satellite III repeats. J Cell Biol 164(1):25–33

Jones LH, Bunnage ME (2017) Applications of chemogenomic library screening in drug discovery. Nat Rev Drug Discov 16:285–296

Khurana V, Tardiff DF, Chung CY, Lindquist S (2015) Toward stem cell-based phenotypic screens for neurodegenerative diseases. Nat Rev Neurol 11(6):339–350

Li L, Zhou Q, Voss TC, Quick KL, LaBarbera DV (2016) High-throughput imaging: Focusing in on drug discovery in 3D. Methods 96:97–102

Lindquist SL, Kelly JW (2011) Chemical and biological approaches for adapting proteostasis to ameliorate protein misfolding and aggregation diseases: progress and prognosis. Cold Spring Harb Perspect Biol **3**(12):a004507

Liu C, Su J, Yang F, Wei K, Ma J, Zhou X (2015) Compound signature detection on LINCS L1000 big data. Mol BioSyst 11(3):714–722

Mattiazzi UM, Styles EB, Verster AJ, Friesen H, Boone C, Andrews BJ (2016) High-content screening for quantitative cell biology. Trends Cell Biol 26(8):598–611

Moffat JG, Rudolph J, Bailey D (2014) Phenotypic screening in cancer drug discovery - past, present and future. Nat Rev Drug Discov 13:588–602

Moffat JG, Vincent F, Lee JA, Eder J, Prunotto M (2017) Opportunities and challenges in phenotypic drug discovery: an industry perspective. Nat Rev Drug Discov 16(8):531–543

Moore JD (2015) The impact of CRISPR–Cas9 on target identification and validation. (2015). Drug Discov Today 20:450–457

Moutsatsos IK, Parker CN (2016) Recent advances in quantitative high throughput and high content data analysis. Expert Opin Drug Discovery 11(4):415–423
Neef DW, Jaeger AM, Thiele DJ (2011) Heat shock transcription factor 1 as a therapeutic target in neurodegenerative diseases. Nat Rev Drug Discov 10(12):930–944
Neudegger T, Verghese J, Hayer-Hartl M, Hartl FU, Bracher A (2016) Structure of human heat-. shock transcription factor 1 in complex with DNA. Nat Struct Mol Biol 23(2):140–146
Nijman SM (2015) Functional genomics to uncover drug mechanism of action. Nat Chem Biol 11:942–948
Nyström A, Thriene K, Mittapalli V, Kern JS, Kiritsi D, Dengjel J, Bruckner-Tuderman L (2015) Losartan ameliorates dystrophic epidermolysis bullosa and uncovers new disease mechanisms. EMBO Mol Med **7**(9):1211–1228
Pegoraro G, Misteli T (2017) High-throughput imaging for the discovery of cellular mechanisms of disease. Trends Genet 33(9):604–615
Persson M, Hornberg JJ (2016) Advances in predictive toxicology for discovery safety through high content screening. Chem Res Toxicol 29(12):1998–2007
Petersen DN, Hawkins J, Ruangsiriluk W, Stevens KA, Maguire BA, O'Connell TN, Rocke BN, Boehm M, Ruggeri RB, Rolph T, Hepworth D, Loria PM, Carpino PA (2016) A small-molecule anti-secretagogue of PCSK9 targets the 80S ribosome to inhibit PCSK9 protein translation. Cell Chem Biol 23(11):1362–1371
Phukan J (2010) Arimoclomol, a coinducer of heat shock proteins for the potential treatment of amyotrophic lateral sclerosis. IDrugs 13(7):482–496
Rye CS, Chessum NE, Lamont S, Pike KG, Faulder P, Demeritt J, Kemmitt P, Tucker J, Zani L, Cheeseman MD, Isaac R, Goodwin L, Boros J, Raynaud F, Hayes A, Henley AT, de Billy E, Lynch CJ, Sharp SY, Te Poele R, Fee LO, Foote KM, Green S, Workman P, Jones K (2016) Discovery of 4,6-disubstituted pyrimidines as potent inhibitors of the heat shock factor 1 (HSF1) stress pathway and CDK9. Medchemcomm **7**(8):1580–1586
Saibil H (2013) Chaperone machines for protein folding, unfolding and disaggregation. Nat Rev Mol Cell Biol 14(10):630–642
Sandqvist A, Sistonen L (2004) Nuclear stress granules: the awakening of a sleeping beauty? J Cell Biol 164(1):15–17
Santagata S, Xu YM, Wijeratne EM, Kontnik R, Rooney C, Perley CC, Kwon H, Clardy J, Kesari S, Whitesell L, Lindquist S, Gunatilaka AA (2012) Using the heat-shock response to discover anticancer compounds that target protein homeostasis. ACS Chem Biol **7**(2):340–349
Santagata S, Mendillo ML, Tang YC, Subramanian A, Perley CC, Roche SP, Wong B, Narayan R, Kwon H, Koeva M, Amon A, Golub TR, Porco JA Jr, Whitesell L, Lindquist S (2013) Tight coordination of protein translation and HSF1 activation supports the anabolic malignant state. Science 341(6143):1238303
Santos R, Ursu O, Gaulton A, Bento AP, Donadi RS, Bologa CG, Karlsson A, Al-Lazikani B, Hersey A, Oprea TI, Overington JP (2017) A comprehensive map of molecular drug targets. Nat Rev Drug Discov 6(1):19–34
Sawyers CL (2005) Making progress through molecular attacks on cancer. Cold Spring Harb Symp Quant Biol 70:479–482
Scannell JW, Bosley J (2016) When quality beats quantity: decision theory, drug discovery, and the reproducibility crisis. PLoS One 11(2):e0147215
Schenone M, Dančík V, Wagner BK, Clemons PA (2013) Target identification and mechanism of action in chemical biology and drug discovery. Nat Chem Biol 9(4):232–240
Schirle M, Jenkins JL (2016) Identifying compound efficacy targets in phenotypic drug discovery. Drug Discov Today 21:82–89
Shi Y, Inoue H, Wu JC, Yamanaka S (2017) Induced pluripotent stem cell technology: a decade of progress. Nat Rev Drug Discov 16:115–130
Swinney DC, Anthony J (2011) How were new medicines discovered? Nat Rev Drug Discov 10:507–519
Thomas N (2010) High-content screening: a decade of evolution. J Biomol Screen 15(1):1–9

Vilaboa N, Boré A, Martin-Saavedra F, Bayford M, Winfield N, Firth-Clark S, Kirton SB, Voellmy R (2017) New inhibitor targeting human transcription factor HSF1: effects on the heat shock response and tumor cell survival. Nucleic Acids Res 45(10):5797–5817

Vydra N, Toma A, Widlak W (2014) Pleiotropic role of HSF1 in neoplastic transformation. Curr Cancer Drug Targets 14(2):144–155

Wagner BK (2016) The resurgence of phenotypic screening in drug discovery and development. Expert Opin Drug Discovery 11(2):121–125

Wagner BK, Schreiber SL (2016) The power of sophisticated phenotypic screening and modern mechanism-of-action methods. Cell Chem Biol 23:3–9

Wassermann AM, Lounkine E, Hoepfner D, Le Goff G, King FJ, Studer C, Peltier JM, Grippo ML, Prindle V, Tao J, Schuffenhauer A, Wallace IM, Chen S, Krastel P, Cobos-Correa A, Parker CN, Davies JW, Glick M (2015) Dark chemical matter as a promising starting point for drug lead discovery. Nat Chem Biol 11(12):958–966

Westerheide SD, Bosman JD, Mbadugha BN, Kawahara TL, Matsumoto G, Kim S, Gu W, Devlin JP, Silverman RB, Morimoto RI (2004) Celastrols as inducers of the heat shock response and cytoprotection. J Biol Chem 279(53):56,053–56,060

Westerheide SD, Kawahara TL, Orton K, Morimoto RI (2006) Triptolide, an inhibitor of the human heat shock response that enhances stress-induced cell death. J Biol Chem 281(14):9616–9622

Westerheide SD, Anckar J, Stevens SM Jr, Sistonen L, Morimoto RI (2009) Stress-inducible regulation of heat shock factor 1 by the deacetylase SIRT1. Science 323(5917):1063–1066

Workman P (2005) Drugging the cancer kinome: progress and challenges in developing personalized molecular cancer therapeutics. Cold Spring Harb Symp Quant Biol 70:499–515

Xu K, Sun X, Erokwu BO, Cernak I, Lamanna JC (2011) A heat-shock protein co-inducer treatment improves behavioral performance in rats exposed to hypoxia. Adv Exp Med Biol 701:313–318

Yoon IS, Au Q, Barber JR, Ng SC, Zhang B (2010) Development of a high-throughput screening assay for cytoprotective agents in rotenone-induced cell death. Anal Biochem 407(2):205–210

Yoon YJ, Kim JA, Shin KD, Shin DS, Han YM, Lee YJ, Lee JS, Kwon BM, Han DC (2011) KRIBB11 inhibits HSP70 synthesis through inhibition of heat shock factor 1 function by impairing the recruitment of positive transcription elongation factor b to the hsp70 promoter. J Biol Chem 286(3):1737–1747

Zeng XC, Bhasin S, Wu X, Lee JG, Maffi S, Nichols CJ, Lee KJ, Taylor JP, Greene LE, Eisenberg E (2004) Hsp70 dynamics *in vivo*: effect of heat shock and protein aggregation. J Cell Sci 117(Pt 21):4991–5000

Zhang D, Zhang B (2016) Selective killing of cancer cells by small molecules targeting heat shock stress response. Biochem Biophys Res Commun 478(4):1509–1514

Zhang JH, Chung TD, Oldenburg KR (1999) A simple statistical parameter for use in evaluation and validation of high throughput screening assays. J Biomol Screen 4(2):67–73

Zhang B, Gu X, Uppalapati U, Ashwell MA, Leggett DS, Li CJ (2008) High-content fluorescent-based assay for screening activators of DNA damage checkpoint pathways. J Biomol Screen 13(6):538–543

Zhang B, Au Q, Yoon IS, Tremblay MH, Yip G, Zhou Y, Barber JR, Ng SC (2009a) Identification of small-molecule HSF1 amplifiers by high content screening in protection of cells from stress induced injury. Biochem Biophys Res Commun 390(3):925–930

Zhang Y, Au Q, Zhang M, Barber JR, Ng SC, Zhang B (2009b) Identification of a small molecule SIRT2 inhibitor with selective tumor cytotoxicity. Biochem Biophys Res Commun 386(4):729–733

Zhang W, Bai Y, Wang Y, Xiao W (2016) Polypharmacology in drug discovery: a review from systems pharmacology perspective. Curr Pharm Des 22:3171–3181

Zhou Y, Liu G, Chen J, Reddy PS, Yoon IS, Zhang M, Zhang B (2009) Barber, J.R, Ng, S.C. (2009a) Pyrimido[5,4-e][1,2,4]triazine-5,7(1H,6H)-dione derivatives: their cytoprotection effect from rotenone toxicity and preliminary DMPK properties. Bioorg Med Chem Lett 19(21):6114–6118

Zhou Y, Vu K, Chen Y, Pham J, Brady T, Liu G, Chen J, Nam J, Murali Mohan Reddy PS, Au Q, Yoon IS, Tremblay MH, Yip G, Cher C, Zhang B, Barber JR, Ng SC (2009b) Chloro-oxime derivatives as novel small molecule chaperone amplifiers. Bioorg Med Chem Lett 19(11):3128–3135

Zhou Y, Wei L, Brady TP, Murali Mohan Redddy PS, Nguyen T, Chen J, Au Q, Yoon IS, Yip G, Zhang B, Barber JR, Ng SC (2009c) Pyrimido [5,4-e][1,2,4]triazine-5,7(1H,6H)-dione derivatives as novel small molecule chaperone amplifiers. Bioorg Med Chem Lett 19(15):4303–4307

Chapter 9
Regulation of Autophagy by the Heat Shock Factor 1-Mediated Stress Response Pathway

Yoshihisa Watanabe and Masaki Tanaka

Abstract Heat shock factor 1 (HSF1) is a master regulator of heat shock response and controls transcription of heat shock proteins, including molecular chaperones. It is recently becoming clear that macroautophagy (hereafter autophagy) is regulated by the HSF1-mediated proteostasis network in cells exposed to stress conditions. HSF1 controls expression of several autophagy related genes and activates the autophagy receptor sequestosome1 (p62/SQSTM1). This chapter focuses on regulation of p62-associated formation of inclusions and selective autophagy clearance of harmful proteins by HSF1. The contributions of autophagy induction by HSF1 activation to lifespan extension are also discussed.

Keywords Autophagy · Heat shock factor 1/HSF1 · Longevity · p62/SQSTM1 · Protesostasis network

Abbreviations

ALP	Macroautophagy lysosome pathway
ATF4	Activating transcription factor 4
ATF6	Activating transcription factor 6
ATFS1	Activating transcription factor associated with stress-1
ATG	Autophagy-related gene
BAG3	Bcl2-associated athanogene 3
C. elegans	Caenorhabditis elegans

Y. Watanabe (✉)
Department of Basic Geriatrics, Graduate School of Medical Science,
Kyoto Prefectural University of Medicine, Kyoto, Japan
e-mail: y-watana@koto.kpu-m.ac.jp

M. Tanaka (✉)
Department of Anatomy and Neurobiology, Graduate School of Medical Science,
Kyoto Prefectural University of Medicine, Kyoto, Japan
e-mail: mtanaka@koto.kpu-m.ac.jp

A. A. A. Asea, P. Kaur (eds.), *Heat Shock Proteins and Stress*,
Heat Shock Proteins 15, https://doi.org/10.1007/978-3-319-90725-3_9

ER	Endoplasmic reticulum
FOXO	Forkhead box protein O
HSC70	Heat shock cognate protein 70
HSE	Heat shock element
HSF1	Heat shock factor 1
HSP	Heat shock protein
HSPB8	Small heat-shock protein-beta 8
IGF-1	Insulin-like growth factor 1
KIR	Keap1 interaction region
NBR1	BRCA1 gene 1
NDP52	Nuclear dot protein 52
OPTN	Optineurin
p62/SQSTM1	Sequestosome1
PE	Phosphatidylethanolamine
PtdIns3K	Phosphatidylinositol 3-kinase
TAX1BP1	Tax1-binding protein 1
TFEB	Transcription factor EB
TOLLIP	Toll-interacting protein
ULK1	Atg1/UNC51-like kinase-1
UPS	Ubiquitin–proteasome system
XBP1	X-box binding protein 1

9.1 Introduction

Accumulation of unfolded or aggregated proteins is very harmful to the cell and causes neurodegenerative diseases, cancer, diabetes and amyloidosis (Stefani and Dobson 2003). Harmful unfolded proteins are generated by genetic mutations or various stresses, such as heat shock and oxidative stress. The proteostasis network promotes protein folding to protect cells from the deleterious effects of unfolded proteins (Sala et al. 2017). It is well known that molecular chaperones, such as heat shock protein (HSP)60, HSP70, HSP90 and HSP100, facilitate refolding of unfolded proteins (Saibil 2013). Another proteostasis event also involves protein degradation processes, including the ubiquitin–proteasome system (UPS) and the macroautophagy (hereafter autophagy) lysosome pathway (ALP). These systems play important roles in degradation of damaging proteins (Wong and Cuervo 2010). Heat shock factor 1 (HSF1), a master regulator of the heat shock response, controls transcription of molecular chaperones, co-chaperones and ubiquitin (Carper et al. 1987; Franceschelli et al. 2008). However, it is not known whether, under stress, autophagy is regulated by the HSF1-associated proteostasis network. It is now becoming clear that HSF1 is also involved in degradation of protein aggregates via autophagy. In this chapter, we describe the basic machinery of autophagy, the mechanisms of its regulation by HSF1 and the role of the HSF1–autophagy axis in longevity.

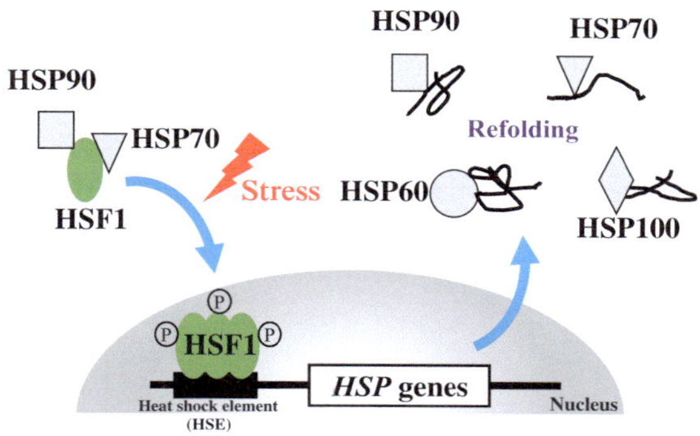

Fig. 9.1 The HSF1-mediated proteostasis network. Monomeric HSF1 interacts with HSP90 and HSP70 in the cytoplasm. When various stresses generate unfolded proteins, HSF1 trimers become localized to the nucleus and bind to the heat shock element (HSE) of the *HSP* genes, followed by upregulation of HSP expression. Unfolded proteins are refolded by molecular chaperones, such as HSP100, HSP90, HSP70 and HSP60, to protect cells from harmful proteins

9.1.1 HSF1-Mediated Proteostasis Network

Unfolded proteins and protein aggregates are generated intra- and extracellularly after exposure to various stresses, such as heat shock and oxidative stress (Morimoto et al. 1997). Accumulation of these proteins adversely affects cellular functions, resulting in cell death. To alleviate this situation, refolding and degradation of harmful proteins are activated through the proteostasis network. In the proteostasis network, distinct surveillance mechanisms exist for proteome function in the cytoplasm, endoplasmic reticulum (ER) and mitochondria. Transcriptional activation of HSP, including molecular chaperones and co-chaperones, is mediated by HSF1 in the cytoplasm (Akerfelt et al. 2010), whereas transcriptional factors, such as X-box binding protein 1 (XBP1), activating transcription factor 4 (ATF4), activating transcription factor 6 (ATF6) and activating transcription factor associated with stress-1 (ATFS1), respond to protein misfolding in the ER and mitochondria (Haynes et al. 2013; Labbadia and Morimoto 2015; Walter and Ron 2011). The HSF1-mediated proteostasis network is described in Fig. 9.1. Under normal conditions, HSF1 exists as an inert monomer, that interact with HSP70 and HSP90 (Morimoto 1998). Under stress conditions, HSF1 forms homotrimers and becomes localized to the nucleus, where it binds to heat shock elements within the promoters of HSP genes, resulting in upregulation of HSP expression (Morimoto 1998). Genome-wide analysis in mammalian cells showed that 46 genes, including those coding for HSP, have HSF1 binding sites in their promoters and that their expression can be induced by heat shock (Trinklein et al. 2004). Cytosolic molecular chaperones, such as HSP60, HSP70, HSP90 and HSP100, refold unfolded proteins, thus protecting cells from stress (Saibil 2013). Moreover, expression of Bcl2-associated athanogene 3 (BAG3), a co-chaperone for HSP70 and heat shock cognate protein 70 (HSC70), is also regulated by HSF1 (Jacobs and Marnett 2009). It was recently shown that the small heat-shock protein-beta 8 (HSPB8)–BAG3–HSP70 chaperone complex was involved in degrading defective ribosomal products within stress granules. Stress granules are membrane-less ribonucleoprotein assemblies that exist under proteotoxic stress (Alberti et al. 2017; Ganassi et al. 2016). This degradation is carried out by either chaperone-mediated autophagy or autophagy-independent clearance (Ganassi et al. 2016). Furthermore, as described further below, BAG3 also has an important role in selective autophagy of misfolded proteins (Gamerdinger et al. 2011).

9.1.2 Autophagy-Lysosome Pathway

The ALP is an intracellular metabolic process in which cytoplasmic proteins and organelles, sequestered by autophagosomes, are degraded in lysosomes or vacuoles (Mizushima 2007). More than 30 autophagy-related genes (*ATG*) were identified by yeast genetic studies and their orthologs in higher eukaryotes were subsequently described in many studies (Feng et al. 2014; Kabeya et al. 2000; Mizushima et al. 1998; Tsukada and Ohsumi 1993). Molecular functions of the Atg core machinery are classified into 4 subgroups: (1) the Atg1/UNC51-like kinase-1 (ULK1) complex; (2) Atg9 and its cycling system; (3) the phosphatidylinositol 3-kinase (PtdIns3K) complex; and (4) two ubiquitin-like conjugation systems (Atg8/LC3 and Atg12) (Feng et al. 2014). In *Saccharomyces cerevisiae*, the Atg1 kinase complex is composed of Atg1, Atg13, Atg 17, Atg29 and Atg31, whereas mammalian Atg1 (ULK1) interacts with FIP200, Atg13 and Atg101 (Mizushima 2010). The ULK1 complex is essential for initiation of autophagy and is directed to the ER, together with Atg9 vesicles and the Beclin1-VPS34 (PtdIns3K) complex, in mammals (Fig. 9.2). The PtdIns3K complex generates PtdIns3P, which recruits PtdIns3P binding proteins, such as DFCP1 and WIPIs, promoting formation of isolation membranes (Fig. 9.2). Atg12 and LC3 are regulated by an ubiquitin-like conjugation system. Atg12 is covalently conjugated to Atg5 by the E1-like enzyme Atg7 and the E2-like enzyme Atg10, sequentially binding to Atg16L1 to form the Atg12–Atg5–Atg16L1 complex. LC3 is conjugated to phosphatidylethanolamine (PE) by the E1-like enzyme Atg7, the E2-like enzyme Atg3 and the E3-like enzyme Atg12–Atg5–Atg16L1 complex. Both Atg12–Atg5–Atg16L1 and LC3–PE localize to the developing autophagosome and LC3–PE is probably a scaffold protein supporting membrane expansion and determining the size of autophagosomes (Geng and Klionsky 2008). Finally, mature autophagosomes are fused with lysosomes, resulting in degradation of cellular components.

Although autophagy was originally considered to be a non-selective bulk degradation pathway, selective autophagy degradation has recently become well known because various autophagy receptor proteins were discovered (Svenning and Johansen 2013). Nutrient starvation induces degradation of cytosolic proteins and

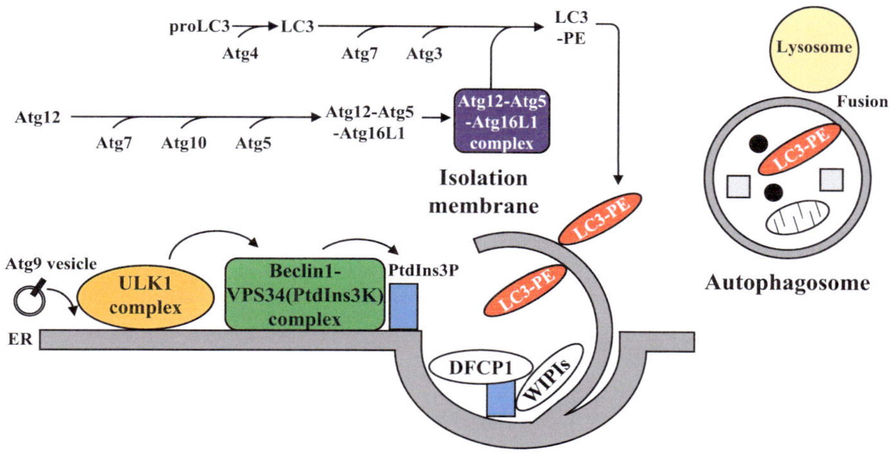

Fig. 9.2 The autophagy signaling pathway in mammals. The activated ULK1 complex translocates to the phagophore formation site and recruits the Beclin1–VPS34 (PtdIns3K) complex. VPS34 generates PtdIns3P, resulting in accumulation of the PtdIns3P binding protein, DFCP1, at the omegasome. These events contribute to generation of an isolation membrane. Then, the isolation membrane is elongated by two ubiquitin-like conjugation systems, the Atg5–Atg12 complex conjugation system and LC3-phosphatidylethanolamine (PE) conjugation system. Finally, autophagosomes fuse with lysosomes, followed by degradation of its intraluminal content for recycling

organelles by nonselective autophagy, whereas selective autophagy is targeted to protein aggregates, damaged organelles, including mitochondria and lysosomes, and intracellular pathogens (Stolz et al. 2014). Autophagy receptors interact with individual substrates in ubiquitin-dependent or ubiquitin-independent mechanisms (Khaminets et al. 2016; Watanabe and Tanaka 2011). For example, protein aggregates are bound to sequestosome1 (p62/SQSTM1), a neighbor of the BRCA1 gene 1 (NBR1), optineurin (OPTN), Cue5 and Toll-interacting protein (TOLLIP), whereas damaged mitochondria are targeted to OPTN, nuclear dot protein 52 (NDP52) and Tax1-binding protein 1 (TAX1BP1) (Khaminets et al. 2016). Autophagy receptors contain a Atg8/LC3-interacting region (LIR; Trp/Phe/Tyr–x–x–Leu/Ile/Val) and selectively engulf cargo into autophagosomes (Khaminets et al. 2016).

Selective autophagy eliminates protein aggregates that accumulate because of proteasomal inhibition or overexpression of aggregation-prone proteins (Ravikumar et al. 2002; Watanabe and Tanaka 2011; Watanabe et al. 2012; Zhu et al. 2010). Autophagy receptors, such as p62, NBR1 and OPTN, are required for this selective elimination of protein aggregates. These receptors have both the LIR domain and the UBA and UBAN domains, which bind to ubiquitinated proteins, resulting in their selective sequestration into autophagosomes (Wild et al. 2014). Furthermore, the co-chaperone BAG3 was recently described as a mediator of the autophagy pathway (Gamerdinger et al. 2011). The HSP70–BAG3 complex interacts with unfolded proteins in an ubiquitin-independent manner and transports them, in the aggresome, along microtubules using the dynein motor complex. This transport is followed by autophagic degradation of the unfolded proteins (Gamerdinger et al. 2011).

Autophagy induction is regulated not only by activation of Atg proteins, but also by coordinated transcriptional activation of the *ATG* and accessory genes. For example, expression of *LC3b*, *ATG3*, *ATG5* and *ATG12* is activated by the forkhead box protein O (FOXO) transcriptional factors (Salih and Brunet 2008; Xu et al. 2011). Transcription factor EB (TFEB) is a master regulator of lysosomal biogenesis and regulates transcription of *ATG9*, *WIPI*s, *ATG18*, *LC3b* and *p62* (Lapierre et al. 2015; Settembre et al. 2011). Under stress conditions, HSF1 transcriptionally induces *ATG18* and *p62* expression in *Caenorhabditis elegans* (*C. elegans*) and *ATG7* and *BAG3* expression in mammalian cells (Desai et al. 2013; Kumsta et al. 2017), suggesting that HSF1 is involved in autophagy induction.

9.1.3 p62-Associated Proteostasis

p62 is a multifunctional protein acting as a signaling hub for several signal transduction pathways. Mutations in the p62 gene result in Paget disease of the bone, amyotrophic lateral sclerosis or frontotemporal lobar degeneration (Kwok et al. 2014). Although p62 functions as a selective autophagy receptor, the protein is also a selective autophagy substrate. Indeed, p62 is markedly accumulated in liver-specific *ATG7* knockout (KO) mice (Komatsu et al. 2007). *ATG7* KO mice also showed accumulation of inclusion bodies of ubiquitinated proteins, whereas ubiquitin-positive inclusions were not detected in *ATG7* and *p62* double KO mice (Komatsu et al. 2007). These results indicated an important role for p62 in formation of inclusion bodies with ubiquitinated proteins. Inclusion bodies and amyloid deposits are considered pathological hallmarks in various neurodegenerative disorders, including Alzheimer's disease, Parkinson's disease and amyotrophic lateral sclerosis. p62 was commonly observed in these inclusion bodies (Kuusisto et al. 2002; Mizuno et al. 2006; Nakano et al. 2004). It is believed that p62-mediated formation of inclusions with ubiquitinated proteins is protective against cell death in neurons and cardiac myocytes (Doi et al. 2013; Su and Wang 2011). Moreover, loss of p62 function caused significant exacerbation of eye degeneration in the polyQ disease model flies (Saitoh et al. 2015). This evidence suggested that p62 is critical for protein quality control in various tissues.

Moreover, p62 is also involved in Nrf2–Keap1 antioxidative signaling pathways, which control gene expression of antioxidant proteins and other detoxification enzymes. Keap1 acts as an oxidative stress sensor and suppresses the transcription activity of Nrf2, the master regulator of antioxidant responses. Under oxidative stress, Nrf2 is dissociated from Keap1, enabling translocation of Nrf2 into the nucleus. p62 competitively interacts with Keap1 and these complexes are degraded by the ALP. However, autophagy deficient mice had abnormal accumulation of p62 in various tissues and developed multiple liver tumors (Komatsu et al. 2005; Takamura et al. 2011). Like in these mice, p62-positive structures called Mallory-Denk bodies and hyaline bodies were observed in several cases of human hepatocellular carcinoma (Katsuragi et al. 2016). It is believed that excess p62 accumulation because of autophagy deficits leads to persistent Nrf2 activation, attributed to competitive binding of p62 to Keap1, resulting in increased liver tumor development (Inami et al. 2011). Indeed, additional loss of p62 suppressed tumorigenesis in the *ATG7*-deficient liver, supporting this hypothesis (Komatsu et al. 2007).

9.1.4 Regulation of Autophagy by a Stress Response Pathway

Heat shock stress at 43 °C for 30 min induces autophagy in various cultured cell lines, such as HeLa and HEK293 cells (Zhao et al. 2009). However, it remains unclear how autophagy machinery is activated during proteostatic stress although several studies recently elucidated this mechanism. HSF1, a master regulator of stress response pathways, plays an important role in autophagy induction under stress conditions. For example, transcription of several ATG and accessory genes is regulated by HSF1 (Fig. 9.3). Furthermore, HSF1 is also involved in activation of p62. p62 is phosphorylated by mTORC1, casein kinase 1 and TBK1 at Ser349 (corresponding to Ser351 in the mouse) and Ser403 (corresponding to Ser405 in the mouse) under various stresses such as oxidative stress, mitochondrial impairment and proteasomal inhibition (Fig. 9.3). However, an HSF1 inhibitor or HSF1 KO suppressed p62 phosphorylation at both sites, suggesting that these phosphorylation events were regulated in an HSF1-dependent manner under stress conditions (Watanabe et al. 2017). Ser349 and Ser403 phosphorylation sites are located in the Keap1 interaction region (KIR) and the UBA domain, respectively (Ichimura et al. 2013; Matsumoto et al. 2011; Pilli et al. 2012). In particular, Ser349 phosphorylation is believed to be required for stable binding to ubiquitinated proteins and Keap1 (Ichimura et al. 2013; Watanabe et al. 2017). However, the factors contributing to p62 phosphorylation downstream of HSF1 are unknown. After binding, the p62–substrate complex was oligmerized via the PB1 domain, resulting in inclusion formation and its autophagic elimination (Komatsu et al. 2007; Lamark et al. 2003). However, HSF1 inhibition suppressed inclusion formation of ubiquitinated proteins, even when they were accumulated intracellularly and, thus, autophagic clearance of protein aggregates was delayed (Watanabe et al. 2017). Collectively, these findings suggested that HSF1 activates the autophagy system not only through induction of autophagy-related gene expression but also through formation of inclusions containing protein aggregates, formed via p62 phosphorylation.

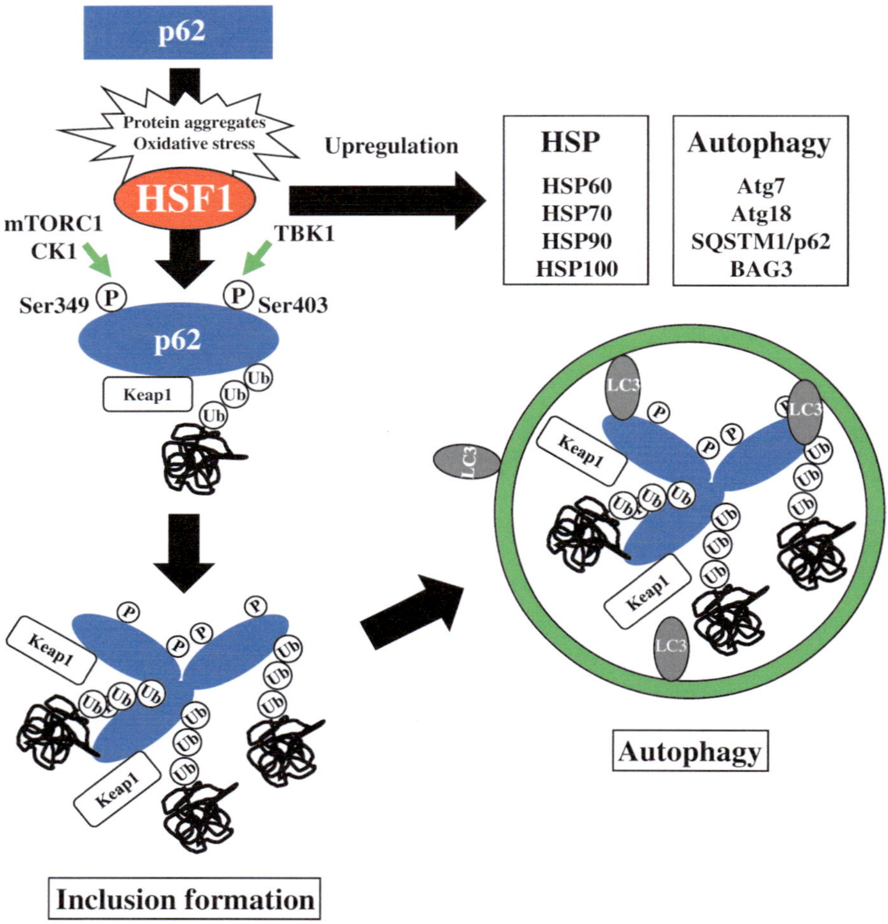

Fig. 9.3 Regulation of p62-associated proteostasis by HSF1. HSF1 activation enhances p62 phosphorylation (at Ser349 and Ser403) under stress conditions, stabilizing its binding to ubiquitinated proteins or Keap1. Subsequently, p62 oligomerizes and forms inclusion bodies with ubiquitinated proteins and Keap1. These inclusion bodies are sequestered into autophagosomes and degraded by lysosomes

9.1.5 *HSF1–Autophagy Axis in Longevity*

Caloric restriction extends the lifespan of many organisms, from yeast to mammals (Bishop and Guarente 2007). Genetic studies of longevity using *C. elegans* demonstrated involvement of the insulin-like growth factor 1 (IGF-1) signaling pathway in lifespan extension. Mutation of the insulin/IGF-1 receptor *daf-2* gene caused constitutive activation of the FOXO transcription factor DAF-16 and significantly increased lifespan (Kenyon et al. 1993). HSF1 cooperatively activated DAF-16 and induced expression of specific genes, including genes encoding small HSP, resulting in lifespan extension (Hsu et al. 2003). In contrast to this result, it was reported elsewhere that enhanced induction of HSP was not necessary for increased lifespan (Baird et al. 2014). Furthermore, autophagy genes are also required for both dauer formation and lifespan extension. The *C. elegans* ortholog of Beclin1 (*bec-1*) is involved in normal dauer morphogenesis and lifespan extension, whereas the *C. elegans* orthologs of *unc-51* (*ATG1*), *atg-7* (*ATG7*), *lgg-2* (*ATG8*) and *atg-18* (*ATG18*) are essential for dauer development (Melendez et al. 2003). The beneficial effects of autophagy on longevity were shown in *Drosophila*. Expression of *Atg2*, *Atg8a*, *Atg18* and *bchs* (*ALFY*) was decreased, in an age-dependent fashion, in fly neural tissues (Simonsen et al. 2008). Moreover, enhanced *Atg8a* expression in older fly brains extended the average adult lifespan by 56% (Simonsen et al. 2008). Recently, exposure of *C. elegans* to hormetic heat shock and HSF1 overexpression early in life induced expression of autophagy-related genes and increased lifespan (Kumsta et al. 2017). However, knockdown of autophagy related genes, such as *unc-51*, *bec-1* and *lgg-2*, prevented the increased lifespan after hormetic heat shock and overexpression of HSF1 (Kumsta et al. 2017). Such evidence suggested that HSF1-mediated autophagy also plays a crucial role in determining lifespan.

9.2 Conclusions

To date, studies on the HSF1-mediated proteostasis network were primarily focused on the regulatory mechanisms of molecular chaperones and UPS. However, it has now been shown that this network is also important for autophagy under stress conditions. In particular, both the HSF1-mediated proteostasis network and autophagy system are very much involved in pathogenesis of neurodegenerative diseases and cancer and they are, therefore, attractive targets for drug development. Further elucidation of links between HSF1 and autophagy will facilitate discovery of more effective drugs for these diseases. Furthermore, constitutive activation of autophagy by drugs, food or dietary supplements may beneficially affect longevity.

Acknowledgements This work was supported by Grants-in-Aid for Scientific Research from the Japan Society for the Promotion of Science (YW: 15 K09320 and MT: 17H03553).

References

Akerfelt M, Morimoto RI et al (2010) Heat shock factors: integrators of cell stress, development and lifespan. Nat Rev Mol Cell Biol 11:545–555

Alberti S, Mateju D et al (2017) Granulostasis: protein quality control of RNP granules. Front Mol Neurosci 10:84

Baird NA, Douglas PM et al (2014) HSF-1-mediated cytoskeletal integrity determines thermotolerance and life span. Science 346:360–363

Bishop NA, Guarente L (2007) Genetic links between diet and lifespan: shared mechanisms from yeast to humans. Nat Rev Genet 8:835–844
Carper SW, Duffy JJ et al (1987) Heat shock proteins in thermotolerance and other cellular processes. Cancer Res 47:5249–5255
Desai S, Liu Z et al (2013) Heat shock factor 1 (HSF1) controls chemoresistance and autophagy through transcriptional regulation of autophagy-related protein 7 (ATG7). J Biol Chem 288:9165–9176
Doi H, Adachi H et al (2013) p62/SQSTM1 differentially removes the toxic mutant androgen receptor via autophagy and inclusion formation in a spinal and bulbar muscular atrophy mouse model. J Neurosci 33:7710–7727
Feng Y, He D et al (2014) The machinery of macroautophagy. Cell Res 24:24–41
Franceschelli S, Rosati A et al (2008) Bag3 gene expression is regulated by heat shock factor 1. J Cell Physiol 215:575–577
Gamerdinger M, Kaya AM et al (2011) BAG3 mediates chaperone-based aggresome-targeting and selective autophagy of misfolded proteins. EMBO Rep 12:149–156
Ganassi M, Mateju D et al (2016) A surveillance function of the HSPB8-BAG3-HSP70 chaperone complex ensures stress granule integrity and dynamism. Mol Cell 63:796–810
Geng J, Klionsky DJ (2008) The Atg8 and Atg12 ubiquitin-like conjugation systems in macroautophagy. 'Protein modifications: beyond the usual suspects' review series. EMBO Rep 9:859–864
Haynes CM, Fiorese CJ et al (2013) Evaluating and responding to mitochondrial dysfunction: the mitochondrial unfolded-protein response and beyond. Trends Cell Biol 23:311–318
Hsu AL, Murphy CT et al (2003) Regulation of aging and age-related disease by DAF-16 and heat-shock factor. Science 300:1142–1145
Ichimura Y, Waguri S et al (2013) Phosphorylation of p62 activates the Keap1-Nrf2 pathway during selective autophagy. Mol Cell 51:618–631
Inami Y, Waguri S et al (2011) Persistent activation of Nrf2 through p62 in hepatocellular carcinoma cells. J Cell Biol 193:275–284
Jacobs AT, Marnett LJ (2009) HSF1-mediated BAG3 expression attenuates apoptosis in 4-hydroxynonenal-treated colon cancer cells via stabilization of anti-apoptotic Bcl-2 proteins. J Biol Chem 284:9176–9183
Kabeya Y, Mizushima N et al (2000) LC3, a mammalian homologue of yeast Apg8p, is localized in autophagosome membranes after processing. EMBO J 19:5720–5728
Katsuragi Y, Ichimura Y et al (2016) Regulation of the Keap1–Nrf2 pathway by p62/SQSTM1. Curr Opin Toxicol 1:54–61
Kenyon C, Chang J et al (1993) A C. elegans mutant that lives twice as long as wild type. Nature 366:461–464
Khaminets A, Behl C et al (2016) Ubiquitin-dependent and independent signals in selective autophagy. Trends Cell Biol 26:6–16
Komatsu M, Waguri S et al (2005) Impairment of starvation-induced and constitutive autophagy in Atg7-deficient mice. J Cell Biol 169:425–434
Komatsu M, Waguri S et al (2007) Homeostatic levels of p62 control cytoplasmic inclusion body formation in autophagy-deficient mice. Cell 131:1149–1163
Kumsta C, Chang JT et al (2017) Hormetic heat stress and HSF-1 induce autophagy to improve survival and proteostasis in C. elegans. Nat Commun 8:14337
Kuusisto E, Salminen A et al (2002) Early accumulation of p62 in neurofibrillary tangles in Alzheimer's disease: possible role in tangle formation. Neuropathol Appl Neurobiol 28:228–237
Kwok CT, Morris A et al (2014) Sequestosome-1 (SQSTM1) sequence variants in ALS cases in the UK: prevalence and coexistence of SQSTM1 mutations in ALS kindred with PDB. Eur J Hum Genet 22:492–496
Labbadia J, Morimoto RI (2015) The biology of proteostasis in aging and disease. Annu Rev Biochem 84:435–464

Lamark T, Perander M et al (2003) Interaction codes within the family of mammalian Phox and Bem1p domain-containing proteins. J Biol Chem 278:34568–34581
Lapierre LR, Kumsta C et al (2015) Transcriptional and epigenetic regulation of autophagy in aging. Autophagy 11:867–880
Matsumoto G, Wada K et al (2011) Serine 403 phosphorylation of p62/SQSTM1 regulates selective autophagic clearance of ubiquitinated proteins. Mol Cell 44:279–289
Melendez A, Talloczy Z et al (2003) Autophagy genes are essential for dauer development and life-span extension in C. elegans. Science 301:1387–1391
Mizuno Y, Amari M et al (2006) Immunoreactivities of p62, an ubiqutin-binding protein, in the spinal anterior horn cells of patients with amyotrophic lateral sclerosis. J Neurol Sci 249:13–18
Mizushima N (2007) Autophagy: process and function. Genes Dev 21:2861–2873
Mizushima N (2010) The role of the Atg1/ULK1 complex in autophagy regulation. Curr Opin Cell Biol 22:132–139
Mizushima N, Sugita H et al (1998) A new protein conjugation system in human. The counterpart of the yeast Apg12p conjugation system essential for autophagy. J Biol Chem 273:33889–33892
Morimoto RI (1998) Regulation of the heat shock transcriptional response: cross talk between a family of heat shock factors, molecular chaperones, and negative regulators. Genes Dev 12:3788–3796
Morimoto RI, Kline MP et al (1997) The heat-shock response: regulation and function of heat-shock proteins and molecular chaperones. Essays Biochem 32:17–29
Nakano T, Nakaso K et al (2004) Expression of ubiquitin-binding protein p62 in ubiquitin-immunoreactive intraneuronal inclusions in amyotrophic lateral sclerosis with dementia: analysis of five autopsy cases with broad clinicopathological spectrum. Acta Neuropathol 107:359–364
Pilli M, Arko-Mensah J et al (2012) TBK-1 promotes autophagy-mediated antimicrobial defense by controlling autophagosome maturation. Immunity 37:223–234
Ravikumar B, Duden R et al (2002) Aggregate-prone proteins with polyglutamine and polyalanine expansions are degraded by autophagy. Hum Mol Genet 11:1107–1117
Saibil H (2013) Chaperone machines for protein folding, unfolding and disaggregation. Nat Rev Mol Cell Biol 14:630–642
Saitoh Y, Fujikake N et al (2015) p62 plays a protective role in the autophagic degradation of polyglutamine protein oligomers in polyglutamine disease model flies. J Biol Chem 290:1442–1453
Sala AJ, Bott LC et al (2017) Shaping proteostasis at the cellular, tissue, and organismal level. J Cell Biol 216:1231–1241
Salih DA, Brunet A (2008) FoxO transcription factors in the maintenance of cellular homeostasis during aging. Curr Opin Cell Biol 20:126–136
Settembre C, Di Malta C et al (2011) TFEB links autophagy to lysosomal biogenesis. Science 332:1429–1433
Simonsen A, Cumming RC et al (2008) Promoting basal levels of autophagy in the nervous system enhances longevity and oxidant resistance in adult drosophila. Autophagy 4:176–184
Stefani M, Dobson CM (2003) Protein aggregation and aggregate toxicity: new insights into protein folding, misfolding diseases and biological evolution. J Mol Med (Berl) 81:678–699
Stolz A, Ernst A et al (2014) Cargo recognition and trafficking in selective autophagy. Nat Cell Biol 16:495–501
Su H, Wang X (2011) Autophagy and p62 in cardiac protein quality control. Autophagy 7:1382–1383
Svenning S, Johansen T (2013) Selective autophagy. Essays Biochem 55:79–92
Takamura A, Komatsu M et al (2011) Autophagy-deficient mice develop multiple liver tumors. Genes Dev 25:795–800
Trinklein ND, Murray JI et al (2004) The role of heat shock transcription factor 1 in the genome-wide regulation of the mammalian heat shock response. Mol Biol Cell 15:1254–1261
Tsukada M, Ohsumi Y (1993) Isolation and characterization of autophagy-defective mutants of Saccharomyces cerevisiae. FEBS Lett 333:169–174

Walter P, Ron D (2011) The unfolded protein response: from stress pathway to homeostatic regulation. Science 334:1081–1086

Watanabe Y, Tanaka M (2011) p62/SQSTM1 in autophagic clearance of a non-ubiquitylated substrate. J Cell Sci 124:2692–2701

Watanabe Y, Tatebe H et al (2012) p62/SQSTM1-dependent autophagy of Lewy body-like alpha-synuclein inclusions. PLoS One 7:e52868

Watanabe Y, Tsujimura A et al (2017) HSF1 stress response pathway regulates autophagy receptor SQSTM1/p62-associated proteostasis. Autophagy 13:133–148

Wild P, McEwan DG et al (2014) The LC3 interactome at a glance. J Cell Sci 127:3–9

Wong E, Cuervo AM (2010) Integration of clearance mechanisms: the proteasome and autophagy. Cold Spring Harb Perspect Biol 2:a006734

Xu P, Das M et al (2011) JNK regulates FoxO-dependent autophagy in neurons. Genes Dev 25:310–322

Zhao Y, Gong S et al (2009) Induction of macroautophagy by heat. Mol Biol Rep 36:2323–2327

Zhu K, Dunner K Jr et al (2010) Proteasome inhibitors activate autophagy as a cytoprotective response in human prostate cancer cells. Oncogene 29:451–462

Chapter 10
Involvement of Heat Shock Proteins in Invertebrate Anhydrobiosis

Alexander Nesmelov, Elena Shagimardanova, Takahiro Kikawada, and Oleg Gusev

Abstract The anhydrobiosis is a unique state notable for a complete lack of detectable metabolic activity and an ability to withstand extreme stresses. It is induced, as its name suggests, by severe water loss, extending up to complete desiccation. Despite of severe stress, organisms entering anhydrobiosis normally retain their viability. This phenomena is ensured by a variety of protective mechanisms, including expression of protective proteins. The latter include Heat shock proteins (Hsp), which have long been recognized for their importance in a wide range of stress protection mechanisms. This chapter summarizes the theory and available experimental data, both suggesting the importance of Hsp in invertebrate anhydrobiosis. However, most of experimental data are the results of expression studies. We show that they are insufficient to make robust conclusions on the role of Hsp in anhydrobiosis. To date, only two robust evidences based on loss of-function-experiments are available, leaving for the future research the complete elucidation of Hsp function in anhydrobiosis.

Keywords Anhydrobiosis · Anhydrobiotic engineering · Desiccation stress · Gene expression · Heat shock proteins · Invertebrates · Molecular chaperons

---

A. Nesmelov · E. Shagimardanova
Institute of Fundamental Medicine and Biology, Kazan Federal University, Kazan, Russia

T. Kikawada
Division of Biotechnology, Institute of Agrobiological Sciences, National Institute of Agriculture and Food Research Organization, Tsukuba, Japan

Graduate School of Frontier Sciences, The University of Tokyo, Chiba, Japan

O. Gusev (✉)
Institute of Fundamental Medicine and Biology, Kazan Federal University, Kazan, Russia

KFU-RIKEN Translational Genomics Unit, RIKEN, Yokohama, Japan
e-mail: oleg.gusev@riken.jp

A. A. A. Asea, P. Kaur (eds.), *Heat Shock Proteins and Stress*,
Heat Shock Proteins 15, https://doi.org/10.1007/978-3-319-90725-3_10

Abbreviations

ArHsp22 and ArHsp21	Small heat shock proteins
ATP	Adenosine triphosphate
CRISPR/Cas9	Protein-9 nuclease (Cas9) associated with clustered regularly interspaced short palindromic repeats (CRISPR)
DNA	Deoxyribonucleic acid
DnaJ (Hsp40)	Chaperone DnaJ, also known as Hsp40
F08H9.3 and *F08H9.4*	Small heat shock protein genes
geLC-MS/MS	In-gel tryptic digestion followed by liquid chromatography-tandem mass spectrometry
Gy	Gray of ionizing radiation
HSC	Heat shock cognate protein
hsp	Heat shock protein genes
Hsp	Heat shock proteins
hsp70	Genes of a heat shock proteins of a particular family
Hsp70	Heat shock proteins of a particular family
IR	Ionizing radiation
LD50	Dose lethal to 50% of treated organisms
LEA	Late embryogenesis abundant proteins
mRNA	Messenger RNA
Mt-shsp17.2	Small heat shock protein gene
p26	Small heat shock protein
PCD	Programmed cell death
qPCR	Quantitative polymerase chain reaction
RNA	Ribonucleic acid
RNAi	RNA interference
SDS-PAGE	Sodium dodecyl sulfate polyacrylamide gel electrophoresis
sHsp	Small heat shock protein
shsp	Small heat shock protein gene
TRET1	Trehalose transporter

10.1 Introduction

Anhydrobiosis is an ametabolic state, inducible by severe desiccation. Unlike death, anhydrobiosis is normally reversible and a completely dried organism without detectable signs of metabolism is able to resume its life when the environment becomes hospitable again due to the return of water (Keilin 1959). This adaptation demonstrates the ability to cope with an extraordinary stress, as illustrated by the massive damage to cell components that extreme dehydration causes in desiccation-sensitive organisms. The main types of drying-related damage are denaturation and aggregation of proteins, leakage and fusion of membranes as a result of phase transitions, and destabilization of RNA, DNA and chromatin (França et al. 2007; Tunnacliffe et al. 2010). Despite the severity of desiccation stress, organisms that are able to enter anhydrobiosis are found in many taxa.

An interesting feature of the anhydrobiotic state is that it is typically associated with extreme tolerance to diverse abiotic stresses such as low or high temperatures and ionizing radiation (IR), among others (Vanhaecke et al. 1989; Ramløv and Westh 2001; Watanabe et al. 2002). In addition, animals in the anhydrobiotic state can often survive for years or even decades longer than their hydrated lifespan, as shown by the successful rehydration of *Polypedilum vanderplanki* larvae after 17 years of dry storage (Adams 1983). These properties of anhydrobiosis have attracted the attention of researchers who aim to adapt them for the artificial preservation of biomaterials. Such a preservation technology, which has been called 'anhydrobiotic engineering' (de Castro et al. 2000), would dramatically decrease both the cost and complexity of the storage and transport of a wide range of biomaterials. Its implementation, however, requires an understanding of the molecular mechanisms of anhydrobiosis, including the identification of all the associated protective molecules. Organisms entering anhydrobiosis are now known to accumulate many protective components, including (in many cases) the non-reducing sugar trehalose and various proteins (Watanabe et al. 2002; Crowe 2014). The latter mainly comprise the late embryogenesis abundant (LEA) proteins, antioxidant proteins and heat shock proteins (Hsp) (Cornette and Kikawada 2001; Wharton 2015). Hsp are likely to be indispensable for anhydrobiosis, as they are for a wide variety of other stress responses (Feder et al. 1999). Indeed, as shown below, they accumulate in the anhydrobiotic state in many desiccation-tolerant organisms. In this chapter, we summarize the theory and experimental data on the functions of Hsp in invertebrate anhydrobiosis, extending our analysis to cover gaps that currently exist in these data.

10.1.1 HSP in Anhydrobiosis: Putative Functions

Hsp are a diverse superfamily of conserved proteins that are found in all living organisms in varying combinations. They differ in molecular size and mode of action, generally falling into two major categories: ATP-dependent large Hsp and ATP-independent small Hsp (sHsp). In turn, the large Hsp can be divided into four ubiquitous conserved families: Hsp100s, Hsp90s, Hsp70s and Hsp60s. The number in each large Hsp family name is approximately indicative of molecular mass; the sHsp are from 12 to 42 kDa in size (Haslbeck et al. 2015). Large Hsp can have both stress-inducible and constitutively expressed versions, whereas sHsp are often expressed only in response to stress (Richter et al. 2010). Constitutively expressed Hsp in are often abbreviated as Hsc. Hsp were first discovered, as their name suggests, as key defensive components of the heat shock response, but were shown subsequently to participate in a wide range of cellular processes relating to proteome maintenance, as part of both stress responses and normal physiology

(Hartl et al. 2011). Nowadays they are often called molecular chaperones. Their main activities include enabling the correct folding of newly synthesized proteins, the refolding of denatured proteins, the assembly of protein oligomers, protein trafficking and assistance in protein degradation (Hartl et al. 2011).

The above activities are not equally distributed among the different categories of Hsp; indeed, there is considerable specialization of function. Thus, the division into the two main categories of Hsp, large Hsp and sHsp, reflects not only a size difference, but also a different mode of action. Large ATP-dependent Hsp work in complexes with other proteins and facilitate the correct folding of newly synthesized proteins and the refolding of misfolded proteins, including decomposition of stable protein aggregates (Richter et al. 2010). These activities are referred to as 'foldase' and 'disaggregase' functions, respectively. Some large Hsp and their partner proteins also participate in protein degradation (Hartl et al. 2011). In contrast, sHsp form oligomeric complexes which bind and passivate misfolded proteins to prevent their further aggregation. This mode of Hsp action is termed 'holdase' activity. Such holdase sHsp work in conjunction with large Hsp, which effect the release and refolding of misfolded proteins (Richter et al. 2010).

All the various types of Hsp activity seem to be important, if not crucial, for anhydrobiosis. Denaturation and aggregation of proteins were shown to occur during desiccation *in vitro* and can lead to the loss of enzyme activity (Carpenter et al. 1987, 1989; Dong et al. 1995; Kikuta et al. 2017). Moreover, these processes are thought to be key factors in desiccation stress (Wolkers et al. 2002). Thus, holdases, foldases and disaggregases are essential for survival of desiccation as they help to prevent and repair the protein damage caused by extreme dehydration. However, are Hsp the only molecules that can perform such functions? It is likely that at least the holdase function of sHsp can be carried out by intrinsically disordered proteins (mainly LEA proteins) and, to some extent, trehalose, both of which can prevent protein aggregation (Crowe 2007; Battaglia et al. 2008; Hand et al. 2011). Intrinsically disordered proteins and trehalose are accumulated by many anhydrobiotic animals, perhaps reflecting the need for additional such activity under the extreme conditions provided by desiccation stress. These components differ from sHsp in their mode of action, preventing protein aggregation without a strong interaction with their targets (Crowe 2007; Chakrabortee et al. 2012). In contrast to sHsp activities, the refolding and disaggregation activities of large Hsp most likely cannot be performed by LEA proteins or other protective components associated with anhydrobiosis (Chakrabortee et al. 2007, 2012). These large Hsp activities are thus important for anhydrobiosis, as suggested by the fact that they are frequently induced as anhydrobiotic animals desiccate. However, the different activities of the large Hsp may be necessary at different stages of anhydrobiosis, as suggested by the often asynchronous expression of Hsp70 and Hsp90 (for example, Gusev et al. 2011; Wang et al. 2014). The constitutive and inducible forms of particular large Hsp might also have different functions during desiccation and rehydration in the anhydrobiotic tardigrade *Milnesium tardigradum* and, similarly, in the flesh fly in response to desiccation (Hayward et al. 2004; Wang et al. 2014). We should note that, in partially desiccated cells, the foldase and disaggregase activities of large Hsp could be restricted due to the limited availability of ATP as metabolism shuts down (Clegg 2001). Thus, the catalytic activity of large Hsp may be postponed until rehydration and subsequent recovery from desiccation. Nevertheless, large Hsp can bind misfolded proteins independently of ATP, and may remain in this state until ATP becomes available once more (Richter et al. 2010). In this respect, large Hsp can also exhibit holdase function over prolonged periods and this likely contributes to the prevention of protein aggregation (King and MacRae 2015).

Another putative function of Hsp in anhydrobiosis is the inhibition of programmed cell death (PCD). Possible triggers of PCD in anhydrobiosis are DNA damage and protein aggregation. Desiccation causes severe DNA fragmentation even in anhydrobiotic animals, despite the presence of protective mechanisms (Neumann et al. 2009). In anhydrobiotic *P. vanderplanki* larvae, DNA fragmentation is similar to that caused by 3000 Gy of IR, whereas the LD50 (dose lethal to 50% of treated organisms) for IR in wild-type *Drosophila melanogaster* is less than 1400 Gy (Parashar et al. 2008; Gusev et al. 2010). However, in anhydrobiotes even such severe DNA damage fails to trigger PCD, because dried *P. vanderplanki* larvae survive and recover full DNA integrity upon rehydration (Gusev et al. 2010); *M. tardigradum* also recovers from the massive DNA damage caused by long-term storage in the dry state (Neumann et al. 2009). This is particularly surprising because the major DNA damage experienced by anhydrobiotes persists for an extended period of time while DNA repair processes take place. For example, *P. vanderplanki* larvae typically restore DNA integrity within 72 to 96 h of rehydration (Gusev et al. 2010), yet PCD is not triggered and is presumably specifically regulated in anhydrobiotes (Menze et al. 2007). Moreover, direct biochemical evidence for the PCD inhibition by the sHsp p26 has been obtained in the anhydrobiotic cysts of *Artemia franciscana* (Villeneuve et al. 2006). This finding suggests that Hsp have an important role in PCD inhibition in anhydrobiotes, as they do in non-anhydrobiotic organisms (Kennedy et al. 2014).

To summarize so far, we have shown how the functions usually attributed to Hsp, such as inhibition of protein aggregation, PCD inhibition, refolding of proteins and assistance in their degradation, are likely to be essential for successful anhydrobiosis. However, as shown in the following sections, the details of the contribution of Hsp to anhydrobiosis, including which proteins are involved and the extent of their roles, are for the most part yet to be uncovered.

10.1.2 HSP in Anhydrobiosis: Experimental Data

Since the initial observation of sHsp accumulation in anhydrobiotic *A. franciscana* cysts in 1994, a number of other experimental reports have also concluded that Hsp are functional in anhydrobiosis. These reports cover all known groups of anhydrobiotic invertebrates, including crustaceans, tardigrades, nematodes, rotifers and insects.

10.1.2.1 Crustaceans

One of the best-studied anhydrobiotic organisms is the brine shrimp, *A. franciscana*. The embryonic cysts of this organism, which are the only developmental stage able to enter anhydrobiosis, can survive desiccation to a water content less than 2% of hydrated levels (Toxopeus et al. 2014). These cysts were the first anhydrobiotic structure in which the accumulation of a specific protein (named p26, due to its 26 kDa molecular mass) was observed; p26 was subsequently identified as a small heat shock/α-crystallin protein (Clegg et al. 1994, 1999). The importance of this protein for tolerance of desiccation and temperature extremes was verified by knockdown experiments (King and MacRae 2012). As mentioned above, this protein was also shown to inhibit apoptosis (Villeneuve et al. 2006). These findings were extended when Hsp70, Hsp90 and two small Hsp named ArHsp22 and ArHsp21 were detected in *A. franciscana* (Clegg et al. 1999; Qiu et al. 2008a, b). Of these, only ArHsp21 accumulates during cyst development. However, despite this, ArHsp21 has, at best, a minor role in cyst tolerance of desiccation and freezing (King et al. 2013). Thus, Hsp accumulation during cyst development can not only be attributed to preparation for anhydrobiosis, but also to regulation of development itself, since Hsp tend to accumulate at stages related to developmental arrest (for example, Rinehart et al. 2007).

10.1.2.2 Tardigrades

Tardigrades are also well-studied anhydrobiotic invertebrates, and there are many reports on the role of Hsp. Possible Hsp overexpression in *Richtersius coronifer* in response to desiccation was first reported in 2001 (Ramløv and Westh 2001). However, the authors did not confirm that the low-molecular-mass protein observed by SDS-PAGE in samples after desiccation was actually an Hsp. Later, Schill and coworkers reported the induction of the mRNA of one of three *hsp70* isoforms during transition into anhydrobiosis in *M. tardigradum* (Schill 2004). At the same time, mRNAs corresponding to all three *hsp70* isoforms were upregulated in response to heat shock. In contrast, Reuner et al. reported the induction in *M. tardigradum* of only one of the three *hsp70* mRNAs during desiccation, as well as one *hsp90* mRNA, in the anhydrobiotic state (Reuner et al. 2010). In the same research, all other *hsp* genes were downregulated at all stages of anhydrobiosis (desiccation, desiccated state, rehydration) and after heat shock, with the sole exception of a small α-crystallin sHsp gene, *Mt-shsp17.2*, which was induced in response to heat shock (Reuner et al. 2010). In another study, the total level of Hsp70 was found to have decreased in anhydrobiotic samples of *R. coronifer*, with some recovery after 1 h rehydration (Jönsson et al. 2007). Other authors showed that desiccated tardigrades of the species *Paramacrobiotus richtersi* did not differ significantly from the hydrated state in the relative levels of Hsp70 and Hsp90 detected by immunoblotting (Rizzo et al. 2010). In a later study in the same species, Hsp90 protein was reported to be induced in desiccated samples (Rebecchi et al. 2011). Despite some discrepancies in the reported data on Hsp expression in tardigrades, they are generally in accordance with the hypothesis that *hsp* mRNAs are accumulated during desiccation, but that Hsp protein synthesis does not occur until after rehydration (Guidetti et al. 2011). However, these data are still fragmentary as they cover different species and time points, which are not necessarily comparable.

10.1.2.3 Nematodes

Many species of these roundworms are known to be capable of entering anhydrobiosis and Hsp function has been studied in this context in some of them. For example, in the Antarctic nematode *Plectus murrayi*, no significant difference in expression of *hsp70* and *hsp90* transcripts was observed between desiccated and control samples (Adhikari et al. 2009). In a distantly related anhydrobiotic nematode, *Aphelenchus avenae*, expression profiling revealed no differentially expressed *hsp* genes upon mild desiccation (Reardon et al. 2010). This mild desiccation is a form of preconditioning that is required to enable some anhydrobiotes to prepare for severe desiccation. In another anhydrobiotic nematode, *Panagrolaimus superbus*, the mRNA of one *shsp* was upregulated in the desiccated state, while other Hsp tested (*shsp*, *DnaJ (hsp40)* and *hsp70*) were downregulated (Tyson et al. 2012). *Caenorhabditis elegans* dauer larvae, which are a desiccation-tolerant developmental stage of this model nematode, have been subjected to a rigorous analysis before and after mild desiccation, which is necessary for the successful induction of anhydrobiosis (Erkut et al. 2013). Five *shsp* genes and one *hsp70* gene were found to be induced at the transcriptional level after 24 h of mild desiccation. Interestingly, of these, only one sHsp (Hsp-12.6 αB-crystallin) was found to be induced at the translational level, as detected by geLC-MS/MS (Erkut et al. 2013). Two *shsp* genes (*F08H9.3* and *F08H9.4*) and an *hsp70* gene were shown to be important for anhydrobiosis after gene silencing produced desiccation-sensitive phenotypes.

10.1.2.4 Rotifers

Similarly to tardigrades, most of these small aquatic invertebrates of the Bdelloidea class are able to enter anhydrobiosis at any stage of the life cycle (Caprioli et al. 2004). For species of the Monogononta class of rotifers, this ability is limited to the specific stage of dormant (resting, diapausing) eggs, which are products of the sexual cycle. In contrast, amictic eggs which are products of the asexual cycle, cannot survive desiccation. *Shsp* mRNA is upregulated in resting eggs of the monogonont rotifer *Brachionus plicatilis* (Denekamp et al. 2009). In another study, *hsp70* with *DnaJ (hsp40)* co-chaperones and *hsc70/hsp90* organising protein gene were upregulated in resting eggs of the same species (Clark et al. 2012). To the best of our knowledge, there are no reports on the role of Hsp in bdelloid rotifer anhydrobiosis.

10.1.2.5 Insects

The African chironomid *P. vanderplanki* was for a long time considered to be the only insect capable of entering anhydrobiosis (Cornette and Kikawada 2001). *Polypedilum ovahimba* had been described as a second candidate anhydrobiote among the insects, but only circumstantial evidence was available (Cranston 2014). However, recently a close relative of *P. vanderplanki*, *Polypedilum pembai*, was unequivocally shown to be anhydrobiotic and therefore represents a second insect species with this property (Cornette et al. 2017). Most probably, these species diverged after the anhydrobiotic phenotype evolved and thus they are likely to employ the same mechanisms of desiccation tolerance. Nevertheless, currently the best studied example of anhydrobiosis in insects remains the *P. vanderplanki* larva, which is the only ontogenetic stage of the midge that is able to enter anhydrobiosis. A wide variety of information has been obtained, including data on the whole genome and whole transcriptome scale (Gusev et al. 2014). More detailed studies have also been performed of the expression and function of a number of proteins, including the trehalose transporter TRET1 (Kikawada et al. 2007), trehalose metabolic enzymes (Mitsumasu et al. 2010), aquaporins (Kikawada et al. 2008), LEA proteins (Kikawada et al. 2006; Hatanaka et al. 2013), Hsp (Gusev et al. 2011; Kozlova et al. 2016) and antioxidant proteins (Nesmelov et al. 2016, 2018). An expressed sequence tag database contains 19 clusters of *hsp*-like genes, including *DnaJ* (*hsp40*), *hsp60, hsp70, hsp82, hsp90* and at least four *shsp* (Cornette et al. 2010). Some of these *hsp* genes are induced by desiccation, especially at the late 36 h time point. Six such genes – *Pv-hsp90, Pv-hsp70, Pv-hsc70, Pv-hsp60, Pv-hsp20,* and *Pv-p23* – have been analyzed by qPCR, which confirmed them to be induced by desiccation (Gusev et al. 2011). Genome and transcriptome sequencing revealed 22 *hsp* genes, 19 of which are induced more than threefold at the mRNA level in larvae after 48 h desiccation (Gusev et al. 2011, supplementary data). The recent development of a CRISPR/Cas9 system and RNAi for the *P. vanderplanki*-derived anhydrobiotic cultured cells Pv11 should provide a powerful tool for the verification of the role of Hsp in *P. vanderplanki* anhydrobiosis (Nakahara et al. 2010; Okada et al. 2015; Watanabe et al. 2016; Sogame et al. 2017).

10.1.3 HSP in Anhydrobiosis: Future Prospects

As described in the above sections, the induction of Hsp at the mRNA or protein levels has been observed in all known groups of anhydrobiotic animals. Different groups of Hsp were found to be upregulated or downregulated at different stages of anhydrobiosis, probably reflecting their different functionality, as suggested previously (Hayward et al. 2004; Wang et al. 2014). In two cases, in *A. franciscana* (*p26 shsp*) and in *C. elegans* (two *shsp* genes, one *hsp70*), the essential role of some Hsp in defining the anhydrobiotic phenotype has been confirmed by gene silencing (King and MacRae 2012; Erkut et al. 2013). Taken together with the well-recognized role of Hsp as important components of various stress responses, these data provide robust evidence that at least some Hsp are required for anhydrobiosis.

However, the details of how Hsp contribute to anhydrobiosis are largely unknown. In particular, it will be crucial to perform loss-of-function experiments to determine the minimum set of Hsp required for a given anhydrobiosis phenotype. As shown in the next paragraph, although data on the expression of the full set of Hsp are available for a number of anhydrobiotes, this is probably insufficient to decipher which of these are necessary for successful anhydrobiosis. Furthermore, the abundance of a given mRNA at a particular stage of anhydrobiosis may not correlate well with the timing and extent of the synthesis of the corresponding protein, nor with its activity (King and MacRae 2015). For example, in *C. elegans* dauer larvae, mild desiccation for 24 h induced three out of four tested *hsp* at the mRNA level, but the cognate proteins were not found to be upregulated at 24 or 96 h (Erkut et al. 2013). Interestingly, transcripts of the F08H9.4 gene, whose putative Hsp-16-like protein product was not detected, were expressed at 1.5-times higher levels than mRNA of the *hsp-12.6* gene, whose corresponding protein was successfully detected: expression levels as a measure of mRNA abundance at the 24 h desiccation time point were 273,108 and 183,969, respectively (dataset S1 in Erkut et al. 2013). Protein activity should also be taken into account, since the activity of large Hsp can be modified by partner proteins (Saibil 2013). Intriguingly, to date, only one such partner protein is reported to be upregulated in anhydrobiosis along with its Hsp partner, i.e. Hsc70/Hsp90 organizing protein in the eggs of *B. plicatilis* rotifers (Clark et al. 2012). However, these proteins are undoubtedly important for the function of large Hsp (Richter et al. 2010). It worth noting that Hsp are just a part (albeit an important one) of a very complex system that maintains proteome integrity (Hartl et al. 2011). In anhydrobiotes, this system might be even more complicated since these organisms often accumulate LEA proteins, other intrinsically disordered proteins and trehalose, which can overlap with Hsp in their functions.

The importance of adopting the correct experimental strategy when attempting to identify the "crucial set" of Hsp (or indeed any other category of molecule) required for anhydrobiosis is illustrated by the case of *Deinococcus radiodurans*. This bacterium is extremely resistant to IR, ultraviolet light and desiccation. Extensive studies resulted in the identification of 57 genes that are essential for *D. radiodurans* radioresistance, since loss of function of any one of these genes produces a sensitive phenotype (reviewed in Makarova et al. 2007). Of these, only 15 genes were reproducibly upregulated more than threefold by IR in two independent microarray studies, while 19 other genes were upregulated in at least one such study (Liu et al. 2003; Tanaka et al. 2004; Makarova et al. 2007). Thus, almost half the genes essential for radioresistance were not upregulated by exposure to IR, and, conversely, many of genes upregulated by IR were not found to be essential (Fig. 10.1). This example is important particularly because many anhydrobiotes are also cross-resistant to IR and therefore these phenotypes have a degree of overlap. Similarly, in *Saccharomyces cerevisiae* 650 genes were identified as implicated in desiccation

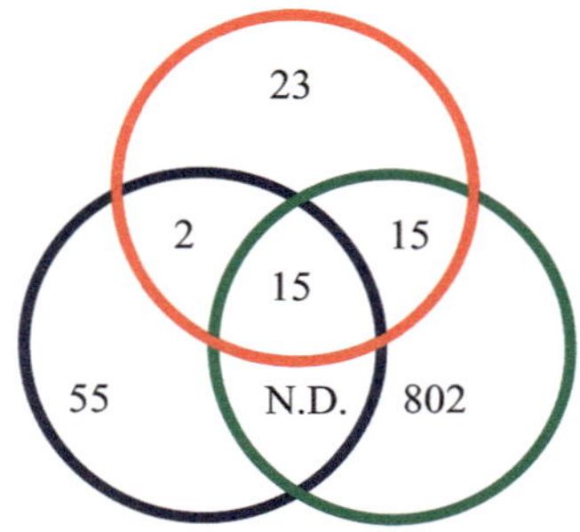

Fig. 10.1 Venn diagram of the number of *D. radiodurans* genes implicated in radioresistance (*red*) and/or upregulated in response to IR, as reported by Liu et al. 2003 (*green*) or Tanaka et al. 2004 (*blue*). N.D. – no data

tolerance and only 48 of them were upregulated by desiccation, comprising less than 1% of upregulated genes (Ratnakumar et al. 2011). These examples show that transcriptome data can be misleading when searching for stress tolerance genes. Thus, loss-of-function experiments seems to be absolutely necessary to determine which Hsp (and other proteins) are essential for anhydrobiosis. Such studies on anhydrobiotic animals from different groups of invertebrates will expand our understanding of the interplay of Hsp with protective components and proteins specific for anhydrobiosis.

10.2 Conclusions

The available data suggest that Hsp are important for desiccation tolerance in all known groups of anhydrobiotic invertebrates. The Hsp studied in these organisms vary from just one example to the full set of Hsp identified at the transcriptome and/or proteome level. Some of the Hsp studied were shown to be upregulated at the mRNA and/or protein level during induction of anhydrobiosis. In a few cases, Hsp have been shown to be essential for anhydrobiosis in the organism investigated, since gene silencing produces a desiccation-sensitive phenotype. Unfortunately, such robust evidence is scarce. To date, the complete set of Hsp that are crucial for anhydrobiosis, as shown by loss-of-function experiments, has not been determined in any organism. Thus, a full picture of Hsp function in anhydrobiosis is yet to be revealed. Further studies involving genome editing and RNA silencing are complicated in part because anhydrobiotes are mainly non-model organisms and the corresponding techniques have not been fully developed. However, Hsp are promising candidates for such studies, which are an essential step in our understanding of the anhydrobiosis phenomenon and its exploitation in the development of nature-inspired technologies for biomaterial preservation.

Acknowledgements We apologize to those researchers who have contributed to the elucidation of the role of Hsp in anhydrobiosis whose work we have not cited due to space limitations. We are grateful to Dr. A. Tunacliffe for critical reading of the manuscript and valuable comments. This work was performed according to the Russian Government Program of Competitive Growth of Kazan Federal University in the analysis of Hsp putative functions, and was supported by the Russian Science Foundation (№ 17-44-07002) in the analysis of Hsp functions in *P. vanderplanki*. The work was also supported by the JSPS KAKENHI (Grant Numbers JP17H01511, JP25128714 and JP16K07308) in the analysis of Hsp in tardigrades and nematodes.

References

Adams A (1983) Cryptobiosis in Chironomidae (Diptera) – two decades on Antenna. Bull R Entomol Soc Lond 8:58–61

Adhikari BN, Wall DH, Adams BJ (2009) Desiccation survival in an Antarctic nematode: molecular analysis using expressed sequenced tags. BMC Genomics 10:69

Battaglia M, Olvera-Carrillo Y, Garciarrubio A, Campos F, Covarrubias AA (2008) The enigmatic LEA proteins and other hydrophilins. Plant Physiol 148:6–24

Caprioli M, Krabbe Katholm A, Melone G et al (2004) Trehalose in desiccated rotifers: a comparison between a bdelloid and a monogonont species. Comp Biochem Physiol Part A Mol Integr Physiol 139:527–532

Carpenter JF, Crowe JH (1989) An infrared spectroscopic study of the interactions of carbohydrates with dried proteins. Biochemistry 28:3916–3922

Carpenter JF, Martin B, Crowe LM, Crowe JH (1987) Stabilization of phosphofructokinase during air-drying with sugars and sugar/transition metal mixtures. Cryobiology 24:455–464

Chakrabortee S, Boschetti C, Walton LJ et al (2007) Hydrophilic protein associated with desiccation tolerance exhibits broad protein stabilization function. Proc Natl Acad Sci U S A 104:18073–18078

Chakrabortee S, Tripathi R, Watson M, Kaminski GS (2012) Intrinsically disordered proteins as molecular shields. Mol BioSyst 8:210–219

Clark MS, Denekamp NY, Thorne MAS et al (2012) Long-term survival of hydrated resting eggs from *Brachionus plicatilis*. PLoS One 7:e29365

Clegg JS (2001) Cryptobiosis – a peculiar state of biological organization. Comp Biochem Physiol Part B Biochem Mol Biol 128:613–624

Clegg J, Jackson SA, Warner AH (1994) Extensive intracellular translocations of a major protein accompany anoxia in embryos of *Artemia franciscana*. Exp Cell Res 212:77–83

Clegg JS, Willsie JK, Jackson SA (1999) Adaptive significance of a small heat shock/α-Crystallin protein (p26) in encysted embryos of the brine shrimp, *Artemia franciscana*. Am Zool 39:836–847

Cornette R, Kikawada T (2011) The induction of anhydrobiosis in the sleeping chironomid: current status of our knowledge. IUBMB Life 63:419–429

Cornette R, Kanamori Y, Watanabe M et al (2010) Identification of anhydrobiosis-related genes from an expressed sequence tag database in the cryptobiotic midge *Polypedilum vanderplanki* (Diptera; Chironomidae). J Biol Chem 285:35889–35899

Cornette R, Yamamoto N, Yamamoto M et al (2017) A new anhydrobiotic midge from Malawi, *Polypedilum pembai* sp.n. (Diptera: Chironomidae), closely related to the desiccation tolerant midge, *Polypedilum vanderplanki* Hinton. Syst Entomol 42:814–825

Cranston PS (2014) A new putatively cryptobiotic midge, *Polypedilum ovahimba* sp. nov. (Diptera: Chironomidae), from southern Africa. Austral Entomol 53:373–379

Crowe JH (2007) Trehalose as a "Chemical chaperone": fact and fantasy. In: Csermely P, Vígh L (eds) Advances in experimental medicine and biology. Springer, New York, pp 143–158

Crowe JH (2014) Anhydrobiosis: an unsolved problem. Plant Cell Environ 37:1491–1493

de Castro AG, Lapinski J, Tunnacliffe A (2000) Anhydrobiotic engineering. Nat Biotechnol 18:473

Denekamp NY, Thorne MAS, Clark MS et al (2009) Discovering genes associated with dormancy in the monogonont rotifer *Brachionus plicatilis*. BMC Genomics 10:108

Dong A, Prestrelski SJ, Allison SD, Carpenter JF (1995) Infrared spectroscopic studies of lyophilization- and temperature-induced protein aggregation. J Pharm Sci 84:415–424

Erkut C, Vasilj A, Boland S et al (2013) Molecular strategies of the *Caenorhabditis elegans* dauer larva to survive extreme desiccation. PLoS One 8:e82473

Feder ME, Hofmann GE (1999) Heat-shock proteins, molecular chaperones, and the stress response: evolutionary and ecological physiology. Annu Rev Physiol 61:243–282

França MB, Panek AD, Eleutherio ECA (2007) Oxidative stress and its effects during dehydration. Comp Biochem Physiol Part A Mol Integr Physiol 146:621–631

Guidetti R, Altiero T, Rebecchi L (2011) On dormancy strategies in tardigrades. J Insect Physiol 57:567–576

Gusev OA, Nakahara Y, Vanyagina V et al (2010) Anhydrobiosis-associated nuclear DNA damage and repair in the sleeping chironomid: linkage with radioresistance. PLoS One 5:e14008

Gusev OA, Cornette R, Kikawada T, Okuda T (2011) Expression of heat shock protein-coding genes associated with anhydrobiosis in an African chironomid *Polypedilum vanderplanki*. Cell Stress Chaperones 16:81–90

Gusev OA, Suetsugu Y, Cornette R et al (2014) Comparative genome sequencing reveals genomic signature of extreme desiccation tolerance in the anhydrobiotic midge. Nat Commun 5:4784

Hand SC, Menze MA, Toner M, Boswell L, Moore D (2011) LEA proteins during water stress: not just for plants anymore. Annu Rev Physiol 73:115–134

Hartl FU, Bracher A, Hayer-Hartl M (2011) Molecular chaperones in protein folding and proteostasis. Nature 475:324–332

Haslbeck M, Vierling E (2015) A first line of stress defense: small heat shock proteins and their function in protein homeostasis. J Mol Biol 427:1537–1548

Hatanaka R, Hagiwara-Komoda Y, Furuki T et al (2013) An abundant LEA protein in the anhydrobiotic midge, PvLEA4, acts as a molecular shield by limiting growth of aggregating protein particles. Insect Biochem Mol Biol 43:1055–1067

Hayward SAL, Rinehart JP, Denlinger DL (2004) Desiccation and rehydration elicit distinct heat shock protein transcript responses in flesh fly pupae. J Exp Biol 207:963–971

Jönsson KI, Schill RO (2007) Induction of Hsp70 by desiccation, ionising radiation and heat-shock in the eutardigrade *Richtersius coronifer*. Comp Biochem Physiol Part B Biochem Mol Biol 146:456–460

Keilin D (1959) The problem of anabiosis or latent life: history and current concept. Proc R Soc London Ser B Biol Sci 150:149–191

Kennedy D, Jäger R, Mosser DD, Samali A (2014) Regulation of apoptosis by heat shock proteins. IUBMB Life 66:327–338

Kikawada T, Nakahara Y, Kanamori Y et al (2006) Dehydration-induced expression of LEA proteins in an anhydrobiotic chironomid. Biochem Biophys Res Commun 348:56–61

Kikawada T, Saito A, Kanamori Y et al (2007) Trehalose transporter 1, a facilitated and high-capacity trehalose transporter, allows exogenous trehalose uptake into cells. Proc Natl Acad Sci U S A 104:11585–11590

Kikawada T, Saito A, Kanamori Y et al (2008) Dehydration-inducible changes in expression of two aquaporins in the sleeping chironomid, *Polypedilum vanderplanki*. Biochim Biophys Acta Biomembr 1778:514–520

Kikuta S, Watanabe SJ, Sato R et al (2017) Towards water-free biobanks: long-term dry-preservation at room temperature of desiccation-sensitive enzyme luciferase in air-dried insect cells. Sci Rep 7:6540

King AM, MacRae TH (2012) The small heat shock protein p26 aids development of encysting *Artemia* embryos, prevents spontaneous diapause termination and protects against stress. PLoS One 7:e43723

King AM, MacRae TH (2015) Insect heat shock proteins during stress and diapause. Annu Rev Entomol 60:59–75

King AM, Toxopeus J, MacRae TH (2013) Functional differentiation of small heat shock proteins in diapause-destined *Artemia* embryos. FEBS J 280:4761–4772

Kozlova O, Cherkasov A, Przhiboro A, Shagimardanova E (2016) Complexity of expression control of HSP70 genes in Extremophilic midges. Bionanoscience 6:388–391

Liu Y, Zhou J, Omelchenko MV et al (2003) Transcriptome dynamics of *Deinococcus radiodurans* recovering from ionizing radiation. Proc Natl Acad Sci U S A 100:4191–4196

Makarova KS, Omelchenko MV, Gaidamakova EK et al (2007) *Deinococcus geothermalis*: the pool of extreme radiation resistance genes shrinks. PLoS One 2:e955

Menze MA, Hand SC (2007) Caspase activity during cell stasis: avoidance of apoptosis in an invertebrate extremophile, *Artemia franciscana*. Am J Physiol Regul Integr Comp Physiol 292:R2039–R2047

Mitsumasu K, Kanamori Y, Fujita M et al (2010) Enzymatic control of anhydrobiosis-related accumulation of trehalose in the sleeping chironomid, *Polypedilum vanderplanki*. FEBS J 277:4215–4228

Nakahara Y, Imanishi S, Mitsumasu K et al (2010) Cells from an anhydrobiotic chironomid survive almost complete desiccation. Cryobiology 60:138–146

Nesmelov AA, Devatiyarov RM, Voronina TA et al (2016) New antioxidant genes from an Anhydrobiotic insect: unique structural features in functional motifs of Thioredoxins. Bionanoscience 6:568–570

Nesmelov AA, Kikawada T, Gusev O (2018) Antioxidant system in the anhydrobiotic midge as an essential adaptative mechanism for desiccation survival. In: Iwaya-Inoue M, Sakurai M, Uemura M (eds) Survival strategies in extreme cold and desiccation: adaptation mechanisms and their application. Springer, Singapore

Neumann S, Reuner A, Brümmer F, Schill RO (2009) DNA damage in storage cells of anhydrobiotic tardigrades. Comp Biochem Physiol Part A Mol Integr Physiol 153:425–429

Okada J, Kikuta S, Gusev O et al (2015) Construction of optimized CRISPR/Cas system to reveal the mechanisms of Anhydrobiosis in the sleeping Chironomid. Cryobiol Cryotechnology 61:69–73

Parashar V, Frankel S, Lurie AG, Rogina B (2008) The effects of age on radiation resistance and oxidative stress in adult *Drosophila melanogaster*. Radiat Res 169:707–711

Qiu Z, MacRae TH (2008a) ArHsp22, a developmentally regulated small heat shock protein produced in diapause-destined *Artemia* embryos, is stress inducible in adults. FEBS J 275:3556–3566

Qiu Z, MacRae TH (2008b) ArHsp21, a developmentally regulated small heat-shock protein synthesized in diapausing embryos of *Artemia franciscana*. Biochem J 411:605–611

Ramløv H, Westh P (2001) Cryptobiosis in the eutardigrade *Adorybiotus* (*Richtersius*) *coronifer*: tolerance to alcohols, temperature and de novo protein synthesis. Zool Anz 240:517–523

Ratnakumar S, Hesketh A, Gkargkas K et al (2011) Phenomic and transcriptomic analyses reveal that autophagy plays a major role in desiccation tolerance in *Saccharomyces cerevisiae*. Mol BioSyst 7:139–149

Reardon W, Chakrabortee S, Pereira T et al (2010) Expression profiling and cross-species RNA interference (RNAi) of desiccation-induced transcripts in the anhydrobiotic nematode *Aphelenchus avenae*. BMC Mol Biol 11:6

Rebecchi L, Altiero T, Cesari M et al (2011) Resistance of the anhydrobiotic eutardigrade *Paramacrobiotus richtersi* to space flight (LIFE-TARSE mission on FOTON-M3). J Zool Syst Evol Res 49:98–103

Reuner A, Hengherr S, Mali B et al (2010) Stress response in tardigrades: differential gene expression of molecular chaperones. Cell Stress Chaperones 15:423–430

Richter K, Haslbeck M, Buchner J (2010) The heat shock response: life on the verge of death. Mol Cell 40:253–266

Rinehart JP, Li A, Yocum GD et al (2007) Up-regulation of heat shock proteins is essential for cold survival during insect diapause. Proc Natl Acad Sci U S A 104:11130–11137

Rizzo AM, Negroni M, Altiero T et al (2010) Antioxidant defences in hydrated and desiccated states of the tardigrade *Paramacrobiotus richtersi*. Comp Biochem Physiol Part B Biochem Mol Biol 156:115–121

Saibil H (2013) Chaperone machines for protein folding, unfolding and disaggregation. Nat Rev Mol Cell Biol 14:630–642

Schill RO (2004) Stress gene (*hsp70*) sequences and quantitative expression in *Milnesium tardigradum* (Tardigrada) during active and cryptobiotic stages. J Exp Biol 207:1607–1613

Sogame Y, Okada J, Kikuta S et al (2017) Establishment of gene transfer and gene silencing methods in a desiccation-tolerant cell line, Pv11. Extremophiles 21:65–72

Tanaka M, Earl AM, Howell HA et al (2004) Analysis of *Deinococcus radiodurans's* transcriptional response to ionizing radiation and desiccation reveals novel proteins that contribute to extreme radioresistance. Genetics 168:21–33

Toxopeus J, Warner AH, MacRae TH (2014) Group 1 LEA proteins contribute to the desiccation and freeze tolerance of *Artemia franciscana* embryos during diapause. Cell Stress Chaperones 19:939–948

Tunnacliffe A, Hincha DK, Leprince O, Macherel D (2010) LEA proteins: versatility of form and function. In: Lubzens E, Cerda J, Clark M (eds) Dormancy and resistance in harsh environments, topics in current genetics. Springer-Verlag, Berlin, Heidelberg, pp 91–108

Tyson T, O'Mahony Zamora G, Wong S et al (2012) A molecular analysis of desiccation tolerance mechanisms in the anhydrobiotic nematode *Panagrolaimus superbus* using expressed sequenced tags. BMC Res Notes 5:68

Vanhaecke P, Sorgeloos P (1989) International study on *Artemia*. XLVII. The effect of temperature on cyst hatching, larval survival and biomass production for different geographical strains of brine shrimp *Artemia* spp. Ann Soc R Zool Belg 118:7–23

Villeneuve TS, Ma X, Sun Y et al (2006) Inhibition of apoptosis by p26: implications for small heat shock protein function during *Artemia* development. Cell Stress Chaperones 11:71–80

Wang C, Grohme MA, Mali B, Schill RO, Frohme M (2014) Towards decrypting cryptobiosis – analyzing anhydrobiosis in the tardigrade *Milnesium tardigradum* using transcriptome sequencing. PLoS One 9:1–15

Watanabe M, Kikawada T, Minagawa N, Yukuhiro F, Okuda T (2002) Mechanism allowing an insect to survive complete dehydration and extreme temperatures. J Exp Biol 205:2799–2802

Watanabe K, Imanishi S, Akiduki G, Cornette R, Okuda T (2016) Air-dried cells from the anhydrobiotic insect, *Polypedilum vanderplanki*, can survive long term preservation at room temperature and retain proliferation potential after rehydration. Cryobiology 73:93–98

Wharton DA (2015) Anhydrobiosis. Curr Biol 25:R1114–R1116

Wolkers WF, Tablin F, Crowe JH (2002) From anhydrobiosis to freeze-drying of eukaryotic cells. Comp Biochem Physiol Part A Mol Integr Physiol 131:535–543

Chapter 11
Molecular Stress Responses against Trace Metal Contamination in Aquatic Invertebrates

Adriano Magesky and Émilien Pelletier

Abstract The highly conserved heat shock proteins have been studied for decades. To protect cell against proteomic dysfunction provoked by metal toxicity this group of proteins is dynamically expressed in aquatic invertebrates from several zoological groups. We herein summarized the main studies regarding Hsp (heat shock proteins) in aquatic invertebrates. Uptake mechanisms of trace metal toxicants by cells are first discussed. The toxicity mechanisms of copper, lead and cadmium in freshwater, brine water and saltwater gave rise to a clearer understanding of the activation of stress-inducible heat shock proteins. The regulation of HSP60, HSP70, HSP90 and small Hsp by some redox-inactive metals occurred as a function of dose concentration and duration of chemical assault. Are also reviewed and evaluated here, the induction time, metal bioaccumulation, toxicity threshold, mRNA-Hsp transcripts expression of different HSP families, and additional factors likely able to modulate Hsp expression such as temperature, age and salinity. Our work finally is pointing to some directions towards the Hsp research in aquatic invertebrates could make some important progresses in the future.

Keywords Cadmium · Copper · Freshwater invertebrates · Heat shock proteins (HSP) · HSP70 · Marine invertebrates · Small Hsp · Trace metal toxicity

Abbreviations

ALAD	Aminolevulinate dehydratase
CAT	Catalase
ER	Endoplasmic reticulum
ERK	Extracellular signal-regulated kinase
G6PD	Glucose-6-phosphate dehydrogenase
GRP78	78 kDa glucose-regulated protein or HSPA5

A. Magesky (✉) · É. Pelletier
Institut de sciences de la mer de Rimouski (ISMER), Université du Québec à Rimouski, Rimouski, Québec, Canada

A. A. A. Asea, P. Kaur (eds.), *Heat Shock Proteins and Stress*,
Heat Shock Proteins 15, https://doi.org/10.1007/978-3-319-90725-3_11

GSH	Glutathione reduced form
GSH-Px	Glutathione peroxidase
HSE	Heat shock elements
HSF	Heat shock factors
JNK	c-Jun N-terminal kinases
MAPK	Mitogen-activated protein kinases
NO	Nitric oxide
NOX	NADPH oxidase
OC	Okadaic acid
PAAm-AgNPs	Poly(allylamine)-coated silver nanoparticles
ROS	Reactive oxygen species
SH	Sulfhydryl groups
SOD	Superoxide dismutase
TBT	Tributyltin

11.1 Introduction

Populations of aquatic invertebrates are naturally under biotic and abiotic stressors such as temperature, food depravity, hypoxia, diseases, space competition, predation, solar radiation etc. Particularly challenging for brine and saltwater invertebrates are salinity changes and desiccation as well. Often, invertebrates are forced to cope with sub-optimal conditions and eventually with severe environmental stresses in natural waters (Holmstrup et al. 2010). Under major physiological disturbances, cells can display a number of sophisticated mechanisms of stress response to rapidly avoid macromolecular damages and keeping cell homeostasis or re-establishing it. The activation of the molecular chaperone system has been considered a very effective and conservative cellular mechanism allowing stress tolerance. Henceforth, cells will rise for survival with adaptation or they will be eliminated following damaging and cellular machinery dysfunction (Dimauro et al. 2016). It happens depending on the type, duration and intensity of the stressing stimuli. It is well established that interactions between natural stressors and contaminants might take place in the environment. As a result, harmful effects may come up worse than previously expected from either stress type taken separately (Holmstrup et al. 2010). Nowadays some few studies have reached such aspects.

Heat shock proteins (Hsp) represent a supergene family divided into several families (Gu et al. 2012). Among these diverse families, the HSP70 is one of the largest and most highly conserved classes with molecular weight ranging from 68 to75 kDa (Gupta et al. 2010). As a general mechanism, HSP70 are rapidly synthesized under hostile conditions, and their units migrate to the nucleus to bind to pre-ribosomes and protein complexes thus avoiding denaturation. Hsp basically prevent the formation of insoluble aggregates that damages the cell. The 70 kDa molecular weight chaperones break up the aggregates in order to repair denatured proteins, refold and recuperate their original structure and biological activity. HSP70 have been found in multiple subcellular compartments including cell membrane, cytosol, mitochondria and endoplasmic reticulum. Moreover, they are the first to be induced under stress conditions in many aquatic invertebrates.

Thereby, HSP70 has been used so far as an important tool in pollution monitoring for years, notably with mussels, sea urchin embryos and some other marine invertebrates. Hsp (mostly inducible-70 kDa) have had some success in environment health assessment, even though the comparison to undisturbed populations (seem as appropriate controls) remains somewhat problematic and subject to controversy. With optimized experimental conditions, the quantification of HSP70 expression also brings mechanistic information about a specific toxicant: induction time, toxicity threshold after low to high contaminant doses, interaction among chaperones and co-chaperones, and investigation of additional factors able to modulate Hsp expression. For instance, by measuring differences in HSP70 expression in controlled assays it was possible to distinguish toxicity levels of a metal (Ag) either in its dissolved or particulate form (Magesky et al. 2017).

The regulation of the Hsp response occurs both transcriptionally and translationally. Transcripts from stress-inducible Hsp genes rise within minutes of the onset of stress suggesting the activation of pre-existing stress-sensitive transcription factors (De Pomerai 1996). Studies progressively focused on the transcripts expression of several Hsp molecules in freshwater and marine invertebrates rather than one type of proteins. Several molecular pathways related to stressful conditions as metal exposure, temperature rising or bacteria challenge have been considered in such analyzes. Transcriptomics techniques offer an accurate and holistic view on molecular mechanisms and signaling pathways controlling the stress response. Therefore, it unravels not only the complexity of activation-inactivation of many Hsp and their co-chaperones, but also apoptotic factors, antioxidant defenses, and growth factors during a chemical insult. Indeed, it has been a more realistic approach to a better understanding of mechanisms through which aquatic invertebrates withdrawn proteomic stress.

Currently, aquatic ecosystems are under growing anthropic pressures which include wastewater discharges, runoff from agricultural and urban areas; water and sediment contamination by mining activities and drilling, eutrophication, etc. Metals are generally released into natural waters by industrial and domestic sewage discharges, mining, farming, and electronic wastes (Gheorghe et al. 2017). Albeit some metals can be considered as essential at lower doses for living organisms (e.g. Cu^{2+}, Mn^{+2}, Se^{-2}, Cr^{+2}, Fe^{+2}, etc.), increasing doses might probably induce toxicity effects and disturb their metabolism homeostasis. Little research has yet been describing such aspects in aquatic invertebrates. In addition, some non-essential metals (Cd^{+2}, Pb^{2+}, Ag^{+}, Hg^{2+}, Ni^{2+}, As^{3+} etc.) can bring intoxication at relatively lower doses (Campbell and Couillard 2004). Most of the studies reviewed hereafter dealt with toxicity effects of some trace metals such as Hg^{2+}, Zn^{2+}, Pb^{2+}, Ag^{+}, As^{3+}, Se^{2-}, with a special attention to Cd^{2+} and Cu^{2+}.

After decades of regulations, cadmium contamination is still a real problem for ecosystems health. This is a common pollutant found in urban runoff, industrial discharges, fungicides, insecticides and mine drainage. Environmental concentrations of dissolved cadmium in some northern coastal areas range between 2.75 ng. L^{-1} and 0.145 μg.L^{-1} (Sinclair et al. 2015). In northern freshwater environments, concentration as high as <0.1 to 112 μg.L^{-1}/Cd^{2+} (0.4 μg.L^{-1} in average) can also be found (CEQG 2014). Some commercial products such as ceramic pigments can release about 83% of hazardous free Cd^{2+} in aqueous suspension (Liu et al. 2017). Similarly, cooper (Cu^{2+}) sources are coal ash effluents, mine drainage, industrial wastewater and urban runoff. Freshwater lakes near pollution sources like mining and smelting operations, for example, can easily reach 1581 mg.kg^{-1} in sediments (CEQG 1999). Likewise, marine sites close to industrial harbors can have copper in sediments as high as 440 mg.kg^{-1}. Benthic organisms are likely exposed to particulate and dissolved cooper in interstitial and overlying waters (CEQG 1999). While low concentrations of cadmium frequently lead to powerful toxic effects and Hsp expression (especially for 70 kDa HSP) in aquatic invertebrates, higher doses of cooper are necessary to sometimes stimulate similar responses.

An exhaustive literature review showed that around 61% of the research papers had HSP70 family as a biomarker for metal contamination for freshwater invertebrates. HSP90 has been found in 17.4% of experimental works, followed by 8.7% for HSP60, 8.7% for small HSP20 family, and 4.3% for HSP40. Either in experimental toxicity studies or *in situ* environmental risk assessments having marine invertebrates as biological models, HSP70 family appeared in approximately 54% of the studies, HSP90 have been found in 16%, HSP60 in 13%; and about 17% studies for the small Hsp (~10–27 kDa) in which 13% corresponded to HSP20 and 4.2% to HSP10. To date, the study of GRP78 (78 kDa glucose-regulated protein GRP78 or HSPA5) in aquatic invertebrates seemed to confirm some mechanisms already described in humans, but further investigations are still needed to fulfill the gap of information about their role in organisms inhabiting different ecosystems.

Several animal models have been used to describe toxicity mechanisms of trace metal contaminants under laboratory conditions and their effects in the environment. Amongst freshwater invertebrates, 26% of the studies were carried out with crustaceans, 22% with insect larvae, 22% with mollusks, 17% with planarians, 8.5% with cnidarians and sponges and 4.4% with rotifers. For marine invertebrates, mollusks (essentially bivalves) are by far the most used models with 56.6% of the studies published, followed by crustaceans with 15.2%, and echinoderms with 13.2%. Some other phyla corresponded approximately to 15% of models which include 11% with sponges and cnidarians, 4% with rotifers. Most of the research papers reviewed hereafter have been carried out in the temperature range of 18 to 25 °C, and rarely at lower temperatures often encountered in northern seas and lakes. A wide range of nominal concentrations have been chosen without direct measurements of metal concentrations during the course of experiments, and only a few studies assessed the metal accumulation in the organisms themselves, which is a major lack of information when comes the time, as an example, to discuss conflicting results in toxicity experiments. Another important point is the number of individuals chemically treated per volume (mL) in each replicate. The toxic effects can be drastically changed by adding more organisms into the exposure media (charge effect), so providing the exact number of organisms exposed is crucial for any experimental repeatability and quality control. Overall, the last two decades have brought increasing knowledge about stress-induced Hsp in aquatic invertebrates contaminated by metals, but there are limited data for organisms from cold waters, and an under-representation of some zoological groups among toxicological models.

11.2 Behavior, Bioavalibility and Cellular Uptake of Trace Metals in Aquatic Ecosystems

After entering natural waters, metals can be dissolved, get adsorbed by particulate organic matter, phytoplankton, other microorganisms. Metals are likely to be uptaken by filtering organisms as dissolved and/or particulate forms, actively ingested and accumulated by other aquatic animals; and they will be at some point settled in bottom sediment. From the sediment compartment, metals might be remobilized and dissolved again in the water column. As part of chemical transformation processes in natural waters, metals may be found as particulate (>200 nm), colloidal (10–200 nm) or dissolved form (<10 nm) (Campbell and Couillard 2004). Once in solution, hydrated ionic metals (*metal*-$(H_2O)^{z+}$) will form hydroxy-complexes (H_2O_{n-1}-metal-OH), inorganic complexes (*metal*-Cl^-, F^-, SO_4^{2-}, HCO_3^-, CO_3^{2-}), organic complexes with monomers (*metal*-ligands as H_2N–HC*–COOH, or (–C(O)OH)$_n$) or organic complexes with natural or man-made polymers (fulvic acids, humic acids, proteins etc.) (Campbell and Couillard 2004). These processes obviously occur in a dynamic way and depend on pH variation and chemical affinity with the ligands.

Several factors are controlling metal bioavailability to living organisms such as physico-chemical factors (temperature, salinity, pH, and ionic strength), redox conditions, concentration of organic matter (dissolved or particulate) and metal speciation (Gheorghe et al. 2017). On biological surface in general, dissolved metals will first reach a protective layer surrounding living organisms like mucus on epithelial surfaces. This layer is mostly composed by polysaccharides and glycoproteins and ions will likely cross this protective layer composed of macromolecules harboring different functional groups. As these groups normally have oxygen as an electron donor (C(O)OH, –P(O)OH ou –C(OH)), many of them may dissociate and generate a negatively charged surface following a pH range found in natural waters and bind trace metals.

Having crossed this barrier, metals must then get through the cellular membrane. Doing so, two groups of ligands will likely interact with them (Campbell and Couillard 2004) (Fig. 11.1): (1) physiologically unreactive groups to which any metal can be connected without causing any metabolic disturbance, and (2) physiologically active sites where metals will probably affect cell metabolism. This interaction of free ions with cell membrane happens as a complexation reaction on cell surface. If the cell ligand corresponds to a physiologically active site, the complex-

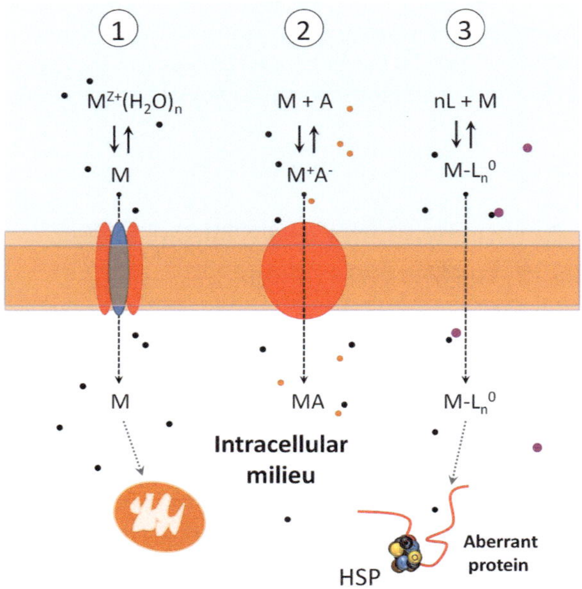

Fig. 11.1 Interactions between metals and biological membranes separating an aquatic organism from surrounding water (modified from Campbell and Couillard 2004). Trace metals reaching intracellular milieu will likely interact with organelles and proteins. Cation facilitated diffusion through the cell (1). Cations randomly transported by anions into the cell (2). Passive diffusion of lipophilic metallic complexes (3). (M = metal; M^{z+} $(H_2O)_n$ = free cation; ML = organic metallic complex in solution; ML_n^0 = hydrophobic metallic complex; A^- = anion; L (forming organic complex with M) = amine acid or polycaboxylic acid, for example; black dots: cationic metals, orange dots: anions; purple dots: organic ligand)

ation will trigger a biological response. Considering the hypothesis of a transport site as a ligand, the complexation reaction will activate the ion transportation into intracellular milieu. Once the metal is internalized, further reactions with metabolically active sites will necessarily occur. When a ligand is crucial to allow nutritional trace elements entry into the cell, the contaminant will compete with essential metals inhibiting their assimilation. For particulate metals, organisms must ingest the particles carrying the metals (like phytoplankton, bacteria, organic matter, etc.), so the solubilization will take place in the gut where the pH range is totally different from external medium. Metals can be also associated with hydrophobic molecules that easily pass through cell membrane before getting into intracellular milieu (Fig. 11.1).

As an example, soluble silver (free Ag^+) is a membrane-disrupter known to have chemical affinity with thiol groups. The Ag^+ interaction with phospholipid bilayer induces a destabilization of the outer cell membrane. It basically occurs through the presence of thiol groups in the cell-wall proteins (Lapresta-Fernández et al. 2012). Then, it gives rise to a decrease in the electrochemical gradient of the transmembrane proton, which collapses the plasma-membrane potential and de-activates energy-dependent reactions. After levels of ATP-ion transport and metabolite sequestration are depleted, cell homeostasis starts to shutdown. With organic thiols, free Ag^+ forms a thermodynamically favorable complex of silver (I) thiolate $(AgSR)_n$ whose $\log K_f$ (constant for the forward reaction) is ~13 mo ^{2} dm^{-6} (Bell and Kramer 1999). With the inorganic ones (HS^-, S^-), many chemical species might be formed such as AgSH, $[Ag(SH)_2]^-$ and $[Ag_2(SH)_2]^{2-}$ (Bell and Kramer 1999; Toh et al. 2014). This process is a function of anion concentrations on the biological surface. In the silver case, disruption of cell membrane functions may occur by electrostatic attraction between Ag^+ and transmembrane proteins or a steric interference caused by nanoparticulate silver (AgNPs) (Su et al. 2009). Then, ROS (reactive oxygen species) may be produced once the transmembrane space proton pool or the respiratory chain electron flow is destabilized.

11.3 Cytotoxicity of Trace Metals in Aquatic Invertebrates: Antioxidant Defense Failure, ROS Overproduction and Cell Death

There are several ways in which metals can disturb the structure and function of organelles and cells. The first one is the physical interaction of metals with cell structures causing metabolic impairment and lately physical harm. There is also a depletion or stimulation of enzymes or metabolites. For instance, the overproduction of ROS disturbs the balance between ROS and antioxidant agents leading to intracellular redox state changes. Thirdly, inhibition or stimulation of specific proteins like Hsp, metallothioneins, etc. (Moore 1985). The mechanism of interaction between metals and cell structures is obviously governed by metal oxidation states. The redox-active metals such as iron ($Fe^{2+/3+}$), copper ($Cu^{1+/2+}$), chromium (Cr^{2+}), vanadium ($V^{2+, 3+, 4+, 5+}$) and cobalt ($Co^{2+/3+}$) undergo redox-cycling reactions sometimes producing reactive free radicals (Jomova et al. 2012). The redox-inactive ones like cadmium (Cd^{2+}), arsenic ($As^{3+/5+}$), mercury (Hg^{2+}), lead (Pb^{2+}) and nickel (Ni^{2+}) have their primary route for toxicity based on (1) depletion of glutathione and some other thiol-containing antioxidants, and (2) bonding to sulfhydryl groups of proteins (Valko et al. 2016). Either redox-active or inactive metals may enhance ROS production.

11.3.1 Copper

Copper ($Cu^{+1/2+}$) is normally required for structural and catalytic properties of many enzymes like cytochrome *c* oxidase, copper-zinc superoxide dismutase (Cu-Zn-SOD), tyrosinase, p-hydroxyphenyl pyruvate hydrolase, and others (Gaetke and Chow 2003). Notwithstanding, exceeding levels of $Cu^{+1/2+}$ caused by environment contamination will provoke cytotoxicity. The basis of cooper-induced cellular toxicity is the propensity of $Cu^{+1/2+}$ ions to induce ROS as a prooxidant agent leading to final oxidative stress. Both Cu^{+} and Cu^{2+} ions can participate in oxidation and reduction reactions as follows:

$$O_2^{-*} + Cu^{2+} \rightarrow O_2 + Cu^{+} \ (1)$$
$$Cu^{+} + H_2O_2 \rightarrow Cu^{2+} + OH^{-} + OH^{*} \ (2)$$

With superoxide (O_2^{-*}) present or some reducing agents like GSH (glutathione reduced form), Cu^{2+} is reduced to Cu^{+} (1) which is capable of catalyzing the formation of hydroxyl radicals (OH^{-}) from hydrogen peroxide (H_2O_2) (Bremner 1998).

Another clear consequence of $Cu^{+/2+}$-induced production of ROS is an increased lipid peroxidation. According to Powell (2000), hydroxyl radicals might react with unsaturated fatty acids to give rise to lipid radicals and amino-bearing carbons from which hydrogen can be taken away. Gilbert and Avenant-Oldewage (2017) investigated the exposure of trace metals to the freshwater fish monogenean parasite *Paradiplozoon ichthyoxanthon* and observed that Cu^{2+} was the main metal sequestered in vitellaria organs (~44.49%) followed by Au^{+} (~5.66%), Cr^{2+} (~1.24%), and $Fe^{2+/3+}$ (~0.61%). Production of ROS was also reported in the integument, but not in the parasite eggs where metals were not detected. On the contrary, metal binding to the protein sclerotin-shells of cestode eggs was described by Degger et al. (2009) as a result of high affinity of egg cases for divalent ions. It was likely due to the tanning oxidative process involved in egg formation.

11.3.2 Lead

The main mechanism explaining Pb^{2+}-induced oxidative stress is related to cell membrane damage by a modification in its fatty acid composition (Patra et al. 2011). Some effects over the antioxidant defense systems have been mentioned as well. *In vitro* studies showed that double bounds in the fatty acid of cell membrane weaken C-H bonds on carbon atoms near the double bonds making H^{+} removal easier (Yiin and Lin 1995). The fatty acid length and unsaturation are key factors for membrane susceptibility to peroxidation, so it is likely that Pb^{2+} effects on arachidonic acid, for example, could be responsible for rising lipid peroxidation on membranes (Halliwell and Gutteridge 1999). Membrane lipids are organized into phase-separated microdomains also called lipid rafts which have a specific composition and a molecular dynamic quite different from the ones of the surrounding liquid crystalline phase (Eeman and Deleu 2010). All toxicity effects driven by Pb^{2+} may culminate in a lateral phase separation of the phospholipid membrane, that is, a segregation of gel domains from liquid crystalline domains (Eze 1991; Patra et al. 2011). Lead is thus able to affect membrane-related processes like enzymatic activity on membranes, endo- and exocytosis, ions transportation, signal transduction processes (Yiin and Lin 1995).

The effect on antioxidant defense systems of cells is the second mechanism for Pb^{2+}-induced oxidative stress. As a rule, the ionic mechanism of Pb^{2+} toxicity is based on the ability of free Pb^{2+} to replace other bivalent cations such as Ca^{2+}, Mg^{2+},

Fe^{2+} and monovalent ones as Na^+ disturbing cell metabolism (Jaishankar et al. 2014). In essence, nonessential metals can compete with essential trace metals modifying enzymatic activation and disabling antioxidant effective responses. Selenium (Se^{2-}), for example, is used as cofactor of GSH-Px (glutathione peroxidase); a well known antioxidant enzyme. Because GSH-Px, CAT (catalase) and SOD (superoxide dismutase) must have trace elements as cofactors for proper molecular conformation and function, their activities can be potentially disturbed by agonistic competition between Pb^{2+} and Se^{2-}. In other words, GSH-Px activity might be reduced after interaction of Pb^{2+} with selenocysteine functional groups of enzyme molecule. Furthermore, Pb^{2+} (as well as Hg^{2+} and Cd^{2+}) can alter enzymatic activity of aminolevulinate dehydratase (ALAD), superoxide dismutase (SOD), catalase (CAT), glutathione peroxidase (GSH-Px), and glucose-6-phosphate dehydrogenase (G6PD) due to lead affinity for sulfhydryl groups (SH).

As an example, the freshwater bivalve *Hyridella australis* exposed to lead contaminated sediments (<0.01, 205 ± 9 and 419 ± 16 μg.g dry weight) throughout 28 days at ~20 °C had total antioxidant activity in hepatopancreas cells significantly suppressed (Wadige et al. 2014). Individuals contaminated by low concentrations showed higher levels of total antioxidant activity compared to those exposed to higher doses. Of the lead accumulated in the hepatopancreas, around 83–91% was detoxified and stored in metal rich granules. Nevertheless, lysosomal stability in hepatopancreas decreased in both treatments. Likewise, blood cockles clams *Anadara granosa* living in coastal mudflats chronically exposed to medium doses of Pb^{2+} (86 $\mu g.L^{-1}$) at ~26 °C for 28 days showed more toxicity effects than at lower and higher doses, 43 and 172 $\mu g.L^{-1}$, respectively (Peng et al. 2015). It appeared that enzymatic activity in gills, gonads and digestive glands mostly increased in 86 $\mu g.L^{-1}$ treated-clams between 21 to 28 days.

11.3.3 Cadmium

Considered as potent cell poison, cadmium is a nonessential metal known to provoke oxidative stress by enhancing lipid peroxidation and/or by changing intracellular glutathione levels. However, this metal is unable to generate ROS directly. As cadmium has a high affinity for thiols, glutathione (a major thiol antioxidant and very abundant in cells) is a primary target for free Cd^{2+}. Once the levels of glutathione pool are depleted, cellular redox balance is disturbed leading to a rising oxidative environment. In that way, Cd^{2+} generates radicals through the displacement of Fenton metals (Pari and Murugavel 2007), and it indirectly contributes to the production of O_2^{-*} and H_2O_2 via activation of NADPH oxidase (NOX) (Nemmiche 2017). ROS production is one of the first steps in Cd^{2+}-mediated cytotoxicity, preceding mitochondrial dysfunction by loss of membrane potential. Hence, the cellular stress continues with the activation of caspases and later apoptosis. Reactive species H_2O_2 and OH^{-*} radical are often accompanied by activation of redox sensitive transcription factors (e.g. NF-kB, AP-1 and Nrf2) and alteration of ROS-related gene expression (Schieber and Chandel 2014). Early on during the development, these mechanisms might be crucial to guarantee physiological homeostasis and survival of aquatic invertebrates exposed to metals. The antioxidant system of one-day old paralarvae of *Octopus vulgaris* represented by CAT, SOD and glutathione S-transferase (GST) gene expression readily responded to 10–100 μg.L^{-1}/Cd^{2+} treatment in a 24 h-exposure at ~19 °C (Nicosia et al. 2015). Notwithstanding larvae experienced decreasing survival rate, morphological alterations in the mantle and arm length. An increase of HSP70 mRNA expression at 3.5-fold (10 μg.L^{-1}) and 5.5-fold (100 μg.L^{-1}) was also detected.

Experiment with freshwater crab *Sinopotamon henanense* showed that Cd^{2+} caused production of H_2O_2 in the gills within 2 h of 58 mg.L^{-1} treatments increasing the caspase-3/8/9 activity at ~20 °C (Wang et al. 2012). This study showed that H_2O_2 level dropped after 8 h of cadmium treatment, but it rose again after 24 h. It was reported that Cd^{2+} disturbance may increase antioxidant enzymes activity in a short-term response of less than 8 h, and then inhibit SOD and GSH-Px activities causing mitochondrial damages within 24 h. Henceforth, H_2O_2 production was involved in the Cd^{2+}-induced cell apoptosis in a time- and concentration-dependent manner. Currently it has been also postulated that cadmium provokes apoptosis in a caspase-dependent pathway in *S. henanense* enhancing hiperpolarization of mitochondria membrane potential ($\Delta\psi_m$) at 3.56, 7.12 and 14.25 mg.L^{-1}/Cd^{2+} (Liu et al. 2013). Lipid metabolism disruption is another effect caused by Cd^{2+} in *S. henanense* (Yang et al. 2013). Hepatopancreas is the main detoxifying organ in crustaceans and the first one to accumulate cadmium (Liu et al. 2013). Following 10 to 20 days of exposure overall lipid content of hepatopancreas and ovary in *S. henanense* was significantly reduced in all doses used (1.45, 2.9 and 5.8 mg.L^{-1}) at ~20 °C (Yang et al. 2013). Lipoprotein lipase, a key enzyme for lipid transport in the heamolymph, was significantly down-regulated after 10 days of Cd^{2+}-exposure. One explanation proposed by the authors is that cadmium-lipid peroxidation provokes decomposition of polyunsaturated fatty acids in membrane systems impacting hepatopancreas cell membrane structure. In a like manner, Liu et al. (2011) stated that the freshwater crab *S. yangtsekiense* went through apoptosis and necropoptosis processes after 48 h of 28.49 mg.L^{-1}/Cd^{2+} treatment. Cadmium induced a mitochondria-dependent apoptosis in the hepatopancreas but how mitochondria injury occurred was not clear. Albeit high dosages might show a fast activation of protective mechanisms in experimentally exposed organisms, when doses are excessively high compared to environmental concentrations (100 to 1000 times lower than dosages) the molecular responses might rather reflect the rise of lethality effects.

11.4 Temperature Factor, Metal Toxicity and Activation of Heat Shock Proteins in Aquatic Invertebrates

During a moderate stress exposure, the metabolic compensation comes up with high protein turnover and an elevated basal maintenance cost to manage the costs of stress proteins upregulation and/or activation of the protein degradation pathways

(Ivanina et al. 2008; Ivanina et al. 2009; Sokolova et al. 2012). Both activities are extremely energy-demanding for cells (Hochachka and Somero 2002). When the stress situation is worsening, the progressive rise in ATP demand for maintenance or the growing impairment of aerobic metabolism strikes down ATP production via aerobic metabolism resulting in a partial metabolic change to anaerobiosis (Sokolova et al. 2012). These changes maintain the essential maintenance costs. Actually, this metabolic shift implies some adjustments in mitochondria, enzymes, ion and gas transport capacities as well as membrane composition. Such physiological strategy of aquatic invertebrates in moderate to high contaminated waters ensures nonetheless a short-term survival. The temperature-dependent stress response in oysters *Crassostrea virginica* contaminated by Cd^{2+} (50 $\mu g.L^{-1}$) at 20, 24 and 28 °C was investigated by Lannig et al. (2006). The authors found that increasing temperature amplified the toxic effects of cadmium. It turned out that a combination of rising temperatures and cadmium treatment are significantly more damaging to oysters than the effects of a single stress by itself. It can be explained by the increasing metabolic costs needed for basal maintenance with activation of detoxification pathways like metallothionein production, turnover of damaged proteins and/or maintenance of intracellular ion equilibrium (Leung and Furness 2001). Beyond the basic metabolic processes maintained by cells, mitochondria seem not to be able to maintain an elevated basal metabolic rate for a long time to guarantee cell ability with both cadmium and rising temperatures operating together (Lannig et al. 2006).

Studying the short-term effects of cadmium exposure within 4 h on glutathione synthesis, Hsp (HSP60, HSP70 and HSP90) and metallothioneins production in *C. virginica* at ~20 °C, Ivanina et al. (2008) remarked contrasting protein levels in gills and hepatopancreas cells. Harboring sulfhydryl groups, metallothioneins can capture free radicals and ROS, acting as antioxidants (Coyle et al. 2002). Whereas in hepatopancreas of *C. virginica* mRNA metallothionein increased 2–8-fold in response to cadmium, no significant levels were found in gill cells (Ivanina et al. 2008). Conversely, HSP60 and HSP70 protein expression was significantly enhanced at 1.5–2-fold in Cd^{2+}-treated gills, but not in hepatopancreas. It indicated that metallothionein induction is not sufficient to prevent cell impairment caused by cadmium in the gills, so Hsp expression emerges as a supplementary protective mechanism for cell defense. It can also be understood that intensity of the Hsp regulation and their relative concentrations are highly tissue-specific for both vertebrates and invertebrates (Sanders 1993).

11.5 General Aspects of HSP70 Modulation in Mussels, Sea Urchins and Daphnids

A common mechanism of action of Hsp is the capacity to track unfolded or misfolded polypeptides in intermediate stages of folding and thus facilitates their refolding by ATP-depend via (for the large Hsp) or ATP-independent manner (for low weight Hsp). Under normal conditions without metal stress, Hsp also (i) participate in vital cellular processes like assembly of multiprotein complexes, (iii) transport proteins across cell membranes, (ii) maintain cell structural integrity by interacting with cytoskeletal components to protect them and change their formation and functions (housekeeping role), and finally (iv) select and direct aberrant proteins to the proteasome or lysosomes for later degradation (Kalmar and Greensmith 2009; Dimauro et al. 2016). Noticeably the heat shock proteins also have anti-apoptotic and pro-apoptotic properties after cellular stress induction by interacting with members of the apoptotic cell death cascade.

When cells are challenged by a natural stressor or an anthropic assault, Hsp are immediately regulated by heat shock factors (HSF) to keep cellular homeostasis. Mainly members of HSP70 family and its co-chaperones work immediately to refold damaged proteins. In a pre-conditioning cellular stress, HSP70 as well as several transcription factors and others Hsp are normally under low upregulation. In this case, HSP70 and HSP90 are bond to monomeric HSF1 in suppressed state. Following a trace metal challenge all cellular defense mechanisms work synergistically against oxidative stress. As emphasized before, with an oxidative intracellular milieu, macromolecules can undergo oxidation. And proteins in particular are very sensitive to ionic, redox and pH changes (Kalmar and Greensmith 2009). Consequently, aberrant proteins induce activation of Hsp through HSF1 activation. As follows, HSP70 and HSP90 release HSF1 to assist protein refolding and trimerize. HSF1 are thus translocated to nucleus in order to activate Hsp transcription by binding to specific regions in the promoter of Hsp genes. There are some evidences that this regulation pathway differs among aquatic invertebrates following their life history and adaptation capacity though (Hofmann and Somero 1995).

In order to broadly assess Hsp plasticity in freshwater and saltwater invertebrates, it becomes important to understand their expression in as many zoological groups as possible. Acanthocephalan parasites *Polymorphus minutes*, for example, can greatly interfere with Hsp expression in freshwater amphipods *Gammarus roeseli* preventing them to express HSP70 after 10 μg.L^{-1}/Pb^{2+} exposures at ~12.9 °C (Sures and Radszuweit 2015). Surprisingly, the parasites themselves can greatly express HSP70 against lead. Especially among marine invertebrates living in-shore and under intertidal fluctuations, adaptation to local environment conditions might be challenging. Intertidal, ecologically similar and phylogenetically close species *Mytilus trossulus* and *Mytilus galloprovincialis* facing the same thermic conditions in the field modulated HSP70 expression by 8-fold and 1.5-fold, respectively (Hoffmann and Somero 1995). In addition, no variations between the endogenous levels of HCP70 (constitutive isoform) and HSF1 in both species were found either. In a more stable environment as marine cold waters, Hsp activation against chemical insult is so far poorly understood. Some Antartic invertebrates like seastar *Odontaster validus* and gammarid *Paraceradocus gibber* are unable to induce HSP70 response probably due to the loss of the regulation pathway (Clark and Peck 2009). How these species would protect cells from proteomic stress caused by trace metals is still an unsolved question.

11.5.1 Mussels (Bivalvia)

By comparing HSP70 amino acid sequences of bivalves to their human homologues, it becomes possible to detect some similarities in both structures (Piano et al. 2004). They display some conserved domain structures which consists of (1) a cleavable signal sequence at the N-terminus; (2) an ATPase domain, (3) a peptide binding domain, (4) a G/P-rich C-terminal domain region allowing HSP70 to bind co-chaperones and contains intracellular localization signal sequences (Fabbri et al. 2008). Notwithstanding, some exclusive traits in bivalves HSP70 are remarkable as well. One of them is the extra NQSQ tetrapeptide within the ATPase domain, which is also a feature in the HSC70 of Ostreidae (Kourtidis et al. 2006).

Some research has been carried out to comprehend the regulation mechanisms of molluscan HSP70. It becomes clear that up- and down-regulation, dose- and time-dependent expression pattern of HSP70 in molluscan hemocytes have a complicated regulation pattern not well understood (Wang et al. 2013). The interaction of HSF and HSE (heat shock elements) in molecular promoters drives the HSP70 expression regulation. The HSF constitute a series of pentameric units arranged as inverted adjacent arrays of the sequence 5′-nGAAn-3′ (Morimoto 1998). Species phylogenetically related with different tolerances for Hsp activation may have different sensitivity to trigger the molecular promoters or the HSF.

As described by Buckley et al. (2001), HSF1 is released from HSP70 in intertidal mussels (*Mytilus* sp.), it goes then to the nucleus in response to small changes in the optimal conditions. Inside the nucleus HSF1 may remain inactive on the promoter until a stressful condition comes up. Additionally, the mitogen-activated protein kinases (MAPK) signaling cascade and the extracellular signal-regulated kinase (ERK) signaling pathway can operate with HSP70 gene regulation after HSF1 and the promoter are linked (Buckley et al. 2001; Anestis et al. 2007; Zahoor et al. 2010). In mussels *Mytilus galloprovincialis* the increasing levels of HSP70 expression occurs concomitantly with the phosphorylation of HSF1, p38-MAPK and c-Jun N-terminal kinases (JNK). Alternatively, the transcriptional factor HP1BP3 seemed to negatively regulate HSP70 transcription in the oyster *Crassostrea hongkongensis* after 48 h-exposure to 300 $\mu g.L^{-1}/Cd^{2+}$ at ~24 °C (Xu et al. 2017). HSP70 mRNA basal transcription was clearly enhanced at 24 and 48 h upon Hp1bp3 mRNA depletion. More than inducing transcriptional factors and heat shock proteins, cadmium has a role in triggering the protein ubiquitination process, that is, the removal of aberrant proteins after the chemical exposure (Meng et al. 2017). This process is discussed in details in the Sect. 11.7.2.

Overall, transcriptomics have revealed that several inducible Hsp, transcription factors, apoptotic factors, enzymes, or proteins related to the cytoskeleton etc. are modulated in mollusks during a trace metal challenge (Zhang et al. 2010b; Wu et al. 2013; Negri et al. 2013; Li et al. 2014, Thompson et al. 2015, Xu et al. 2017). Nonetheless, hostile conditions like prolonged starvation, parasitism, desiccation, and UV radiation combined with metals (in single or mixture exposures) still need more investigation. Recently, González-Fernández et al. (2017) showed that different reproductive and nutritive states in mussel *M. galloprovincialis* might influence Hsp response once chronically exposed to 3 and 60 μg.L^{-1} /fluoranthene. The authors found that both nutrition and reproductive states affected the mRNA HSP22, HSP70–2, HSP70–3 and HSP70–4 baseline levels and modulated the transcriptional responses to the fluoranthene contamination.

11.5.2 Sea Urchins (Echinoidea)

During embryogenesis of some species of sea urchin under heat shock, Hsp production only occurs after hatching (Giudice et al. 1999). Nonetheless species like *Arbacia lixula* and *Arbacia punctulata* are enabled to express Hsp at the stage of approximately 64 to 128 cells (Roccheri et al. 1982; Maglott 1982; Howlett et al. 1983). Notably, the cell ability to synthesize Hsp already exists during oogenesis in *Paracentrotus lividus*, but it gets turned off throughout egg maturation and comes back again after hatching. It seems in this case that induced-Hsp are mainly found as constitutive forms (HSC) before hatching. In cells of sea urchins under stress, HSP70 is detected in the soluble cytoplasm where it is kept for around 22 h, then it appears in the ribosomal and subribosomal structures as small fractions and finally in the nucleus where it remains for at least 5 h in large amounts (Roccheri et al. 1982). More recently, it has been found that nitric oxide (NO) functions as a physiological messenger involved in the stress response in sea urchin embryos (Romano et al. 2011). At low dosage (0.25 μg.mL) of aldehyde decadienal produced by diatoms, NO protects embryos from teratogenic effects of decadienal by activating HSP70 gene expression. At higher doses (>2.5 μg.mL^{-1}) NO can mediate pre-apoptotic events leading to the loss of mitochondrial functionality through the generation of peroxynitrite. With Cd^{2+}, NO directly activated HSP70 and indirectly regulated HSP60 and skeletogenic genes. With Mn^{2+}, NO indirectly interfered with HSP60, skeletogenic, multidrug efflux and detoxification genes (Migliaccio et al. 2014).

Actually, 70 kDa heat shock protein also has additional functions in coelomocytes of sea urchins. Extracellular HSP70 mixtures of constitutive (HSC70) and inducible forms (HSP70) produced by coelomocytes of *Lytechinus variegatus* can inhibit a spreading reaction of the clotting response (Browne et al. 2007). The presence of Hsp and/or HSP70 in the extracellular fluid may also induce mitosis of dividing cells and shut-down the immune cell reaction. These preliminary results suggest that HSP70 and perhaps other Hsp might be directly involved in cell immune system modulation and repair mechanisms in sea urchin.

The 60 kDa and 70 kDa heat shock proteins are efficiently induced by 3-month-old juveniles of sea urchin *Strongylocentrotus droebachinesis* to endure dissolved silver (Ag^+) and poly(allylamine)-coated silver nanoparticles (PAAm-AgNPs) effects at 8 °C (Magesky et al. 2017) (Fig. 11.2). In soluble Ag-media (100 μg.L^{-1}), necrosis overwhelmingly began at 72 h. With fast dissolution of Ag^+ in saltwater, toxicity quickly emerged. Swelling was the most evident morphological abnormality in coelomocytes chemically injured by dissolved silver which clearly indicated the silver effects. Hence, HSP70 expression was slightly enhanced from 12 to 24 h

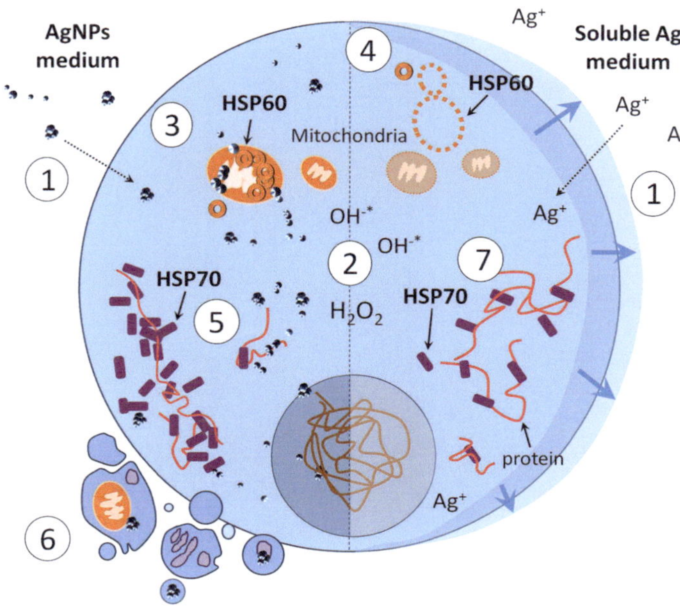

Fig. 11.2 Effects of soluble Ag and nanoAg in sea urchin coelomocyte (Magesky et al. 2017). After dissolved silver exposure, cells undergo a swelling process likely due to disturbance of cell membrane functions; in nanoAg exposure, interactions of silver nanoparticles (AgNPs) with plasmatic membrane occurred with final cell internalization of AgNPs (1); an overproduction of ROS (reactive oxygen species) is detected in cells of sea urchin juveniles exposed to silver (2); while in Ag-treated organisms nanoAg elevated HSP60 expression within 48 h (3), in Ag^+-treated ones HSP60 is progressively downregulated (4); HSP70 is upregulated in 100 $\mu g.L^{-1}$/AgNP-treated juveniles whose cells were strongly burden by nanoparticles (5), an apoptotic-like process was also detected (6); HSP70 is induced in Ag^+ media, but in a less extent compared to AgNPs media (7)

in large Ag^+-treated sea urchins (~1.63 cm Θ), and it followed with the highest peak at 48 h. Meanwhile clots were gradually formed by red spherulocytes and amoebocyte-cells all over the epidermic wounds (Fig. 11.3). More bands of approximately 70 kDa found in Ag^+-treated urchins suggested that repair mechanisms would also operate from 12 to 48 h. Considering the strong toxicity of soluble silver, HSP70 expression seem to represent both cellular protective action against silver and the rise of reparatory processes to cope with tissue injury.

The toxicity of PAAm-AgNPs is based on physical interactions with cells, fast release of Ag^+ forms within 48 h and slower nano-core dissolution over time (Magesky et al. 2017). As a result, 100 $\mu g.L^{-1}$ nanoAg-treated urchins showed the strongest HSP60 and HSP70 overexpression at 48 h. It can be also stated that decreasing levels of chaperonine HSP60 in Ag^+ media was probably due to a drastic mitochondrial membrane impairment or degradation of HSP60 units, their cofactors, and/or coworking chaperones. With nanoparticles of titanium oxide (TiO_2NPs, 10-65 nm) uptaken by coelomocytes of adults of *P. lividus*, no stimulation of HSP70 was detected, but signal transduction downstream to p38 MARK (mitogen-activated protein kinases) was affected (Pinsino et al. 2015). The p38 MAP kinases are

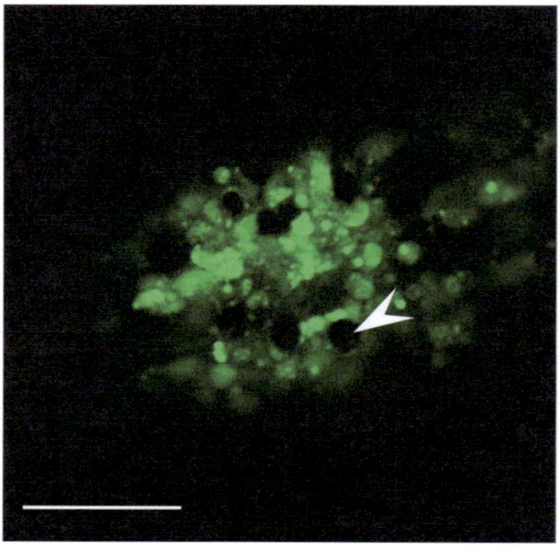

Fig. 11.3 Clotting formation in integument of sea urchins exposed to 100 μg.L^{-1}/Ag^{+} seen by confocal microscopy (Magesky and Pelletier, unpublished material). Cooperation between amoebocyte-cells (cells with SYBT green I-marked nucleus) and red spherule cells emerged throughout time-exposure (48-96 h). White arrowhead indicates one red spherule cell. White bar: 50 μm in laser microscopy

involved in multiple cellular processes like proliferation, differentiation, cell death and migration (Kim and Choi 2010).

11.5.3 Daphnia spp. *(Cladocera)*

Water fleas (*Daphnia* spp.) are small freshwater crustaceans with a long history of use in toxicological assays. Interesting findings published by Schumpert et al. (2014) showed the role of HSP70 in regulating life span of *Daphnia pulex* and *Daphnia pulicaria*. In brief, the authors found that (i) long-lived *D. pulicaria* have a more robust HSP70 response to heat shock than short lived *D. pulex*, (ii) young *Daphnia* display a stronger HSP70 expression than the older ones, (iii) HSP70 response to heat decreases with age in both species (and particularly in short-lived *D. pulex*), and the main raison seems to be the decreasing ability of HSF-1 to bind DNA, which is reduced in old daphnids. It seems that for closely related species in Cladocera HSP70 expression patterns might be divergent enough to eventually have some influence over cell sensitivity to metal exposures. A hypothesis to be tested would be on whether or not antioxidant defenses or other chaperone proteins may eventually compensate lower inducible HSP70 thresholds in older individuals of each species. In addition, comparing HSP70 activation in young and older *Daphnia* spp. challenged by trace metals would clarify how much time is necessary for HSP70 to efficiently cope with toxicity in each case. Furthermore, the detection of HSC70 expression after chemical stress would provide more information regarding wound repair mechanisms in *Daphnia* spp..

Currently it is known that H_2O_2 can directly affect HSF-1 (heat shock factor 1) expression in mammal cells (Nishizawa et al. 1999). In *Daphnia pulex*, both acute heat stress and starvation affected ROS levels (Klumpen et al. 2017). Both nHIF-1α and nHSF-1 fluctuated in a faster way in starved animals. In 24 h-starved daphnids upon acute heat stress, nHIF-1α (nuclear hypoxia-inducible factor 1) fluctuations had some similarity with corresponding ROS fluctuations. Moreover, it seems that nHIF has a critical role to trigger HSP90 mRNA expression in starved individuals. In fact, ROS acts as signaling molecules affecting nHSF-1 levels (nuclear heat shock factor 1) in *Daphnia pulex*. The cHSF-1 (cytoplasmic heat shock factor 1) expression is steadily elevated under heat stress probably by protein expression and/or HSF-1 release from its complex after protein damage (Klumpen et al. 2017). An interesting question to address is how metals would differently affect HSF-1 expression in *Daphnia* spp. by provoking ROS overproduction in stressful conditions like food depravity or heat shock. Some mechanisms of Hsp induction in *Daphnia* spp. discussed in this section are illustrated in Fig. 11.4.

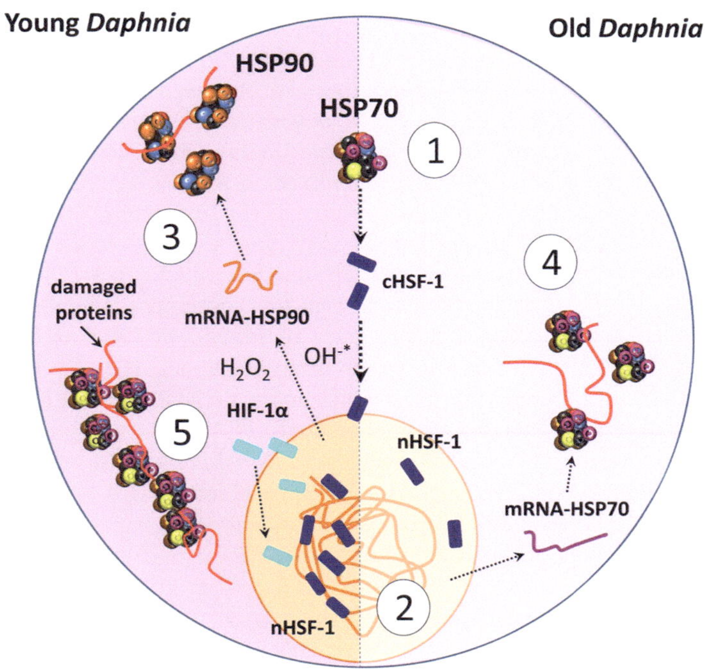

Fig. 11.4 Representation of some regulatory pathways of HSP70 and HSP90 in young and old clones of *Daphnia* spp., after heat stress and starvation (Haap and Köhler 2009; Schumpert et al. 2014; Klumpen et al. 2017). The mechanisms on the left represent HSP activation in young individuals, those on the right refer to old ones. HSF-1 is released from HSP70 as a first step to induce protein expression (1). A lack of DNA-binding by HSF-1 itself is compromised in older individuals compared to younger ones (2). ROS (reactive oxygen species) induce mobilization and/or induction of HIF-1α and expression of mRNA HSP90 after heat shock and starvation (3). Under heat stress, young daphnids may display a more robust HSP70 expression than old specimens (for *D. pulex* and *D. pulicaria*) (4–5). Among 5-day old clones of *D. magna*, another situation occurred when exposed to Cd^{2+}: the sensitive ones produced more HSP70 while the tolerant ones had a lower baseline of HSP70 expression (see main text for more details)

11.6 Stress-Induced Genes and Proteins in Freshwater Invertebrates Following Trace Metals Contamination

11.6.1 Daphnia spp. *(Cladocera)*

Parents and first generation of *Daphnia magna* exposed to Cd^{2+} for 24 h and 21 days at 20 °C showed increasing levels of HSP70 under 80 μg.L^{-1}-treatment while concentrations around 100 μg.L^{-1} resulted in lower levels of 70 kDa proteins (Li et al. 2014). HSP70 expression in parents was however significantly higher than the one from neonates at 80 and 100 μg.L^{-1}. During 21 days, parent generation was exposed to 6 h-pulsed contaminations. With 60 and 80 μg.L^{-1} treatments, HSP70 expression arose immediately until day 4, but went down from 14 to 21th day. As a result, a trans-generation effect arose with high mortality rates detected in first offspring generation in clean water (Li et al. 2014). Comparatively, Haap and Köhler (2009) studied seven clones of *D. magna* 5 days-old from different geographical regions using high dosages of Cd^{2+} (200–800 μg.L^{-1}) in a 24 h-experiment at 24 °C. A correlation between sensitive clones of *D. magna* and HSP70 upregulation was observed with high values of EC50 being related to insensitive clones and the lowest HSP70 expression. It is important to keep in mind that intraspecific tolerance to Cd^{2+} in *D. magna* has greatly to do with reduced HSP70 baseline levels and induction (Haap and Köhler 2009). As another example, 5-day-old clones of *D. magna* pre-exposed and unexposed to 5 μg.L^{-1}/Cd^{2+} upregulated HSP70 at 20° within 24 h (Haap et al. 2016). Then, a second treatment induced high constitutive levels of 70 kDa proteins and stronger upregulation of HSP70 in lower doses of Cd^{2+} (5–300 μg.L^{-1}) in sensitive clones but not in the tolerant ones. These findings clearly demonstrated the problem of having just one clone of cladocerans for toxicological tests with mid to high metal dosages. See Table 11.1 for more details about the experimental set-ups and compared results with *Daphnia* spp.

Testing nominal concentrations close to environmental data (~0.5 μg.L^{-1}/Cd^{+2}), De Coninck et al. (2014) reported no variation in gene expression of both sensitive and tolerant clones of *Daphnia pulex* after a 16 day-chronic exposure. Likewise, 5–7 day-old neonates of *D. pulex* exposed 48 h to Cd^{2+} (0.05 to 5 μM or 5.62–562 μg.L^{-1}) and Zn^{2+} (0.5 to 50 μM or 32.7–3270 μg.L^{-1}) did not show any expression of HSP70–1 gene at 20 °C, a putative orthologue of *D. magna* HSP70 gene (Tang et al. 2015). By contrast, HSP70–2 gene activity was significantly raised at 30.35 μg.L^{-1} (0.27 μM/Cd^{2+}) and 61.8 μg.L^{-1} (0.55 μM/Cd^{2+}) as well as HSP70–3 at 61.82 μg.L^{-1} (0.55 μM/Cd^{2+}). Daphnids accumulated 120 and 340 μg.g at 0.27 and 0.55 μM/Cd^{2+}, respectively. Zinc did not induce HSP70 transcription. Adults and neonates <24 h-old of *D. pulex* also have different HSP83 levels when exposed to arsenite (AsO_3^{3-}) and arsenate (AsO^{3-}) (10–3000 μg.L^{-1}) up to 6 days at 20 °C (Chen et al. 1999). Neonates exposed to AsO_3^{3-} exhibited single bands similar to adults and higher intensity double bands in response to AsO^{3-}. It might represent some additional transcriptional start sites or alternate splicing in young individuals in accordance with the recent findings described by Schumpert et al. (2014).

Table 11.1 Literature review regarding stress-inducible heat shock proteins expression in freshwater invertebrates after trace metal contamination

Biological models tested and authors	Contaminant	Nominal concentration	Number of individuals exposed per replicate	Time exposure and temperature	Level of HSP or HSP-genes expression	Other physiological effects observed
Sponge *Baikalospongia intermedia* (Efremova et al. 2002)	Pb^{2+}, Cu^{2+} and Zn^{2+}	40–400 µg.L^{-1} (Pb^{2+}), 4 and 40 µg.L^{-1} (Cu^{2+}) and 10 and 100 µg.L^{-1} (Zn^{2+})	2 individuals	16 h, 8 °C	Pb^{2+} and Zn^{2+} caused strong induction of HSP70, no expression was observed for Cu^{2+}.	DNA single-strand breaks/alkali-labile sites.
Freshwater polyp *Hydra magnipapillata* (Zeeshan et al. 2017)	Co^{+2}	8, 16 and 60 mg.L^{-1}	Groups of 15 polyps into 6 mL	24, 48, 72, 96 h; _	HSP70 mRNA transcripts were significantly up-regulated at 16 mg.L^{-1}.	Alteration of histology, morphology and regeneration, feeding impairment, reproduction disrupted. ROS formation overwhelming cellular defense mechanisms, DNA damage.
Planarian *Dugesia* (*Girardia*) *schubarti*, (Temenouga et al. 2003)	Cu^{2+}	0–960 µg.L^{-1}	8 individuals (dilution per mL not provided)	24 h to 7 days, 19 °C	Tissue levels of HSP60 remained unaltered.	Catalase activity induced at 40, 80 and 160 µg.L^{-1} concentrations. 360 µg.L^{-1} treatments had an opposite effect. LC_{50} 24 h = 1230 µg. L^{-1}; LC_{50} 7 days = 480 µg.L^{-1}.

(continued)

Table 11.1 (continued)

Biological models tested and authors	Contaminant	Nominal concentration	Number of individuals exposed per replicate	Time exposure and temperature	Level of HSP or HSP-genes expression	Other physiological effects observed
Planarian *Schmidtea mediterranea* (Plusquin et al. 2012)	Cd^{2+}	0.56, 1.12, 2.25, 5.62, 7.9, 10.1, 11.2 and 16.9 mg.L^{-1}	Worms were exposed in 20 mL (number of individuals not given)	2–21 days, ~20 °C	HSP60 and HSP70 genes showed an initial inhibition at the beginning (4-8 h), but returned to normal levels afterwards. HSP70 gene expression increased after 2 days.	Bioaccumulation of cadmium in 2 days and 2 weeks strongly related to exposure concentrations. Body size decreased significantly with high treatments (2 week-exposure). Motility was also affected. Increasing mitotic division in neoblasts (1.12 mg.L^{-1}, 2 weeks; 0.28 to 0.56 mg.L^{-1}, 3 weeks). Antioxidant enzymes activity disturbed throughout 3 weeks.

Planarian *Dugesia japonica* (Ma et al. 2014)	Hg^{2+}, Pb^{2+}	1, 10, 50, 200 μg.L^{-1} (Hg^{2+}); 10, 100, 250, 500 μg.L^{-1} (Pb^{2+})	20 individuals (dilution per mL not provided)	48 h, ~18 °C	10 μg.L^{-1}/Hg^{2+} led to a slight increase of GRP78 gene expression (~1.5-fold); 50 μg.L^{-1}/Hg^{2+} significantly induced GRP78 expression, while 200 μg.L^{-1} led to a 50% of downregulation over controls. 100 μg.L^{-1} and 250 μg.L^{-1}/Pb^{2+} elevated transcripts levels (~1.6- and 2.2-fold, respectively), but reduced gene expression to 0.55-fold in 500 μg.L^{-1}.	–
Planarian *Dugesia japonica* (Zhang et al. 2014)	Cd^{2+}	9.4, 18.8, 37.5, 75, and 150 μg.L^{-1}	10 individuals (1.0–2.0 body length) (dilution per mL not provided)	24 h, 22 °C	No significant differences between control and HSP90 gene expression in exposed organisms; HSP70 transcript levels increase was dose dependent.	Significant higher catalase activity in moderate Cd^{2+} treatments. After 150 μg.L^{-1}/Cd^{2+} exposure, catalase activity decreased.

(continued)

Table 11.1 (continued)

Biological models tested and authors	Contaminant	Nominal concentration	Number of individuals exposed per replicate	Time exposure and temperature	Level of HSP or HSP-genes expression	Other physiological effects observed
Rotifer *Plationus patulus* (Rios-Arana et al. 2005)	Single exposures and mixtures of as^{+}, Cr^{+}, Cu^{2+}, Ni^{2+}, Pb^{2+}, and Zn^{2+}	For lower single metal concentration: 10 μg.L^{-1}, except for Zn^{2+}: 20 μg.L^{-1}; for higher single metal concentration 50 μg.L^{-1} of as^{+3}, Cr^{+}, Cu^{2+} and Ni^{2+}; and 100 μg.L^{-1} of Pb^{2+}; 50 μg.L^{-1} of Zn^{2+}. For mixtures: As^{+3} (50 μg.L^{-1}), Cr^{+}(50 μg.L^{-1}) and Zn^{2+} (150 μg.L^{-1}); 40 μg.L^{-1} for As^{3+}, 50 μg.L^{-1} for Cr^{2+} and Cu^{2+}, and 100 μg.L^{-1} for Pb^{2+} (higher concentrations); 2, 3, 4, 5 and 6 elements combined at low concentrations (10–20 μg.L^{-1}) and all elements at high concentrations (50–150 μg.L^{-1})	200 rotifers per treatment (dilution not provided)	1 h to low single metal concentration, 20 min to high single metal concentration and 20 min for combined elements, 25 °C	The lowest levels of HSP60 were found in 20 μg.L^{-1} zinc-media. Strongest concentrations of cooper increased protein expression but no differences of HSP60 compared to controls were detected. Combined elements like zinc and arsenic increased HSP60 expression. As^{3+}+Cr^{2+}+Cu^{2+}+Pb^{2+} and As $^{3+}$+Cr^{2+}+Cu^{2+}+Ni^{2+}+Pb^{2+} treatments raised HSP60 levels.	–

Aquatic midge *Chironomus tentans* (Lee et al. 2006)	Cd^{2+}, Pb^{2+}, Cr^{+2}	0.2, 2, 20 μg.L^{-1} (Cd^{2+}); 0.05, 0.5, 5 μg.L^{-1} (Pb^{2+}); 0.075, 0.75, 7.5 μg.L^{-1} (Cr^{+2})	10 of the fourth instar larvae into 200 mL beakers containing 100 mL of dechlorinated tap and 0.1 mL of test solution	24 h, 20 ± 1 °C	The HSP70 gene expression increased after metal exposure (Cd^{2+}, Pb^{2+}, Cr^{+2}).	Low levels of Pb^{2+} and Cr^{+2} raised hemoglobin genes expression; both metals decreased hemoglobin genes expression at higher concentrations. Cr^{+2} also had a significant impact on the body weight of exposed larvae after 48 h.
Aquatic midge *Chironomus yoshimatsui* (Yoshimi et al. 2009)	Cd^{2+} and Cu^{2+}	6 mg.L (Cd^{2+}); 3, 6, 9 mg.L (Cu^{2+})	10–15 larvae into 100 mL	1.5 to 24 h, 23 °C ± 0.5 °C	After 1.5 h exposure, HSP70 was activated in a dose dependent-manner by Cd^{2+}, but not Cu^{2+}. An increase of HSC70 happened after 24 h of Cd^{2+} exposure.	–
Aquatic midge *Chironomus riparius* (Planelló et al. 2010)	Cd^{2+}	1124 mg.L^{-1}	No information on number of larvae exposed	12 and 24, 18 °C	12 h after exposure, HSP70 levels were raised throughout 24 h. Constitutive forms (HSC70) remained stable. HSP90 and HSP40 were not altered.	Cd^{2+} altered the expression of the ecdysone receptor gene (EcR) whose mRNA levels were raised.
Aquatic midge *Chironomus riparius* (Martínez-Paz et al. 2013)	Cd^{2+}, tributyltin	0.1, 1, 10 ng.L^{-1} TBTO (tributyltin oxide); 112.4 and 1124 mg.L^{-1} (Cd^{2+})	10 larvae into 500 ml	24 h, 20 °C	Cd^{2+} had a significant effect on HSP27 mRNA upregulation. TBT did not alter HSP27 mRNA levels.	–

(continued)

Table 11.1 (continued)

Biological models tested and authors	Contaminant	Nominal concentration	Number of individuals exposed per replicate	Time exposure and temperature	Level of HSP or HSP-genes expression	Other physiological effects observed
Aquatic midge *Chironomus riparius* (Martín-Folgar and Martínez-Guitarte 2017)	Cd^{2+}	183.3 $mg.L^{-1}$ and 18.3 $mg.L^{-1}$	10 larvae for each treatment time point (n = 6 larvae for 2-6 h exposure; n = 9 larvae for 24 h exposure) (dilution not provided)	2, 4, 6, 24 h; 20 °C	Rapid response by HSP27 gene, whose expression was down-regulated while HSP17 was up-regulated at 2 h. The other genes analyzed remained unaltered. At 24 h, whereas high levels of HSP23, HSP24, HSP27 and HSP34 transcription were detected, HSP22 mRNA levels were down-regulated.	–
Amphipod *Gammarus fossarum* (Schill et al. 2006)	Cd^{2+}	8 to 2000 $\mu g.L^{-1}$	40 adults per tank (~5 L each one), 10 replicates	5 days, 7.5 °C	At 5th day had decreasing levels of HSC/HSP70 in males and females was found; lowest Cd^{2+} concentrations increased HSC/HSP levels in females at day 4; individuals exposed to the strongest concentrations exhibited low stress protein level.	Mortality within the first 5 days of recovery (for both males and females).

Water flea *Daphnia pulex* (Chen et al. 1999)	As^{+3} (arsenite and arsenate)	10, 100, 1000 and 3000 μg.L^{-1}	30 mL of pond water with arsenic (for juveniles, with no number of individuals given), single individuals were placed in 200 ml of filtered pond water containing growing algae and as^{+}	6 days, 20 °C	In arsenate exposures, HSP83 response in adults appeared greater than that of juveniles, while the adult and juvenile responses to arsenite were of similar intensity and each consisted of a single band. Juveniles exposed to arsenite exhibited single bands similar to adults but juveniles exposed to arsenate exhibited higher intensity double bands.	Significant declines in reproduction were observed with 1000 μg.L^{-1} arsenite and 10, 100 and 1000 μg.L^{-1} arsenate. Arsenate was more toxic than arsenite in treatments with equal concentrations.
Water flea *Daphnia magna* (Haap and Köhler 2009)	Cd^{2+}	200, 400, 500, 600 and 800 μg.L^{-1}	5 day-old daphnids into 50 ml for each replicate (4 in total)	24 h, 20 °C	A correlation between sensitive clones and high HSP70 levels was established.	Accumulation in tissues: 250–400 μg.L^{-1} and 500 μg.L^{-1}
Water flea *Daphnia magna* (Li et al. 2014)	Cd^{2+}	10, 20, 40, 60, 80 and 100 μg.L^{-1}	5 young daphnids (<24 h and 3 day-old) into 50 mL (4 replicates) for acute tests (acute experiment), 5 daphnids into 400 mL solutions for HSP induction and pulsed exposures (20 replicates for treatment)	24 h and 6 h (for pulsed exposures over 21 days), 20 ± 1 °C	All individuals tested increased HSP70 levels after 80 μg.L^{-1} Cd^{2+} exposure. HSP70 levels of parents were significantly higher than those from first generation at 80 and 100 μg.L^{-1}.	Increasing the pulsed concentrations of parent generation caused increased 21th day mortality of the first generation of offspring. The mortality rate of first generation was higher than parental generation experiencing both the same pulse.

(continued)

Table 11.1 (continued)

Biological models tested and authors	Contaminant	Nominal concentration	Number of individuals exposed per replicate	Time exposure and temperature	Level of HSP or HSP-genes expression	Other physiological effects observed
Water flea *Daphnia pulex* (Tang et al. 2015)	Cd^{2+}, Zn^{2+} and quantum dots (CdTe, CdSe/ZnS)	From 5.6 to 562 µg.L^{-1} (Cd^{2+}); 32.7 to 3269 µg. L^{-1}(Zn^{2+}); 1, 2, 5, 10, and 20 nM CdTe QDs, and 0.1, 0.5, 1, 2, 5, and 10 nM CdSe/ZnS QDs	10 neonates (5–7 day-old) placed in 30 mL test media	48 h, 20 ± 1 °C	No induction of HSP70–1 gene expression. HSP70–2 mRNA was significantly induced by cd^{+2} at 30.3 and 61.8 µg.L^{-1}. CdSe/ZnS raised HSP70–2 gene expression 1.3-fold at same concentrations. Cd^{2+} (61.82 µg.L^{-1}) also significantly induced HSP70–3 gene. Zn and CdTe did not change HSP70 expression.	A dose-response after Cd^{2+} and quantum dots exposure appeared with metallothionein-1 (MT-1) gene expression increasing levels. No significant inductions of MT-1 with Zn^{2+} were detected. Total Cd^{2+} accumulated after exposure to 25% (0.27 µM) and 50% (0.55 µM) LC50 concentrations were 0.12 ± 0.04 and 0.34 ± 0.2 µg.Mg respectively.

Water flea *Daphnia magna* (Haap et al. 2016)	Cd^{2+}	5, 20, 50, 100 and 300 μg. L^{-1}	Number of 5-day-old daphnids exposed not given, 21 L of medium	24 h, 20 °C	HSP70 was induced in both pre-exposed and non pre-exposed daphnids clones; higher HSC70 levels and stronger upregulation of HSP70 at lower concentrations of cd^{2+} appeared in for S clones (sensitive ones to Cd).	Daphnids without pre-exposure showed higher levels of metallothionein (tolerant clone T); the same effects were less pronounced in the pre-exposed ones. Four generations of Cd^{2+} pre-exposured of clone S had elevated MT expression compared to non-acclimated ones. Clones S also exhibited higher reproductive output.
Zebra mussels *Dreissena polymorpha* (Clayton et al. 2000)	Cu^{2+}, tributyltin	30 to 500 mg.L^{-1}/Cu^{2+}; 1 to 75 mg.L^{-1}/TBT	4 adult mussels in 750 mL (for the TBT) and 4 mussels in 1 L (for Cu^{2+})	24 h, 15 °C	Significant expression of HSP60 at 24 h, highest HSP60 expression at 22 μg cu.L^{-1}; HSP70 elevated expression continued to the highest tested doses (2.5 times control expression); all tested doses of TBT increased HSP60 and HSP70 expression.	–

(continued)

Table 11.1 (continued)

Biological models tested and authors	Contaminant	Nominal concentration	Number of individuals exposed per replicate	Time exposure and temperature	Level of HSP or HSP-genes expression	Other physiological effects observed
Zebra mussels *Dreissena polymorpha* (Singer et al. 2005)	Cd^{2+}, Pb^{2+}, Rh^{3+}, Pt^{4+}, Pd^{2+}	500 μg.L^{-1}	1200 adults per condition into 20 L	~12 °C, 10 weeks	HSP70 induction began around 25–32 days for Pt^{4+} and 18–25 for Pd^{2+}, Rh^{3+}, Pb^{2+} and Cd^{2+}. The highest increase of HSP70 was found with Pd while the lowest was found with Pb^{2+} and Cd^{2+}.	Metal uptake rates (μg.g day^{-1}): Pt^{4+} (2.05), Pd^{2+} (0.51), Rh^{3+} (0.24), Pb^{2+} (0.55), Cd^{2+} (7.73). Accumulation rates were as follows: Pt^{4+} (43–58 μg.g), Pd^{2+} (6–12 μg.g), Rh^{3+} (3 μg.g), Pb^{2+} (10–11 μg.g), Cd^{2+} (91–173 μg.g).
Clam *Corbicula fluminea* (Rocha and Souza 2012)	Pb^{2+}	5 mg.L^{-1} (*in vivo* experiment), 0.2, 2, 20 mg.L^{-1} (*in vitro* experiment)	4 adult clams (in 700–1000 mL) per replicate (3 to 4 per condition)	96 h; 19–22 °C	HSP70 expression increased *in vivo* exposure and *in vitro* experiment with gill cells.	Cell viability reduced by 25%, Na+/K+ ATPase activity inhibited by 70%, expression of MXR phenotype (multixenobiotic resistance) increased (transporter P-glycoprotein increased in cell membranes).

Mussel *Hyriopsis cumingii* (Wang et al. 2017b)	Cd^{2+}	50, 100, 200 μg.L^{-1}	15 adults per replicate (45 in total)	0, 5, 10, 15 and 20 days; 15 ± 2 °C	mRNA expression of HSP90 in the haemolymph was up-regulated under 50 μg.L^{-1} dose at the 10th day. Up-regulation occurred in the gills and haemolymph on the 5th day (for 100 μg.L^{-1} Cd^{2+}).	–
Freshwater snail *Physa acuta* (Martínez-Paz et al. 2016)	Cd^{2+}	18.3 and 183.3 μg.L^{-1}	5 adult snails per treatment (3 replicates each)	5, 24, and 48 h, 20 °C	For both Cd^{2+} concentrations, HSP70 showed a transient upregulation at 24 h; Cd^{2+} did not affect HSP90 transcription.	A slight tendency for upregulation of metallothionein, with no statistical significance, was observed at 24 h with 0.1 and 1 mM Cd^{2+}.

11.6.2 Aquatic Midges Chironomus spp. *(Diptera)*

Aquatic midges have been increasingly used as alternative models to understand Hsp activation after trace metal contamination in freshwater ecosystems (Table 11.1). Four instar larvae of *Chironomus riparius* exposed to very strong Cd^{2+} dosages (18.3 $mg.L^{-1}$ for 24 h and 183.3 $mg.L^{-1}$ for 2, 4 and 6 h) showed a rapid response of small Hsp by lower expression of HSP27 gene (2-6 h) and upregulation of HSP17 (2 h) at 20 °C (Martín-Folgar and Martínez-Guitarte 2017). Following 6 h, mRNA-HSP70 expression was lower or similar to those from the controls. At 24 h, strong transcription activity of HSP23, HSP24, HSP27 and HSP34 was also noted. Although the fast induction of chaperones might represent a clear response to proteomic stress, strong Cd^{2+} concentrations hardly found in the environment might rather bring the rise of a physiological stress closer to lethal effects. In another study using very high concentrations of Cd, Martínez-Paz et al. (2013) demonstrated a significant effect on HSP27 mRNA expression at 15- (112.4 $mg.L^{-1}/Cd^{2+}$) and 35-fold (1124 $mg.L^{-1}/Cd^{2+}$) in larvae of *C. riparius* in a 24 h-treatment. In general, small Hsp have several key functions to maintain cellular integrity such as (i) strong anti-aggregation properties without ATP activation, (ii) cell protection through triage of aberrant and refolded proteins, (iii) interaction with cytoskeleton, (iv) modulation of cell ability to respond to oxidative stress and (v) interference with key apoptotic proteins (Garrido et al. 2012).

The constitutive member of the 70 kDa HSP family is crucial for the chaperoning function of unstressed cells regulating several physiological functions. The overexpression of housekeeping proteins like HCS70 may indicate some activity associated with other molecular chaperones regulating cellular signaling and functions for a re-establishment of cell homeostasis (Liu et al. 2012). It is still less clear however through which pathways a rapid activation of constitutive forms of 70 kDa would emerge in aquatic invertebrates exposed to trace metals. After 12 h-exposure, HSP70 levels were raised 3-fold throughout 24 h while HCS70 and HSP90 levels remained unaltered in 4-instar larvae of *C. riparius* exposed to a very high Cd^{2+} concentration (10 mM or 1124 $mg.L^{-1}$) at 18 °C (Planelló et al. 2010). In *Chironomus yoshimatsui* larvae, HSP70 expression was enhanced by Cd^{2+} (6 and 9 $mg.L^{-1}$ in 24 h) at 23 °C (Yoshimi et al. 2009). After 1.5 h-exposure, HSP70 was activated in a dose dependent-manner by Cd^{2+}, but not by Cu^{2+}. Copper (2 $mg.L^{-1}$) raised HSP70 levels in a dose-dependent manner after 4-7 h. Besides, HSC70 expression was also enhanced after 24 h of Cd^{2+} exposure. Finally, *Chironomus tentans* larvae also experienced HSP70 gene expression upregulation after metal exposure (Cr^{2+}, Pb^{2+} and Cd^{2+}) in a 24 h-treatment at 20 °C (Lee et al. 2006). Both Cr^{2+} and Cd^{2+} increased HSC70 mRNA levels in a concentration-dependant trend.

11.6.3 Mussels and Clams (Veneroidea and Unionoida)

Bivalves are often the largest group of filter-feeding organisms in terms of biomass in several freshwater ecosystems. Table 11.1 summarizes some results and experimental conditions with mussels and clams as models for trace metal toxicity studies in freshwater ecosystems. Because of its high filtration capacity, the zebra mussel *Dreissena polymorpha* (Veneroidea, Dreissenidae) is considered as an accurate monitor for water quality management (Reeders and Bij de Vaate 1990). As a unique trait, *D. polymorpha* can live in both lentic and lotic waters, and estuarine environment tolerating salinity up to 5‰ (Binelli et al. 2015). After a threshold of 19 μg.L^{-1}/Cu^{2+} concentration is reached in adults of D. *polymorpha,* a significant level of HSP60 emerged in 24 h at 15 °C (Clayton et al. 2000). An overexpression of HSP60 arose at 22 μg.L^{-1}/ Cu^{2+} beyond which 60 kDa Hsp levels dropped. Meanwhile, HSP70 elevated expression continued to the highest tested doses though. In this study, tributyltin (TBT) enabled both HSP60 and HSP70 protein expression at the highest levels. In a previous study, adults of *D. polymorpha* were submitted to 500 μg.L^{-1} treatments with Cd^{2+}, Pb^{2+}, Rh^{3+}, Pt^{4+}, Pd^{2+} at ~12 °C over 10 weeks (Singer et al. 2005). The starting period for HSP70 induction was at approximately 25–32 days for Pt^{4+}, and 18–25 days for Pd^{2+}, Rh^{3+}, Pb^{2+} and Cd^{2+}. Palladium enhanced HSP70 to the strongest level while the lowest inductions were found with Pb^{2+} and Cd^{2+}. Furthermore, the lowest rates of metal uptake occurred with Pd^{2+} (0.51 μg.g per day^{-1}) and Rh^{3+} (0.24 μg.g per day^{-1}) which resulted in a lower bioaccumulation by mussels (6–12 μg.g and 3 μg.g, respectively). Both *in vivo* and *in vitro* Pb^{2+} exposures of clam *Corbicula fluminea* (Veneroidea, Corbiculidae) within 96 h revealed that cell viability significantly decreased to ~25% at 19–22 °C. It was observed that lead can inhibit Na^+/K^+ ATPase activity by 70% while HSP70 expression is raised to 2.23-fold for *in vivo* exposure and 1.63-fold (0.2 mg.L^{-1}), 1.09-fold (2 mg.L^{-1}) and 1.2-fold (20 mg.L^{-1}) for *in vitro* experiment with gill cells (Rocha and Souza 2012).

Unionoidean mussels are a monophyletic group composed of large and long-lived organisms (10–40 years or more) that burrow in the sediment of rivers, streams, and lakes. This feature makes them an excellent model to backtrack trace metal pollution in the environment. Wang et al. (2017b) found that 50 μg.L^{-1} of Cd^{2+} gradually raised HSP90 mRNA expression in the haemolymph of unionoidean mussels *Hyriopsis cumingii* (Unionoida, Unionidae) reaching a maximum after 10 days at 15 °C. With 100 μg.L^{-1}/Cd^{2+}, HSP90 rapidly went up to 3.15-fold in haemolymph and 3.36-fold in the gills at 5th day. Finally under 200 μg.L^{-1}/Cd^{2+}, HSP90 mRNA was enhanced in the gills in a time-dependent fashion up to 15 days while HSP90 mRNA had a peak of expression on the 10th day. HSP90 is a chaperone required for the activation and stabilization of several client-proteins (± 200) in which some participate in signal transduction pathways (Li et al. 2012). In collaboration with HSP70, HSP90 promotes recruitment and assembly with client-proteins. Both Hsp act together directing misfolded proteins for degradation.

In comparison to bivalves, adult freshwater snails *Physa acuta* (Gastropoda, Basommatophora) contaminated by Cd^{2+} (18.33 and 183.32 μg.L^{-1}) along 5, 24 and 48 h displayed a HSP70 transcriptional response with a transient upregulation up to 24 h, but returning to basal levels within 48 h at 20 °C (Martínez-Paz et al. 2016). HSP90 expression was not significantly activated though. The pathways involving HSP70 and HSP90 collaboration in terms of protein expression and genes activation still need further research in many freshwater invertebrates.

11.6.4 Rotifers (Plationus patulus)

Rotifers often play an important role in the dynamics of freshwater and marine environments. Due to their grazing capacity, they may affect algal population and water quality (Snell and Janssen 1995). Assessing metabolic defenses of rotifers against metal toxicity can show how species mitigate sublethal effects before reaching the lethal threshold. Species *Plationus patulus* (Monogononta) briefly treated with single and mixtures of As^{+}, Cr^{+}, Hg^{2+}, Zn^{2+} and Cu^{2+} at several nominal concentrations had different patterns of HSP60 expression at 25 °C (Rios-Arana et al. 2005) (Table 11.1). In this study, HSP60 appeared to be sometimes induced by lower nominal concentrations (<50 μg.L^{-1}), or stronger doses of metals. Chromium at low dosage (10 μg.L^{-1}) induced HSP60 by 4.0-fold. Lower doses of Cu^{2+} seemed to change HSP60 expression as follows: 63 μg.L^{-1}/Cu^{2+} declined HSP60 synthesis after 30 min of exposure, but 50 μg.L^{-1}/Cu^{2+} elevated it to 2.0-fold and 20 μg.L^{-1}/Cu^{2+} to 2.0–4.0-fold. Conversely, the lowest level of HSP60 was found with 20 μg.L^{-1}/Zn^{2+}-treated rotifers. Zinc doubled HSP60 levels at 150 μg.L^{-1} though. For 100 μg.L^{-1}/Cu^{2+} condition, HSP60 level was higher, but not statistically significant from others. These conflicting results need to be assessed.

Combined treatments of Zn^{2+}+As^{+} induced HSP60 expression with Zn^{2+} effects predominating by 2.0-fold. Conversely, Zn^{2+}+Cu^{2+} media seemed to decrease 60 kDa HSP each one by 4.0-and 2.0-fold, respectively. For $As^{3+/5+}$+Cr^{+2} + Cu^{2+}+Pb^{2+} mixture treatments, HSP60 were 2.0-fold lower than controls. The mixture of $As^{3+/5+}$+Cr^{+2} + Cu^{2+}+ Ni^{2+}+Pb^{2+}+Zn^{2+} inhibited HSP60 response at low or high concentrations, but $As^{3+/5+}$+Cr^{+2} + Cu^{2+}+Ni^{2+}+Pb^{2+} raised HSP60 levels by 2.0-fold. The toxicity mechanisms explaining cellular effects in terms of anti-oxidant responses and sub-cellular compartmentalization of each metal in combined exposures would better clarify these results. To our knowledge, there is still very little research using freshwater rotifers to address inducible-heat shock proteins modulation after metal contamination.

11.6.5 Planarians (Tricladida)

Freshwater planarians are free-living bilateral flatworms found in lakes, ponds, streams, and springs sometimes hiding under rocks or over aquatic plants. They normally live on the bottom where they can be exposed to potentially harmful substances of dietary origin or even to polluted sediments. Under natural stress such as long-term starvation or an eventual tissue trauma, HSP70 can be induced in the planarian *Polycelis* sp. (Planariidae) (Cheng et al. 2015). In this species, HSP70 is also inhibited by high or low temperatures. In comparison, constitutive HSP90 gene expression is greatly detected in gastrodermis of *Dugesia japonica* (Dugesiidae) undergoing regeneration process (Conte et al. 2011). It appears that this gene is sensitive at the HSP90 transcriptional level to detrimental conditions like heat shock, starvation, and UV radiation.

Although HSP90 gene is not expressed in response to 9.375–150 $\mu g.L^{-1}/Cd^{2+}$ treatments, HSP70 gene response to cadmium is dose-dependent throughout a 24 h-long experiment with *D. japonica* at 22 °C (Zhang et al. 2014). A 21 day-lethality experiment with high doses of cadmium (5–150 μM, or 0.562 to 16.860 mg.L^{-1}) led to oscillating patterns of HSP60 and HSP70 gene protein expression in the planarian *Schmidtea mediterranea* (Dugesiidae) at 20 °C (Plusquin et al. 2012). In the first 8 h, both genes had an initial inhibition but went back to normal levels. After 2-day exposure, only HSP70 gene activity was increased but not steadily persisted throughout 21 days. Another effect observed was the body size reduction exclusively due to cadmium intoxication. Comparatively, increasing doses of cooper (0–960 $\mu g.L^{-1}$) cannot induce HSP60 expression in *Dugesia* (*Girardia*) *schubarti* (Dugesiidae) from 24 h to 7 days at 19 °C (Temenouga et al. 2003).

Contrarily to HSP90 gene, the 78 kDa glucose-regulated protein (GRP78) is not stimulated during starvation in *D. japonica* (Ma et al. 2014). Instead, 100 $\mu g.L^{-1}$ and 250 $\mu g.L^{-1}/Pb^{2+}$ 48 h-exposures elevated GRP78 mRNA levels to ~1.6- and 2.2-fold at 18 °C. The GRP78 is thought to be transiently associated with unfolded, unassembled or aberrant proteins to keep them in a folding-competent state while they pass through endoplasmic reticulum (ER). However, an abnormal dissociation between 78 kDa proteins from unfolded proteins can take place as a consequence of Pb^{2+} direct binding to GPR78, which blocks protein trafficking (Qian and Tiffany-Castiglioni 2003). Moreover, 50 $\mu g.L^{-1}/Hg^{2+}$ dose significantly induced GRP78 expression. Higher doses of both metals (200 $\mu g.L^{-1}/Hg^{2+}$, 500 $\mu g.L^{-1}/Pb^{2+}$) reduced GPR78 gene expression (Ma et al. 2014). Figure 11.6 summarizes the main mechanisms explained in this section. See Table 11.1 for more details about the experimental set-ups and compared results with freshwater planarians.

*11.6.6 Cnidarians (*Hydra spp.*)*

Mismatching capacities for thermotolerance have been found amongst species of freshwater coelenterate *Hydra* spp. (Hydrozoa) (Bosch et al. 1988). Contrarily to *Hydra attenuata*, *Hydra oligactis* did not express detectable levels of HSP60 or any other Hsp in response to thermic shock. Some other species in *Hydra* group share this same feature. To date, it is also difficult to understand how different species of *Hydra* would mitigate metal toxicity with Hsp. Some evidences have shown that the major heat shock protein involved in fast responses to stress in *Hydra* sp. is a 60 kDa

HSP rather than HSP70 (Bosch and Praetzel 1991). Nevertheless, Zeeshan et al. (2017) tested sublethal doses of Co^{2+} (8, 16 and 60 $mg.L^{-1}$) in polyps of *Hydra magnipapillata* in 24, 48, 72 and 96 h exposures and demonstrated that HSP70 mRNA transcripts were significantly up-regulated within 24 h using 16 $mg.L^{-1}$-media. From this dose onward polyps suffered severe morphological anomalies. Strikingly sublethal doses impaired feeding by disorganizing gastroderm tissue, affecting nematocytes and in turn reproduction. An overproduction of ROS was detected as well, which primarily occurred in the lysosomes. ROS liberation was followed by the expression of SOD, GST, GPx and G6PD genes. In this work, 24 h was a turning point to *H. magnipapillata* to undergo Co^{2+} effects: toxicity (LC_{50} = ~66.19 $mg.L^{-1}$) and antioxidant defenses were found both at the highest levels. The total and dissolved cobalt concentrations normally found in pristine freshwater environments are low (<5 $\mu g.L^{-1}$) (Nagpal 2004). However, higher levels can be detected in streams close to populated areas (1–10 $\mu g.L^{-1}$) and near mining districts or regions with intensive agricultural land use (~11–50 $\mu g.L^{-1}$). Hence, the toxic effects caused by Co^{2+} concentrations in the $mg.L^{-1}$ are extremely high and unlikely to happen in a real case scenario.

11.6.7 Sponges (Haplosclerida)

Studying Hsp induction in different animal models is without a doubt another important step to improve our understanding on chaperone regulation pathways, their functions and the constitutive molecular differences through the zoological scale. To date, HSP70 expression have a potential to be an excellent tool for metal toxicity investigation with some less conventional ecotoxicological models such as sponges, *Hydra* sp., rotifers etc. (Efremova et al. 2002; Rios-Arana et al. 2005; Zeeshan et al. 2017) (Table 11.1).

Freshwater sponges can be found in many types of freshwater habitats worldwide. They occur at shallow or deeper environments, settling on aquatic plans, rocks etc. However, little information about their cellular protective mechanisms against xenobiotics is available so far. Cold water sponge *Lubomirskia baicalensis* (Haplosclerida, Lubomirskiidae) strongly expressed HSP70 at low temperatures as 4 °C (Müller et al. 2007). This synthesis is enhanced after incubation with toxin okadaic acid (OC) normally produced by dinoflagellates living in *L. baicalensis* tissues in summer at 17 °C. Another interesting aspect of HSP70 in sponges is their expression in vegetative bodies. HSP70 stabilize their proteins and membranes during dormancy from autumn to spring when temperature is changing (Schill et al. 2006). The cytoprotective role against metal stress in sponges has been investigated by Efremova et al. (2002). The authors demonstrated that tissues of freshwater sponge *Baikalospongia intermedia* (Haplosclerida, Lubomirskiidae) treated with increasing doses of Pb^{2+}, Cu^{2+} and Zn^{2+} for 16 h at 8 °C can show DNA single-strand breaks and HSP70 strong activity, except for Cu^{2+}. It is likely that similar mechanisms take place in other freshwater sponges living in a wide range of temperatures, but such mechanisms have not been revealed yet.

11.7 Stress-Induced Genes and Proteins in Estuarine and Saltwater Invertebrates Contaminated by Metals

11.7.1 Starfishes and Sea Urchins (Asteroidea and Echinoidea)

As echinoderms are phylogenetically related to vertebrates, they are expected to have a complex detoxifying/defensive system against xenobiotics. Living exclusively on the seafloor, sea urchins and starfishes can serve as sentinel organisms to provide information about water and sediment contamination (Table 11.2). In natural populations of *Asterias rubens* (Asteriidae) living in contaminated sites, HSC70 levels found in coelomocytes were significantly correlated to Cd^{2+} and Cu^{2+} concentrations in the tegument of ~1.52–2.89 µg.g dw and ~1.70–2.48 µg.g dw, respectively (Matranga et al. 2012). In laboratory, adults of starfish *A. rubens* showed a tendency having higher levels of HSP70 (50–60 density units. mm^{-1}) after 10-day exposure to 15 mg.L^{-1}/Mn^{2+}, and an overexpression when exposed to 5.5 mg.L^{-1}/ Mn^{2+} for all sampling occasions especially after 5 days (Oweson et al. 2008). As an additional effect of Mn^{2+} exposure, the number of coelomocytes increased by proliferation of cells from the coelomic epithelium. As Mn^{2+} is an essential trace element for cell metabolism, toxicity will likely rise at the highest doses. Dissolved concentrations of manganese in natural waters void of anthropic imputs range from <0.01 mg.L^{-1} to >10 mg.L^{-1} (Reimer 1999). Total Mn concentration (mg.L^{-1}) in sea surface water is around 0.01–1.70 mg.L^{-1} while seawater typically contains ~2 µg. L^{-1}/Mn^{2+}. A typical seasonal trend in total manganese concentration in surface water is observed during high and low runoff periods.

In the life cycle of marine invertebrates, it is generally accepted that early developmental stages are the most sensitive to environmental stressors. For that reason, most of the research on metal toxicity in echinoderms has been concentrated on investigating their effects on the development, recording embryogenesis, larval morphogenic processes, survival and metamorphosis. Perturbation of skeletogenesis and aberrant development have been recorded with *Paracentrotus lividus* (Echinidae) embryos in Cd^{2+} (10^{-5}, 10^{-4} and 10^{-3} M doses) media up to 24 h at 18 °C (Roccheri et al. 2004). Between 15 and 24 h exposure, a set of stress proteins such as small Hsp (25, 28 kDa), 56 kDa, 70–72 and 90 kDa was induced. An increase of Hsp synthesis was observed from 15 to 21 h with 70–72 kDa, 56 kDa and 25 kDa. The 90 kDa, 70–72 HSP and 56 HSP continued to be synthesized even at lowest Cd^{2+} levels. Similarly, Migliaccio et al. (2014) found genes encoding upregulation of HSP60 and HSP70 at pluteus stage by 2.5- and 2.4-fold after Cd^{2+} and Mn^{+2} exposures respectively. Indeed, Mn^{2+} (1 to 62 mg.L^{-1}) led to constitutive protein

Table 11.2 Physiological effects and HSP induction observed in marine invertebrates after trace metal contamination

Biological models tested and authors	Contaminant	Nominal concentration	Number of individuals exposed per replicate	Time of exposure and temperature	Level of HSP or HSP-genes expression	Other physiological effects observed
Sponge *Suberites domuncula* (Schröder et al. 1999)	Cd^{2+}	0.01, 0.1, 1 mg. L^{-1}	Tissues exposed	0.5, 1, 3 or 6 days; 16 °C	After 12 h, high levels of HSP70 at 1 mg.L^{-1}. Long incubation periods: Maximum levels also with lower concentrations (~001 mg.L^{-1}).	Throughout 6 days, 1 mg. L^{-1}-treated tissues accumulated 7.7 μg.g/ww. DNA damage reached the maximum at 1 mg.L^{-1}.
Sponge *Chondrosia reniformis* (Cebrian et al. 2006)	Cu^{2+}	30 and 100 μg. L^{-1}	15 sponges in total, 1 L for each treatment	5 days, 17 °C	HSP70 cross-reacted with 60 kDa (the only band showing up).	Average of metal accumulation: ~6 to 27 μg. g^{-1}. Decreasing clearance rates, increasing collagen/cell rate by death of cells, lower survival rate in transplanted sponges in polluted sites.
Sponge *Crambe crambe* (Agell et al. 2001)	Cu^{2+}	30 and 100 μg. L^{-1}	15 sponges in total, 2 L for each treatment	5 days, 17 °C	HSP54 and HSP72 were expressed. Significant differences found with HSP72 at 30 μg.L^{-1}. Both were inhibited at 100 μg.L^{-1}. HSP72 mobilized in short-term while HSP52 was accumulated in long-term.	~120 μg.g was the average for Cu^{2+} accumulation in tissues of native or transplanted sponges in polluted zones.
Sea anemone *Anemonia viridis* (Nicosia et al. 2014)	Cd^{2+}, Pb^{2+}	2, 10 and 50 μg. L^{-1}	3 anemones per treatment	12 h, 18 °C	Expression of HSP28.6 gene increased. Transcript up-regulated by Pb^{2+}. Cd^{2+} induced HSP28.6 gene by 8-fold at 50 μg.L^{-1}.	–

Sea anemone *Nematostella vectensis* (Elran et al. 2014)	Hg^{+2}, Cu^{2+}, Cd^{2+}, Zn^{2+}	0.3–0.7 μM following each metal	5 anemones per treatment in a triplicate treatment (dilution not provided)	24 h, T° not given	Transcripts mostly expressed in hg-treated anemones with small increase in cu-treated anemones. The highest regulated genes were those of the HSP20 group after Hg^{2+} contamination.	Up-regulation of transcription factors such as Egr1, AP1 and NF-kB one hour after contamination.
Reef coral *Montastraea franksi* (Venn et al. 2009)	Cu^{2+}	3, 10, 30 and 100 μg.L^{-1}	3 individuals per treatment	4 and 8 h; 27 ± 1 °C	HSP70 gene expression significantly upregulated at: 100 μg.L^{-1} (4 h), 30 and 100 μg.L^{-1} (8 h); HSP90 gene expression increased following 8 h exposure at 30 and 100 μg.L^{-1}.	Corals exposed to Cu^{2+} had a significant increase in multi-xenobiotic resistance protein (P-gp) gene expression at 100 μg.L^{-1} (4 and 8 h) and 30 μg.L^{-1} (8 h).
Rotifer *Brachionus koreanus* (Jung and Lee 2012)	Cu^{2+}, Cd^{2+}	0.125, 0.25, 0.5, 1 mg.L^{-1} (Cu^{2+}) and 31, 63, 125, 250 and 500 mg.L^{-1} (Cd^{2+})	10 individuals/well, 3 replicates (6-well plate) for the acute tests and 2000 individuals/mL, 3 replicates for gene expression study	48 h, 25 °C	HSP20, HSP70 and HSP90 mRNA were enhanced at 24 h by cu^{2+}. Cd^{2+} modulated higher and fast responses if compared to Cu^{+2}.	–
Copepod *Tigriopus japonicus* (Rhee et al. 2009)	Cu^{2+}, ag^{+}, Zn^{2+}	0.1, 0.2, 0.5 μg.L^{-1} (Cu^{2+}); 0.01, 0.05, 0.1 μg.L^{-1} (ag^{+}); 0.1, 0.5 and 1.0 μg.L^{-1} (Zn^{+})	200 copepods (dilution not provided)	96 h, 20 ± 1 °C	All metals and concentrations tested caused upregulation of the HSP70 gene. The highest concentration of ag^{+} (0.1 μg.L^{-1}) did not change HSP levels.	

(continued)

Table 11.2 (continued)

Biological models tested and authors	Contaminant	Nominal concentration	Number of individuals exposed per replicate	Time of exposure and temperature	Level of HSP or HSP-genes expression	Other physiological effects observed
Copepod *Tigriopus japonicus* (Kim et al. 2014b)	Ag^{+}, as^{+3}, Cd^{2+}, Cu^{2+}, Zn^{2+}	5, 10, 50, and 100 μg.L^{-1}	300 adult copepods in each container (dilution not provided)	96 h, 20 °C	HSP20 and HSP70 genes were sensitive to all metals within 96 h. Most of the mRNA HSP genes analyzed were up-regulated by Cu^{2+}, but HSP40 mRNA expression was downregulated at 100 μg.L^{-1}. Transcriptional levels of HSP10 gene were downregulated by 100 μg.L^{-1} (as^{+3}, Cd^{2+}, Zn^{2+}). HSP60 mRNA expression was also downregulated in 100 μg.L^{-1} (ag^{+}, as^{+3}).	Intracellular ROS and glutathione (GSH) were significantly induced in ag^{+}, as^{+}, and Cu^{2+} exposures.
Copepods *Tigriopus japonicus* and *Pseudodiaptomus annandalei* (Wang et al. 2017a)	Hg^{2+}	10 μg.L^{-1}	150 copepods into 500 mL	5 h, 22 °C	In *T. japonicus* HSP70 and HSP20/α-cristallin were upregulated. In *P. annandalei* HSP90, HSP70 and 82-kDa HSP3 were upregulated.	In 10 μg.L^{-1} exposures, *T. japonicus* accumulated ~3500 ng.g^{-1}. *P. annandalei* accumulated ~2500 ng.g^{-1}.
Crab *Portunus trituberculatus* (Zhang et al. 2009)	Cu^{2+}	58 mg.L^{-1}	6 crabs/treatment (control and exposed ones) (dilution not provided)	1 h, 20 °C	HSP90-type 1 transcript was mainly expressed in gills, HSP90-type 2 was significantly upregulated in hepatopancreas and gills.	–

Chinese shrimp *Fenneropenaeus chinensis* (Luan et al. 2010)	Cu^{2+}, Cd^{2+}	3177 μg.L^{-1} (Cu^{2+}), 281 μg.L^{-1} (Cd^{2+})	30 juveniles into 600 L for each toxicant (2 replicates each)	6, 12, 24, 72 h; 25 °C	HSP70 up-regulated by Cu^{2+}; the highest expression was detected at 12 h; HSP70 down-regulated after 72 h. Cu^{2+} not causing significant changes in HSC70 expression. HSC expression inhibited by Cd^{2+}.	–
White leg shrimp *Penaeus vannamei* (Sung et al. 2017)	Cu^{2+}, Zn^{2+}	6.9 mg.L^{-1} (Cu^{2+}), 3.5 mg.L^{-1} (Zn^{2+})	30 shrimps into 2 tanks of 2 L	24 h, 28 °C	Heat shock protected shrimps against metal toxicity by previous HSP70 upregulation.	–
Black tiger shrimp *Penaeus monodon* (Shi et al. 2015)	Cu^{2+}, Zn^{2+}, Cd^{2+}	0.18, 0.506 and 7.73 mg.L^{-1}	Not given	3 individuals taken at 6, 12, 24, 48 and 96 h; 25 °C	In the gills, HSP60 transcript level was significantly raised at 12 h after Cu^{2+} exposure, but decreased up to 48 h. HSP10 reached maximum levels at 96 h. For Zn^{2+} similar tendencies were reported. For Cd^{2+}, HSP60 transcript reached a high expression at 24 h. mRNA maximum expression level for HSP10 was at 12 h.	–
Brine shrimp *Artemia franciscana* (Pestana et al. 2016)	Cd^{2+}, Zn^{2+}	Cd^{2+} (0–10–25-50-100-150-200-300 mg.L^{-1}) and Zn^{2+} (0–0.5-1-4-10-25-60-150 mg.L^{-1})	10 organisms/treatment into 20 mL (3 replicates each)	48 h, 28 °C	The individuals exposed to heat shock and metals had HSP levels increased 6 h after contamination period. HSP levels decreased over time, except under Cd^{2+}. Zn^{2+} exposure had no significant impacts on HSP expression.	No effects were observed after 12 h of exposure to Cd^{2+} or Zn^{2+}, either in terms of DNA methylation or histone acetylation (H3 and H4).

(continued)

Table 11.2 (continued)

Biological models tested and authors	Contaminant	Nominal concentration	Number of individuals exposed per replicate	Time of exposure and temperature	Level of HSP or HSP-genes expression	Other physiological effects observed
Purple Sea urchin *Paracentrotus lividus* (Roccheri et al. 2004)	Cd^{2+}	10^{-5}, 10^{-4}, 10^{-3} M Cd^{2+}	5000 mL^{-1} embryos	15, 18, 21 and 24 h; 18 °C	Stress proteins 25, 28, 56, 70–72 and 90 kDa were induced. From 15 to 21 h 70–72 HSP, 56 kDa (HSP) and 25 kDa were enhanced. 25 kDa proteins continued to be expressed at high levels, the 90 kDa, the 70–72 HSP and the 56 HSP continued to be synthesized even at lower levels.	Perturbation of skeletogenesis and aberrant development.
Purple Sea urchin *Paracentrotus lividus* (Pinsino et al. 2010)	Mn^{2+}	1.0, 7.7, 15.4, 30.8, 61.6 mg.L^{-1}	100 mL/beaker; 4000 embryos/mL	48 h, 18–19 °C	After 48 h, levels of HSC60 were 3- (15.4 mg.L^{-1}), 6- (30.8 mg.L^{-1}), 6-fold (61.6 mg.L^{-1}) higher than controls while HCS70 were 3- (15.4 mg.L^{-1}), 4- (30.8 mg.L^{-1}), 5-fold (61.6 mg.L^{-1}) higher than controls.	Mn^{2+} was rapidly accumulated in embryos in a proportional way regarding the concentrations used, except for the highest concentration (61.6 mg.L^{-1}).
Purple Sea urchin *Paracentrotus lividus* (Migliaccio et al. 2014)	Cd^{2+}, Mn^{2+}	0.09, 0.18, 0.96, 1.83, 3.66, 5.49 mg.L^{-1}/Cd^{2+} and 3.6, 7.2, 15.4, 30.8, 61.6, 123.2 mg.L^{-1}/ Mn^{2+}	150 embryos/mL in each well (triplicate experiment)	48 h, 18 ± 2 °C	Genes encoding HSP60 and HSP70 were up-regulated at pluteus stage.	With Cd^{2+}, nitric oxide directly activated HSP70 and indirectly regulated HSP60 genes. With Mn^{2+}, nitric oxide indirectly affected HSP60.

Purple Sea urchin *Paracentrotus lividus* (Pinsino et al. 2015)	Bare TiO_2NPs with 10–65 nm	1 and 5 μg.L^{-1} (500 μl injected into body cavity)	3 sea urchins for each concentrations (20 L in each aquarium, 3 urchins/tank)	24 h, 18 °C	No stimulation of HSP70.	TiO_2NPs did not affect internal membrane polarization of the trans-Golgi/endoplasmatic reticulum (ER) compartments or other vesicle membranes. TiO_2NPs stimulated phagocytic activity of immune cells. They also affected signal transduction downstream to p38 MAPK.
Green Sea urchin *Strongylocentrotus droebachiensis* (Magesky et al. 2017)	Polymer-coated AgNPs, ag^+	100 μg.L^{-1}	5 individuals into 5.5 mL (6 replicates/ condition, 30 individuals, 240 in total for each contaminant)	48 h, 8 °C	Large urchins had significant HSP70 up-regulation at 48 h of exposure for both silver forms while small ones showed increasing levels of HSP70 from 24 h to 48 h. HSP60s were increasingly expressed throughout 48 h under nano-ag treatment, the opposite effect occurred with soluble ag.	Pigment production against silver was much more evident in soluble silver conditions. Transcytosis of nano-ag through coelomic sinuses was observed in 1-month-old juveniles.

(continued)

Table 11.2 (continued)

Biological models tested and authors	Contaminant	Nominal concentration	Number of individuals exposed per replicate	Time of exposure and temperature	Level of HSP or HSP-genes expression	Other physiological effects observed
Seastar *Asterias rubens* (Oweson et al. 2008)	Mn^{2+}	5.5 and 15 mg. L^{-1}	1 specimen per container (31 in total) (dilution not provided)	5, 10, 15 and 25 days; 12 °C	Higher levels of HSP70 after exposure to 15 mg. L after 10 days (50–60 density units mm^{-1}); overexpression when exposed to 5.5 mg.L^{-1} especially after 5 days.	Accumulation of Mn^{2+} was ~6–8.26 ± 0.58 mg.L^{-1}. Number of coelomocytes increased; haematopoietic tissue and coelomocytes showed stress by pronounced carbonylation.
Seastar *Asterias rubens* (Matranga et al. 2012)	Cu^{2+}, Cd^{2+}, Pb^{2+}, Zn^{2+}	0–60 µg.g dw (Cd^{2+}), 100–500 µg.g dw (Zn^{2+}), 0–800 µg.g dw (Pb^{2+}), 0–100 µg.g dw (Cu^{2+})	10 individuals per site	–	HSC70 levels found in coelomocytes were significantly correlated to Cd^{2+} and Cu^{2+} concentrations in the tegument.	Different levels of metals were found in tegument and pyloric caeca.

expression (HSC60 and HSC70) as a function of Mn^{2+} concentrations in 48 h period at ~19 °C. NO (nitric oxide) was produced in sea urchin embryos in response to both treatments and modulated the expression of some metal-induced genes involved in stress response, skeletogenesis, detoxification and multidrug efflux processes. Additional studies about expression and transcription processes of Hsp among echinoderms, hemichordates and groups of chordates should clarify how much these groups share in terms of cellular strategies to tolerate metal toxicity (Agell et al. 2004; Satoh et al. 2014).

11.7.2 Bivalvia

11.7.2.1 Mussels (Mytiloidea)

Amongst marine bivalves, *Mytilus* spp. have been greatly used as standard models for toxicological studies either for laboratory experiments or costal pollution assessments (Beyer et al. 2017) (see Table 11.3). HSP70 and HSP90 expression is a common mechanism employed by some nearshore sea mussels to endure rising temperatures during tidal variations (Buckley et al. 2001; Anestis et al. 2007). In a similar way, contamination with trace metals has shown the expression of some HSP families in mussels. A time-dependent pattern of HSP70 expression was observed when adults of the hard-shelled *Mytilus coruscus* (Mytilidae) were exposed to 20 $\mu g.L^{-1}/Cu^{2+}$ and 200 $\mu g.L^{-1}/Cd^{2+}$ throughout 30 days at ~25 °C (Li et al. 2014). HSP70 mRNA expression increased as time went on with a maximum peak at 15 days in Cu^{2+} exposure (9.41-fold). In Cd^{2+} media the highest expression level was reached with 10.82-fold at day 9. After exposure to rising concentrations of $NaAsO_2$ (arsenite, As^{III}) and Na_2HAsO_4 (arsenate, As^{V}), the species *M. trossulus* (Mytilidae) showed that As^{III} induced HSP70 in a concentration-dependent manner within 24 h at 13 °C (La Porte 2005). Strong signals of HSP70 expression were given by doses above 1000 $\mu g.L^{-1}/As^{III}$. For both As^{III} and As^{V}, 1000 $\mu g.L^{-1}$ dosage significantly raised HSP70 compared to controls but HSP70 levels were not different from each contaminant. The authors stated that the potential induction to HSP70 elevation corresponded to $As^{III} > As^{V} > (CH_3)_2As^{V}$ trend. Despite arsenic concentrations found in some industrially contaminated sites can be limited to 13.8 $\mu g.L^{-1}$, a significant HSP70 elevation only happens in *M. trossulus* at 100 $\mu g.L^{-1}/As^{III}$ (La Porte 2005).

Following a 15-day exposure, nanoparticulate cooper oxide (CuONPs <50 nm, 10 $\mu g.L^{-1}$) caused an overexpression of HSP71 by more than 20-fold at ~17.8 °C in the gills and digestive gland of *M. galloprovincialis* (Mytilidae) (Gomes et al. 2014). As in other species of suspension-feeding bivalves, the gill tissue in *Mytilus* spp. is a core interface for uptake of dissolved metals, for binding them to metallothionein, incorporation into lysosomes, and further transport in blood plasma and circulating hemocytes (Marigómez et al. 2002). Hence, the CuONPs were accumulated in the gills (at ~12.51 $\mu g.g^{-1}$) and hepatopancreas (at ~26.9 $\mu g.g^{-1}$). Overall,

Table 11.3 Physiological effects and HSP induction found in clams, oyster and mussels after trace metal exposure

Biological models tested and authors	Contaminant	Nominal concentration	Number of individuals exposed per replicate	Time of exposure and temperature	HSP or HSP-genes expression	Other physiological effects observed
Blue mussel *Mytilus edulis* (Sanders et al. 1991)	Cu^{2+}	1, 3.2, 10, 32 and 100 μg.L^{-1}	4 mussels into 750 mL per replicate	7 days, 15 °C	Upregulation of HSP60 was detected with 3.2 μg. L^{-1} or more. The mass of total protein tended to decrease with increasing doses of Cu^{2+}.	Clearance rates were significantly reduced at 32 and 100 μg.L^{-1}, assimilation efficiency was altered at 32 μg.L^{-1}, feeding was reduced by higher doses, SFG (scope for growth: Clearance rate, respiration rate and assimilation efficiency) was significantly reduced at 32 and 100 μg. L^{-1}.
Blue mussels *Mytilus edulis* (Tedengren et al. 1999)	Cd^{2+} and Cu^{2+}	10 μg.L^{-1} (Cu^{2+}) and 100 μg.L^{-1} (Cd^{2+})	30 mussels (first experiment), 50 mussels/5 L (2 replicates for treatment in the second experiment)	1-week (first experiment: 5, 10, 24, 48, 96 and 168 h) and 12.5 h (second experiment); 16 °C	Cadmium induced HSP70 expression. Copper seemed to induce higher levels of HSP71 and HSP73 in both populations tested (from Baltic and North Sea) but the differences were not significant	Baltic mussels accumulated 150 to 200 μg.g^{-1} at 100 h and 1-week respectively. North Sea population accumulated less ~100 μg. g^{-1} from 100 h to 1-week of exposure. Mortality also raised after 7 d of Cd^{2+} exposure particularly for Baltic population.

Blue mussels *Mytilus edulis* (Radlowska and Pempkowiak 2002)	Cd^{2+}, Pb^{2+}, Cu^{2+}	10 and 200 μg.L^{-1} (Cd^{2+}), 200 μg.L^{-1} (Cu^{2+}), 200 μg.L^{-1} (Pb^{2+}) or a mixture of Cu^{2+} (100 μg.L^{-1}) and Cd^{2+} (100 μg.L^{-1})	20 mussels into 3 separate tanks (dilution not provided)	8 days, 10 °C	With 200 μg.L^{-1}/Cd^{2+}, elevated levels of HSP70 appeared after 2 and 8 days. A significant increase in HSP70 levels happened with Pb^{2+} the 8th day. Cu^{2+} also induced HSP70 at 200 μg.L^{-1}. The highest expression of HSP70 occurred with long exposure to mixtures.	Mussels accumulated around 100 μg.g^{-1} d.w. Of Cd^{2+} and Cu^{2+} at day 1. At day 7 around 400 μg.g^{-1} d.w. (Cu^{2+}) and 500 μg/g^{-1} d.w. (Cd^{2+}) were accumulated.
Mussel *Mytilus galloprovincialis* (Franzellitti and Fabbri 2005)	Hg^{2+} and Cr^{2+}	150 μg.L^{-1} (Hg^{2+}) and 1, 10, and 50 ng.L^{-1} (Cr^{2+})	10 mussels per treatment (dilution not provided)	8 h to 6 days (Hg^{2+}) and 1-week (Cr^{2+}), 16 °C	Exposure to Cr^{2+} affected HSP70 gene expression by inducing HSC70 and inhibiting HSP70 transcription levels; HSC70 induction at 1 ng/L^{-1}/Cr^{2+} was the highest detected. Hg^{2+} provoked a significant induction of HSP70 expression in the early 8 h.	–

(continued)

Table 11.3 (continued)

Biological models tested and authors	Contaminant	Nominal concentration	Number of individuals exposed per replicate	Time of exposure and temperature	HSP or HSP-genes expression	Other physiological effects observed
Mussel *Mytilus galloprovincialis* (Franzellitti and Fabbri 2006)	Hg^{2+} and CH_3Hg^+	150.4 $\mu g.L^{-1}$	10 mussels per treatment (dilution not provided)	8 h to 6 days, 16 °C	Changes in HSP70 and HSC70 appeared after 6d. HSP70 expression induced at 8 h by Hg^{2+} reaching a maximum after 24 h and returning to basal levels after 6 days. CH_3Hg^+ inhibited HSP70 expression but induced HSC70 expression after 8 h up to the end.	MXR-related gene products Pgp, Mrp2 and Mvp showed some changes during exposures.
Mussel *Mytilus galloprovincialis* (Varotto et al. 2013)	Cd^{2+}, Cu^{2+}, Hg^{2+}	Equimolar mixtures of 5.6–22.5 $\mu g.L^{-1}$ (Cd^{2+}), 3.2–12.7 $\mu g.L^{-1}$ (Cu^{2+}), 10.0–40.1 $\mu g.L^{-1}$ (hg^{+2})	5 mussels/tank (1 L seawater/mussel)	48 h, 18 °C	Transcripts of HSP90 and HSP24 were significantly modulated at all treatments. At the lowest doses the most overexpressed transcripts were HSP90, small HSP24. The most over-expressed at higher dose was also HSP24.	Transcriptional changes related to protein synthesis and turnover, ion homeostasis, cell cycle regulation and apoptosis and intracellular trafficking.
Mussel *Mytilus galloprovincialis* (Gomes et al. 2013)	Bare AgNPs with <100 nm	10 $\mu g.L^{-1}$	5–6 mussels for each treatment (2–3 mussels/L)	15 days, not given	HSP70 down-regulated at 16-fold in the gills	Silver accumulation was ~5.9 ± 1.4 $\mu g.g^{-1}$ dw. Alteration of proteins associated to stress response and involved in cytoskeleton and cell structure (actin and α-tubulin)

Mussel *Mytilus galloprovincialis* (Negri et al. 2013)	Cu^{2+}	2.5, 5, 10 and 40 μg.L^{-1}	Groups of mussels (number of individuals not provided)	4 days; 16 °C, 20 °C and 24 °C	Log 2-fold change for HSP expressed in gills with a temperature gradient corresponded to: HSP70 (0.82), sHSPp26 (0.63), HSP90 (1.19), HSP27 (0.93) under 20 °C + Cu^{2+} treatment. sHSP26 (0.78) and HSP27 (2.00) under 24 °C + Cu^{2+}.	Accumulation in hepatopancreas was ~2–4 μg.Kg^{-1} ww at 24 °C while ~10 μg.Kg dw accumulated at the same temperature. A strong dose-dependent response at 24 °C and 40 μg.L^{-1} on lysosomal membrane stability.
Mussel *Mytilus galloprovincialis* (Gomes et al. 2014)	Bare CuONPs with <50 nm	10 μg.L^{-1}	2–3 mussels/L (50 in 20 L in a triplicate)	15 days, 17.8 ± 1.1 °C	Over-expressed in the gills and digestive gland (>20-fold)	Cooper accumulated at 12.5 μg.g^{-1} in the gills and 26.9 μg.g^{-1} in the digestive gland. Slow elimination rate of Cu^{2+}.
Mussel *Mytilus galloprovincialis* (Liping et al. 2014)	Cd^{2+}	5 and 50 μg.L^{-1}	25 individuals in 30 L (9 groups)	24, 72, and 96 h, 17 °C	HSP22 mRNA moderately expressed in the mantle and gonads, much lower expression in the gills, hepatopancreas and hemocytes. HSP24 highest expression level in the gonad tissue, and the lowest in hemocytes.	Glutathione peroxidase and catalase activities increased under 5 and 50 μg.L^{-1} treatments.

(continued)

Table 11.3 (continued)

Biological models tested and authors	Contaminant	Nominal concentration	Number of individuals exposed per replicate	Time of exposure and temperature	HSP or HSP-genes expression	Other physiological effects observed
Mussel *Mytilus trossulus* (La Porte 2005)	Sodium arsenite (As^{III}), sodium arsenate (As^{V})	0.010 to 1000 $mg.L^{-1}$ (As^{III}), 1 to 10 $mg.L^{-1}$ (As^{V})	280 mussels into 14 tanks (dilution not provided)	24 h, 13 °C	As^{III} induced HSP70 in a concentration-dependent manner. Equivalent induction was given by 1 $mg.L^{-1}$ As^{V}.	–
Hard-shelled mussel *Mytilus coruscus* (Li et al. 2014)	Cu^{2+}, Cd^{2+}	20 $\mu g.L^{-1}$ (Cu^{2+}) and 200 $\mu g.L^{-1}$ (Cd^{2+})	50 mussels/10 L per treatment	3, 5, 10, 15, 20, 25 and 30 days, 25 °C	A time-dependent pattern observed. HSP70 mRNA expression increased with maximum peak at 15 d in Cu^{2+} exposure (9.41-fold). For Cd^{2+} the highest expression level reached with 10.82-fold at day 9.	–
Brown mussels *Perna perna* (Franco et al. 2006)	Zn^{2+}	0.654 to 6.538 $mg.L^{-1}$	4 mussels for each group (dilution not provided)	48 h, not given	HSP60 protein levels were significantly induced by Zn^{2+} (30 $\mu g.L^{-1}$) in gills	Antioxidant defenses up-regulated at lower levels of Zn^{2+} (0.654 $\mu g.L^{-1}$); high lipid peroxidation end products and low protein thiol and glutathione levels at the highest concentrations.

Manila clam *Venerupis philippinarum* (Li et al. 2010)	Cd^{2+}	10 and 40 $\mu g.L^{-1}$	50 clams per condition (dilution not provided)	24, 48 and 96 h; 20–22 °C	HSP-1 (20.29 kDa) and HSP-2 (24.66 kDa) were identified. HSP-1 was down-regulated by 10 $\mu g.L^{-1}/Cd^{2+}$ but at 96 h HSP-1 levels went back to the original levels., HSP-2 expression downregulated from 24 h to 48 h by 40 $\mu g.L^{-1}$, but got recovered to 96 h.	–
Manila clam *Venerupis philippinarum* (syn. *Ruditapes philippinarum*) (Wu et al. 2013)	Na_2HAsO_4 (As^V)	20 $\mu g.L^{-1}$	72 specimens in total (12 in each replicate/20 L)	3 days, 22 °C	Down-regulation of HSP27 at 1.5-fold.	At different salinities (31.1, 23.3 and 15.6 psu), accumulation rates 0.44, 0.46, and 0.51 $\mu g.g^{-1}$ dw found in gills, respectively. Under normal salinity, cytoskeletal actin and alpha-1 tubulin were down-regulated (cell injury), disturbance in energy metabolism and/or osmotic regulation, apoptosis, oxidative stress occurred

(continued)

Table 11.3 (continued)

Biological models tested and authors	Contaminant	Nominal concentration	Number of individuals exposed per replicate	Time of exposure and temperature	HSP or HSP-genes expression	Other physiological effects observed
Manila clam *Venerupis philippinarum* (Boscolo-Papo et al. 2014)	Cd^{2+}, Cu^{2+}, hg^{+2} and Pb^{2+}	Not provided	50 animals were caught per site (10 for each one of the 4 sites), not given	Specimens were collected from contaminated sites, temperature not given	Tissues analyzed showed immunopositivity to HSP70, levels of protein expression were not assessed.	Metal accumulation (mg. Kg^{-1}) was recorded as follows: Cd^{2+}:0.09 to 0.19; Cu^{2+}: 1.25 to 2.06; hg^{+2}: 0.05 to 0.07; Pb^{2+}: 0.11 to 0.57. Haemocytes exhibited an immunopositivity to cytochrome P4501A (CYP1A).
Asian marine razor-clam *Sinonovacula constricta* (Zhang et al. 2013)	Cd^{2+} and Pb^{2+}	1 and 10 mg.L^{-1}	60 clams per conditions (dilution not provided)	36 h, 21 ± 1 °C	Small Hsp were progressively downregulated in hemocytes and gill under Cd^{2+} exposure.	–
Hard clam *Meretrix meretrix* (Li et al. 2013)	Cd^{2+}	40 μg.L^{-1}	30 clams per condition (dilution not provided)	24, 48 and 96 h; 20 °C	Expression level of HSP20 was elevated in hepatopancreas, gills, and mantle tissues. Lower response occurred in hemocytes, gonad and adductor muscle.	–

Zhikong scallop *Chlamys farreri* (Gao et al. 2007)	Cd^{2+}, Pb^{2+}, Cu^{2+}	50, 100 and 200 μg.L^{-1} (Cd^{2+}), 200, 300 and 500 μg.L^{-1} (Pb^{2+}), 5, 10 and 15 μg.L^{-1} (Cu^{2+})	25 scallops in each tank (dilution not provided)	10 to 20 days (Cu^{2+}), 18 °C	All metals at different concentrations induced HSP90 gene expression.	_
Bay scallop *Argopecten irradians* (Zhang et al. 2010b)	Cu^{2+}, Pb^{2+}, Cd^{2+}	5,10 and 15 μg.L^{-1} (Cu^{2+}), 200, 300 and 500 μg.L^{-1} (Pb^{2+}), 50, 100, 200 μg.L^{-1} (Cd^{2+})	250 scallops separated in 10 groups (dilution not provided)	10 days, 18 °C	Cu^{2+}, Pb^{2+} and Cd^{2+} induced the expression of HSP22 gene, but not in a concentration-dependent manner.	_
European flat oyster *Ostrea edulis* (Piano et al. 2004)	Cd^{2+} and Zn^{2+}	100–500 μg.L^{-1}	10 oysters per treatment (dilution not provided)	1 week, _	A dose-dependent enhancement of HSP70 in gills within 7 d and overexpression in hepatopancreas observed. Zn^{2+} not inducing HSP70 expression. Combined metals at the highest concentrations induced an over-expression of HSP70.	Expression of metallothioneins by 500 μg.L^{-1} of Cd^{2+} in gills and hepatopancreas had a 208% and 160% increase, respectively. Zn^{2+} did not cause any expression. Combined exposures also induced the metallothioneins, but less than Cd^{2+} single exposure.
Eastern oysters *Crassostrea virginica* (Ivanina et al. 2008)	Cd^{2+}	5.62, 56.2 and 224.8 mg.L^{-1}	Not given (cells from 2–5 individuals were analyzed)	4 h, 20 °C	Contrasting levels of HSP60 and HSP70 in gills and hepatopancreas. No change in HSP90 observed.	Metallothioneins mRNA expression increased 2–8-fold only in the hepatopancreas.

(continued)

Table 11.3 (continued)

Biological models tested and authors	Contaminant	Nominal concentration	Number of individuals exposed per replicate	Time of exposure and temperature	HSP or HSP-genes expression	Other physiological effects observed
Pacific oyster *Crassostrea gigas* (Moraga et al. 2005)	Cd^{2+} and Cu^{2+}	25.4 μg.L^{-1} Cu^{2+} and 450 μg.L^{-1} Cd^{2+}	20 oysters (dilution not provided)	3, 7 and 15 days; 15 °C	HSP70 detected in organisms exposed 3 and 15 days in gut epithelia close and some cells at the gill filaments.	Metallothioneins detected in the apical part of cells localized in the stomach and gut epithelia; some cells in gill filaments and oocytes.
Pacific oyster *Crassostrea gigas* (Choi et al. 2008)	Cd^{2+}	10, 50, or 100 μg.L^{-1}	25 oysters into 25 L plastic aquaria filled with filtered natural seawater (control) or Cd^{2+}-test solution	1, 3, 7 and 11 days of exposure; 20 ± 1 °C	With 0.01 ppm, HSP90 mRNA reached the highest level at day 11. With 0.05 or 100 μg.L^{-1}/Cd^{2+} HSP90 mRNA expression increased until day 7 and decreased up to day 11.	Levels of glutamate oxaloacetate transaminase and glutamate pyruvate transaminase in the hemolymph increased by day 7 at 0.05 and 0.1 ppm. Metallothionein mRNA expression levels increased in the gill and digestive gland in a dose and time-dependent manner.
Pacific oyster *Crassostrea gigas* (Ivanina et al. 2009)	Cd^{2+}	50 μg.L^{-1}	Total number of individuals not provided, 2 ml per oyster	45–50 days, 12 °C	Interactions between Cd^{2+} effects and temperature on HSP69 levels were significant. Exposure to high temperatures did not further induce HSP69 in Cd^{2+}-treated oysters.	Oysters from Washington and North Carolina had about 80 μg.g^{-1} dry mass of Cd^{2+} while those from Texas had ~40 μg.g^{-1} in the gills. Metallothioneins were induced to 38–94-fold in Cd^{2+} treatments.

Pacific oyster *Crassostrea gigas* (Meng et al. 2017)	Cd^{2+}	6 μg.L^{-1}	20 oysters into 20 L tanks	9 days, 18 ± 2 °C	Induction of proteins HSP70, HSP60, HSP60 and GRP78 and the protein ubiquitination process.	Induction of anti-oxidant responses (like glutathione metabolism), inhibition of DNA glycosylase and gap-filling and ligation enzymes expressions, inhibiting DNA repair capacity.
Oyster *Saccostrea glomerata* (Thompson et al. 2015)	Cd^{2+}, Cu^{2+}, Pb^{2+} and Zn^{2+}	Range in 3 sediment sites: 2.75–15.28 mg. L^{-1} (Cd^{2+}), 144.2–2238 mg.L^{-1} (Cu^{2+}), 649.7–8810 mg.L^{-1} (Pb^{2+}), 1114–5405 mg.L^{-1} (Zn^{2+}) - dissolved: 30–50 μg.L^{-1} (Cd^{2+}), 130–370 μg.L^{-1} (Cu^{2+}), 110–290 μg. L^{-1} (Pb^{2+}), 970–4300 μg.L^{-1} (Zn^{2+})	7 oysters transplanted in each site	4 days after 10 days of acclimation in sites with low metal concentration	Most of the proteins expressed were related to stress responses (11%) and cytoskeleton like actin (45%).	–

(continued)

Table 11.3 (continued)

Biological models tested and authors	Contaminant	Nominal concentration	Number of individuals exposed per replicate	Time of exposure and temperature	HSP or HSP-genes expression	Other physiological effects observed
Oyster *Crassostrea hongkongensis* (Luo et al. 2014)	Cu^{2+}, Zn^{2+}, Pb^{2+}, Cd^{2+} and Ni^{2+}	0.8 to 17. μg.L^{-1} (Cu^{2+}), 3.4–30.1 μg. L^{-1} (Zn^{2+}), 0.24–1.77 μg.L^{-1} (Pb^{2+}), 0,05–0.14 μg.L^{-1} (Cd^{2+}), 2.0–31 μg.L^{-1} (Ni^{2+})	30 individuals from each site	–	70 kDa heat shock cognate protein 2 was significantly up-regulated in the gills of oysters long-term exposed to relatively slight metal contaminations, but it was down-regulated under long term severe metal contamination.	Mantle, gills and digestive gland were the tissues the most targeted by metals. Up-regulation of metallothionein expression and more granular cells were detected in the gills.
Pearl oyster *Pinctada margaritifera* (Yannick et al. 2017)	Cd^{2+}, Cr^{+2}	1.02 and 10.2 μg.L^{-1} (Cd^{2+}), 1.15 and 11 μg.L^{-1} (Cr^{+2})	124 individuals monitored *in situ* after experimental contamination	1, 4 and 17 days for lower concentrations and 1, 4 and 8 days for higher concentrations; 27,9 °C	With 1 μg.L^{-1} of Cd^{2+} and Cr^{+2} only 3 genes out of 12 had significant expression, one of them was HSP90 gene. At higher doses of Cd^{2+} (10 μg.L^{-1}) great expression of HSP70 and HSP90 arose 1 day after the contamination, but it did not persist.	Accumulation of 10 to 60 mg.Kg^{-1} dw (for 1 μg. L^{-1}) after 17 d of contamination and 18 to 85 mg.Kg^{-1} (for 10 μg.L^{-1}) after 7 days of contamination by Cd^{2+}. Metallothionein was up-regulated by Cd^{2+} and Cr^{+2} while antioxidant enzymes were down-regulated.

there was a low elimination rate of Cu^{2+} and 17 overexpressed proteins involved with cytoskeleton and cell structure, transcription regulation, stress response, oxidative stress, energy metabolism, apoptosis, proteolysis, adhesion and mobility. Another metal to provoke similar effects in *M. galloprovincialis* along 15 days was silver in nanoparticulate form. Silver nanoparticles (AgNPs <100 nm, 10 μg.L^{-1}) led to an alteration of proteins associated to stress response and involved in cytoskeleton and cell structure. In this case, HSP70 was found to be down-regulated at 16-fold in the gills (Gomes et al. 2013).

In shorter exposures (8 h to 6 days), mercury in both Hg^{2+} and CH_3Hg^+ forms (150.45 μg.L^{-1}) provoked some changes in HSP70 and HSC70 in *M. galloprovincialis* at 16 °C (Franzellitti and Fabbri 2005, 2006). HSP70 expression was induced at 8 h by Hg^{2+} reaching a maximum after 24 h and returning back to basal levels after 6 days. Comparatively, CH_3Hg^+ inhibited HSP70 expression, but induced HSC70 expression after 8 h up to the end. It turns out that CH_3Hg^+ was the main chemical form to cause overexpression of MXR (multixenobiotic resistance genes)-related gene products Pgp (p-glycoprotein), Mrp2 (multidrug resistance-associated protein 2 and Mvp (major vault protein). Likewise, very much lower doses of Cr^{2+} (1, 10 and 50 ng.L^{-1}) in a one-week exposure affected HSP70 gene expression by inducing HSC70 and inhibiting HSP70 transcription levels. It is important to keep in mind that methylmercury (CH_3Hg^+) is very toxic (more than ten times as toxic as inorganic Hg^{2+}) and its toxicity emerges in lower doses (~μg.L^{-1}) for aquatic biota. It has been found that even some cellular proteins can intensify CH_3Hg^+ toxicity in cells (Lee et al. 2016). Total mercury in natural surface waters (marine and freshwater) ranges from <1 to 20 ng.L^{-1} while methylmercury concentrations are usually less than 1 ng.L^{-1} (CEQG 2003). For aquatic invertebrates, the chronic toxicity rises in a concentration range of 0.04–1.14 μg.L^{-1}. Hence, gene induction at higher concentrations of CH_3Hg^+ might likely indicate lethal pathways rather than protective mechanisms.

Varotto et al. (2013) reported that after 48 h, equimolar mixtures of Cd^{2+}, Cu^{2+} and Hg^{2+} led to a significant modulation of transcription of HSP90 and HSP24 in gills of *M. galloprovincialis* at 18 °C. With relatively high dosages tested (~50 nM, or 5.62 μg.L^{-1}/Cd^{2+}; 3.17 μg.L^{-1}/Cu^{2+} and 10.0 μg.L^{-1}/Hg^{2+}) the most overexpressed transcripts were HSP90, HSP24.1, and the translation initiation factor 5A. The highest doses as 200 nM led to an overexpression of both genotoxic effects and apoptotic cell death transcripts. Indeed, the bioconcentration factor for Cd^{2+} (69.3) was lower than Cu^{2+} (242.0) and Hg^{2+} (561.4), which highlights cadmium toxicity for this species in lower concentration uptaken. Along similar lines, Liping et al. (2014) proved that HSP22 mRNA expression *M. galloprovincialis* was rather detected in muscle tissue; moderately found in the mantle and gonad, and barely expressed in the gills, hepatopancreas and hemocytes after low to moderate doses in short-term exposures (5 and 50 μg.L^{-1} of Cd^{2+}) at 17 °C. Otherwise, HSP24 highest expression level occurred in the gonad tissue, and the lowest in hemocytes. HSP22 mRNA expression was significantly down-regulated to 0.2-fold at 5 μg.L^{-1} at 24 h. After this, expression levels got recovered. No difference in HSP22 expression appeared at 50 μg.L^{-1}. HSP24 levels also significantly decreased about 0.5–0.6-fold at 24 h and 48 h in 5 μg.L^{-1} treated groups. After exposure period, expression level got recovered to original levels at 96 h. Finally with 50 μg.L^{-1} HSP24 levels decreased to 0.4-fold at 48 h and slightly rose again at 96 h.

Some experiments using blue mussels *Mytilus edulis* (Mytilidae) have similarities with the HSP70 stress-related mechanisms described for *M. galloprovincialis*. With 200 μg.L^{-1}/Cd^{2+}, blue mussels had around 175% and 350% of HSP70 raised levels in gills over the controls after 2 and 8 days at 10 °C, respectively (Radlowska and Pempkowiak (2002). *M. edulis* also reacted with a significant increase in HSP70 levels once exposed to Pb^{2+} (200 μg.L^{-1}) or 260% over the controls on the 8th day. In turn, Cu^{2+} (200 μg.L^{-1}) induced HSP70 at 240% compared to controls. Cooper was the least potent stressor compared to cadmium and lead. As depicted by the authors, Cd^{2+} accumulation was higher than Cu^{2+} in the gills of *M. edulis* at 7th day even though the accumulation started with around 100 μg.g^{-1} of Cd^{2+} and Cu^{2+} (both at 200 μg.L^{-1}) at day 1. After 3 and 5 weeks of depuration, mussels exhibited a high degree of inter-individual variability in HSP70 levels. Similar results were found by Tedengren et al. (1999) with Baltic and North Sea blue mussels exposed to Cd^{2+} (100 μg.L^{-1}) and Cu^{2+} (10 μg.L^{-1}) at 16 °C. In fact, Baltic mussels had slower levels of HSP70 after 1 week-exposure. They reached the highest tissue concentrations of Cd^{2+} showing twice as high ammonia excretion rate, respiration rates and mortality compared to North Sea blue mussels. A significant increase of Cd^{2+} concentration in mussel tissues was observed for both populations. Baltic mussels accumulated 75–100% more Cd^{2+} though (from 150 to 200 μg.g^{-1} at 100 h and 1-week respectively). North Sea population accumulated around 100 μg.g from 100 h to 1-week of exposure. A minor increase of Cu^{2+} in tissue concentration during short exposures was detected without any differences between populations.

Only a few studies have discussed the role of HSP60 against metal stress in mussels. It has been demonstrated that in *M. edulis* a significant increase of HSP60 occurred from the threshold 3.2 μg.L^{-1}/Cu^{2+} to higher doses throughout 7 days at 15 °C (Sanders et al. 1991). As an additional effect, the mass of total proteins tended to decrease with increasing Cu^{2+} concentration. The brown mussel *Perna perna* (Mytilidae) also had HSP60 protein levels significantly induced in gills by Zn^{2+} (30 μg.L^{-1}) within 48 h. High concentrations (653.8 μg.L^{-1}) activated antioxidant defenses, doses even much higher (6538 μg.L^{-1}) increased lipid peroxidation end products and reduced protein thiol and glutathione levels (Franco et al. 2006).

11.7.2.2 Clams (Veneroidea and Tellinoidea)

Differently from marine mussels who get attached to consolidated substratum like rocks, clams inhabit non-consolidated substratum such as mud, porous sand, sand-gravel beaches, etc. Living burrowed in soft sediments as mudflats or mangrove swamps where anoxic conditions are often present, some clams are continuously exposed to hydrogen sulfide at levels up to 384 mg.L^{-1} (Joyner-Matos et al. 2006). Hydrogen sulfide is known to be a highly reactive toxin able to diffuse freely across respiratory surfaces inhibiting cytochrome *c* oxidase and many other enzymes (Bagarinao 1992). In such conditions, small Hsp can be rather produced than HSP70 to handle the chemical stress in some clam species (Joyner-Matos et al. 2006).

In Manila clams *Venerupis philippinarum* (Veneroidea, Veneridae) collected from metal-polluted areas, for example, tissues analyzed showed immunopositivity to HSP70 (Boscolo-Papo et al. 2014). In addition, small Hsp expression was detected in *V. philippinarum* Cd^{2+}-treated throughout 96 h at ~22 °C (Li et al. 2010). Two ~20 kDa forms were identified responding to cadmium toxicity in a short time: HSP-1 (20.29 kDa) and HSP-2 (24.66 kDa). After 24 h the expression level of HSP-1 was down-regulated to 0.4-fold with 10 μg.L^{-1}/Cd^{2+}, but at 96 h HSP-1 levels went back to the original level. With 40 μg.L^{-1}/Cd^{2+}, no changes in Hsp level were observed during 24 h. Nonetheless, it decreased significantly up to 96 h. Regarding HSP-2, the expression level was downregulated from 24 h to 48 h under 40 μg.L^{-1}/Cd^{2+} treatment, getting recovered later at 96 h. With 10 μg.L^{-1}, mRNA-Hsp expression levels were raised to 2.7-fold only after 48 h compared to controls. On the contrary, the same species showed down regulation of HSP27 in 3 days once exposed to 20 μg.L^{-1}/Na_2HAsO_4 at 22 °C (Wu et al. 2013). Rising levels of arsenic accumulation in gills occurred as a function of decreasing salinities.

In hard clam *Meretrix meretrix* (Veneroidea, Veneridae), a 40 μg.L^{-1}/Cd^{2+} 96 h-exposure made the expression level of HSP20 be the strongest in hepatopancreas, gills, mantle tissues, and to a lesser degree in hemocytes, gonad and adductor muscle at 20 °C (Li et al. 2013). Unlike *M. meretrix*, the intertidal clams *Sinonovacula constricta* (Tellinoidea, Solecurtidae) progressively downregulated small Hsp transcripts (~15–30 kDa) in hemocytes and gills after 1 and 10 mg.L^{-1}/Cd^{2+} treatments. It occurred by more than 20-fold at ~17.8 °C throughout 36 h and at ~21 °C (Zhang et al. 2013). As an alternative response, hemocytes and gill cells challenged with 1 and 10 mg.L^{-1}/Pb^{2+} had an overexpression of 15–30 kDa mRNA at 12 h, which dropped drastically at 36 h.

11.7.2.3 Scallops (Pectinoidea)

Scallops *Chlamys farreri* and *Argopecten irradians* can display some different life habits even if they belong to the same clade (Pectinoidea, Pectinidae). While *C. farreri* can be temporarily attached to a hard substratum by byssus threads and later release them to get reoriented again, *A. irradians* are free-living and they normally rest above soft sediment or hard substratum (Alejandrino et al. 2011). Small HSP22 has been suggested to play a critical role in response to bacterial challenge in haemocytes of *C. farreri* (Zhang et al. 2010a). Comparatively, *A. irradians* have a clearly time-dependent expression pattern of HSP70 gene expression when infected with bacteria reaching a peak at 8 h but dropping at 16 h (Song et al. 2006). For this species HSP70 seems to be important to control the effects of a biological agent infection. Rising concentrations of Cd^{2+} (50–200 μg.L^{-1}), Cu^{2+} (5–15 μg.L^{-1}) and Pb^{2+} (500 μg.L^{-1}) induced HSP22 gene in *A. irradians* along 10 days at 18 °C as well, but not in a concentration-dependent way (Zhang et al. 2010b). With Cu^{2+}, expression level in haemolymph increased from 5.8-fold (50 μg.L^{-1}) to 48-fold (100 μg.L^{-1}) and dropped to 24.3-fold (150 μg.L^{-1}). In 200 μg.L^{-1}/Pb^{2+} treatments, HSP22 mRNA was 50 times higher than that of 200 μg.L^{-1}/Cd^{2+}. For a concentration of 100 μg.L^{-1}/Cd^{2+} HSP22 mRNA expression level was 67.2-fold and decreased to 6.0-fold with 200 μg.L^{-1}. In a like manner, all these metals at different doses (μg. L^{-1}) induced HSP90 gene expression in *C. farreri* from 10 days to 20 days at 18 °C (Gao et al. 2007). As expected, a concentration-dependent tendency was established between HSP90 mRNA and Cd^{2+} after 10 days. For Pb^{2+}-treated organisms, HSP90 mRNA steadily increased but a much lower upregulation appeared in 200 and 300 μg.L^{-1} groups (Gao et al. 2007). The strongest mRNA level was showed at 10 μg.L^{-1}/Cu^{2+}-treated groups.

11.7.2.4 Oysters (Ostreoidea and Pterioidea)

Differently from scallops, oysters are sessile animals living permanently attached to solid substratum and must cope with harsh and drastically changing conditions like temperature and salinity variations, and dissecation following the tidal cycles. As discussed by Zhang et al. (2012), an expansion of genes coding for HSP70 and inhibitors of apoptosis exist in some oysters like *Crassostrea gigas* (Ostreoidea, Ostreidae) as a main apparatus to adapt them to common stressors inherent to sessile life in exposed environment, which eventually includes metal pollution discharges.

Using *C. gigas* exposed to 25.4 μg.L^{-1}/Cu^{2+} and 450 μg.L^{-1}/Cd^{2+} at 15 °C, HSP70 can be detected in individuals exposed 3 and 15 days in epithelia close to membrane of the basal portion of intestine, stomach, esophagus cells and some cells at the gill filaments (Moraga et al. 2005). In turn, metallothioneins were detected in the apical part of cells localized in the stomach and gut epithelia, in some gill filament cells and oocytes. The HSP70 regulation in *C. gigas* seems to be severely affected when temperature and contamination by Cd^{2+} come together. Along 45–50 day exposure, HSP69 (an inducible isoform of HSP70) can be upregulated by ~50 μg.L^{-1}/Cd^{+2} in *C.gigas* at 12 °C (Ivanina et al. 2009). Notwithstanding interactions between Cd^{2+} effects and rising temperature (20–40 °C) on HSP69 provoked detrimental effects in *C. gigas* (Choi et al. 2008). In general, exposing Cd^{2+}-treated oysters from three different populations to high temperature did not further induce HSP69 in gill and hepatopancreas. Both stressors have the potential to simultaneously suppress the cytoprotective upregulation of molecular chaperones. Correspondingly, HSC72 and HSC77, HSP90 and HSP60 were not significantly upregulated either. In this same species, Cd^{2+} can also significantly increase HSP90 mRNA expression in the gill and digestive gland in a dose- and time-dependent manner from 1 to 11 days at ~20 °C (Choi et al. 2008).

The impairment of cadmium in protein metabolism and DNA in *C. gigas* has received some attention. Cd^{2+} (6 μg.L^{-1}) induced several Hsp gene family-related proteins (with HSP70, HSP90, HSP60 and GRP78, a 78-kDa glucose-regulated protein) but also the protein ubiquitination process within 9 days at 19 °C (Meng et al. 2017). Protein ubiquitination is a removal of unrepaired proteins through synthesis of ubiquitin-conjugating proteins (E1, E2 and E3). Proteins E1 and E2 were upregulated by Cd^{2+} (6 μg.L^{-1}) whereas mRNA and protein abundance of E3 significantly decreased. It has been suggested that as E3 proteins have sulfhydryl groups in active sites, the molecules might be oxidized by Cd^{2+}. Another possibility is the ATP depletion in oxidative milieu leading to declining levels of ubiquitin conjugates (Meng et al. 2017) (Fig. 11.5).

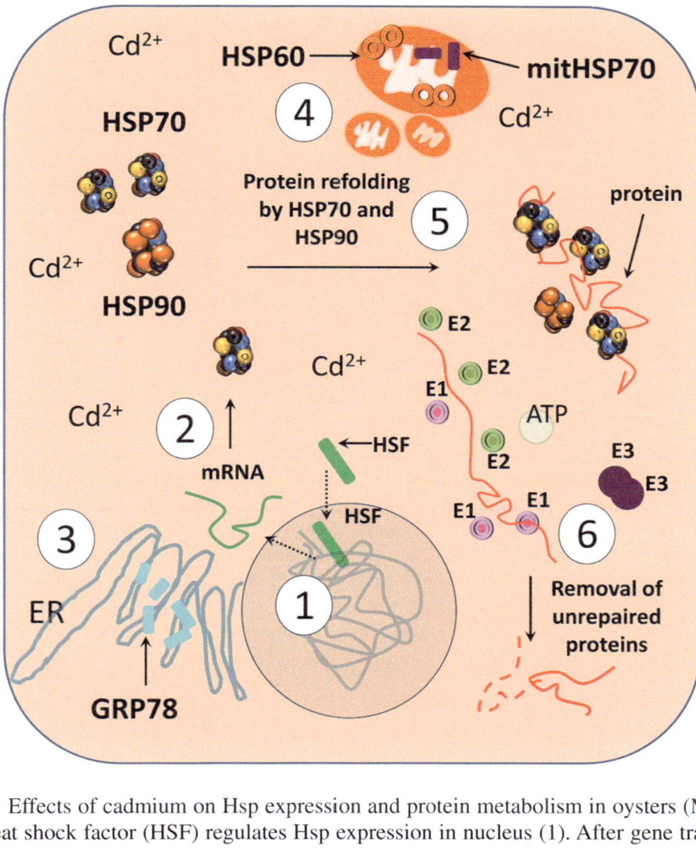

Fig. 11.5 Effects of cadmium on Hsp expression and protein metabolism in oysters (Meng et al. 2017). Heat shock factor (HSF) regulates Hsp expression in nucleus (1). After gene transcription of mRNA-HSP, expression of 70 kDa and 90 kDa proteins begins in cytoplasm (2). While GRP78 (glucose-regulated protein 78) is produced in endoplasmatic reticulum (ER), HSP60 and mitHSP70 (mitochondrial HSP70) are upregulated in mitochondria (3–4). Both HSP70 and HSP90 work together to promote refolding of damaged proteins (5). The initiation of degradation process by ubiquitin proteosome pathway requires ATP (6) (E1: E1 ubiquitin-conjugating enzymes; E2: E2 ubiquitin-conjugating enzymes; E3: E3 ubiquitin-conjugating enzymes). Both mRNA expression and protein abundance of E3 decreased after cadmium treatment. E1 and E3 concentration was increased

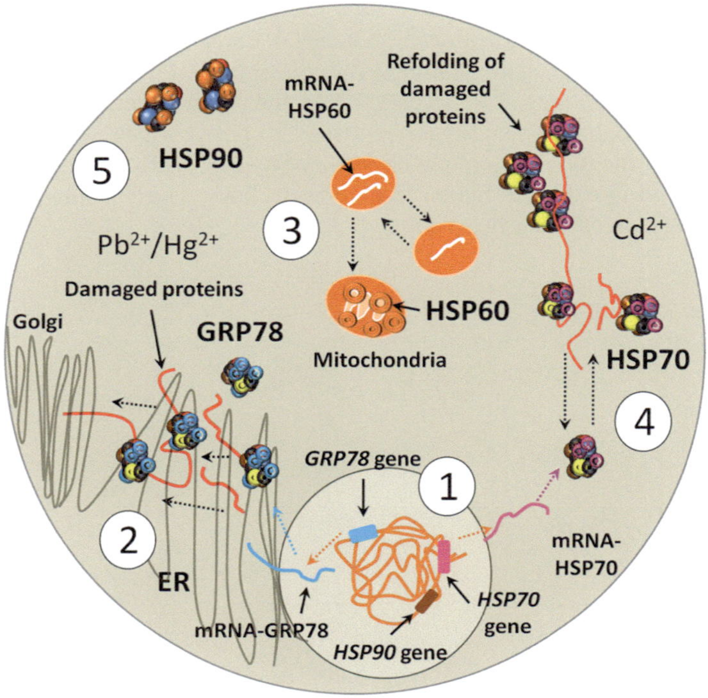

Fig. 11.6 Some mechanisms of HSP activation in planarians facing environmental stresses (starvation and physical harm) and metal contamination (Conte et al. 2011; Zhang et al. 2014; Ma et al. 2014; Plusquin et al. 2012; Cheng et al. 2015). While HSP90 gene is not expressed upon low to high doses of cadmium in *Dugesia japonica* (1), the 78 kDa glucose-regulated protein (GRP78) is upregulated by Pb^{2+} and Hg^{2+} but not by depravity of food (1, 2). Oscillating patterns of HSP60 and HSP70 gene protein expression can be found in *Schmidtea mediterranea* with only HSP70 gene expression persisting through time after cadmium contamination (1, 3–4). Expression of constitutive HSP90 gene can be highly detected in gastrodermis of *D. japonica* in regeneration (5), as well as HSP70 (for both tissue trauma and starvation)

Other species in Ostreoidea have also exhibited interesting mechanisms to mitigate proteomic stress due to trace metal contamination at doses excessively high. Cadmium accumulation in *Crassostrea virginica* (Ostreoidea, Ostreidae) reached around 0.4 mM (45 mg.L^{-1}) to 1 mM (112.4 mg.L^{-1}) in isolated hepatopancreas cells and 200 to 400 μmol.L^{-1} in gill cells following a Cd^{2+} treatment (50, 500 and 2000 μmol·L^{-1} or 5.62; 56.2; 224.8 mg.L^{-1}) for 4 h at 20 °C (Ivanina et al. 2008). Accordingly, contrasting levels of HSP60 and HSP70 arose in gills and hepatopancreas. HSP60 and HSP70 protein expression were significantly raised in Cd^{2+}-exposed gill cells but not in hepatopancreas. Though, metallothioneins mRNA expression increased only in the hepatopancreas. It clearly indicated that metallothionein may provide sufficient protection against Cd^{2+} proteotoxicity in hepatopancreas cells alleviating the need for upregulation of Hsp. Surprisingly, *Ostrea edulis*

(Ostreoidea, Ostreidae) had a dose-dependent enhancement of HSP70 in gills in 100–500 μg.L^{-1}/Cd^{2+} media over 7 days, as well as an over-expression in hepatopancreas (Piano et al. 2004). The expression of metallothioneins caused by 500 μg.L^{-1}/Cd^{2+} in gills and hepatopancreas were 208% and 160% over untreated oysters, respectively.

The species *Crassostrea hongkongensis* (Ostreoidea, Ostreidae) taken from natural locations severely contaminated by Zn^{2+} and Cu^{2+} exhibited Hsp proteins being differently regulated in the gills. 70 kDa heat shock cognate protein 2 (constitutive HSP) was significantly up-regulated in the gills of oysters long-term exposed to relatively slight metal contaminations, but it was down-regulated under a long-term severe metal exposure (Luo et al. 2014). Similarly, HSP70, heat shock cognate protein 5 and HSP60 were detected down-regulated once exposed to long-term metal contaminations. In addition to Hsp levels, there was an appreciable metallothionein expression and more granular cells observed in the gills exposed to long-term and highly heavy-metal contaminated estuary. Metal-rich granules are known to store metals with insoluble forms to detoxify intracellular milieu in mollusks (Marigómez et al. 2002).

It is interesting to note that in the case of Pacific pearl-oyster *Pinctada margaritifera* (Pterioidea, Pteriidae) a great expression of HSP70 and HSP90 rapidly arose 1 day after 10 μg.L^{-1}/Cd^{2+} treatments at 27.9 °C, but it did not persist (Yannick et al. 2017). In a general sense, it can be stated that Hsp expression in oysters is a highly dynamic mechanism involving several factors and chaperones being continuously up- and down-regulated throughout the exposure period and the depuration period. Higher doses might gradually shut-down Hsp expression whereas lower to mid doses will likely stimulate their transcription. In fact, high dosages likely trigger lethal pathways within a few days by disturbing most living mechanisms in cells and tissues which include HSP expression.

11.7.3 Crabs, Shrimps and Small Crustaceans (Decapoda, Anostraca and Copepoda)

Brine shrimps (*Artemia* spp.) cysts normally endure repeated dehydration/rehydration with low mortality rates and no adverse effects on transcription (De Chaffoy et al. 1978). Amongst other physiological and morphological mechanisms, cysts also contain large amounts of small HSPp26 (~20.7 kDa) to tolerate environmental stress (MacRae 2015). Previously heat-shocked (~30 min) swimming nauplii of *Artemia franciscana* at instar II stage (Anostraca) exhibited increased acute tolerance to Cd^{2+} (10–300 mg.L^{-1}) and Zn^{2+} (0.5–150 mg.L^{-1}) at 37 °C (Pestana et al. 2016). After recovering for 6 h at 28 °C from heat shock stress and 48 h-contamination, appreciable levels of HSP70 were still detected after chemical treatment finished. Except for Cd^{2+}, protein levels were gradually reduced over time. In turn, Zn^{2+} exposure had no significant impacts on Hsp expression.

Under natural conditions, many aquatic organisms can be negatively impacted by ultraviolet B radiation (UV-B) which can cause harmful effects in their metabolism including DNA damages and ROS overproduction (Lister et al. 2010). Intertidal copepod *Tigriopus japonicus* (Copepoda, Harpacticidae) once exposed to UV-B radiation are able to subsequently express HSP20 mRNA, HSP70 and HSP90α in a dose-dependent way (Kim et al. 2014a). Conversely, HSP10 and HSP40 mRNA get downregulated at 18 kj.m^{-2} and up. These intriguing results appear compatible with the patterns outlined hereafter. Once 96 h-tested with low doses of Cu^{2+} (0.1–0.5 μg.L^{-1}), Ag^{+} (0.01–0.1 μg.L^{-1}) and Zn^{2+} (0.1–1.0 μg.L^{-1}), HSP70 gene of *T. japonicus* was greatly upregulated at 20 °C (Rhee et al. 2009). With low doses of Hg^{+2} (~10 μg.L^{-1}), *T. japonicus* can also express HSP70 mRNA and HSP20/α-crystalline in 48 h at 22 °C (Wang et al. 2017a). This species experienced different responses under stronger doses of Ag^{+}, As^{+}, Cd^{2+}, Cu^{2+}, Zn^{2+} within 96 h at 20 °C tough (Kim et al. 2014b). Because HSP20 and HSP70 genes were sensitive to all metals, transcriptional levels got elevated. While most of the mRNA Hsp genes analyzed were upregulated in Cu^{2+} media, HSP40 mRNA expression was reduced with 100 μg.L^{-1}. Likewise, transcriptional levels of HSP10 gene were downregulated by 100 μg.L^{-1} (of As^{+}, Cd^{2+} and Zn^{2+}).

Activation of small Hsp and HSP60 transcripts in black tiger shrimp *Penaeus monodon* exposed to Cu^{2+}, Zn^{2+} and Cd^{2+} (0.18, 0.50 and 7.73 mg.L^{-1}) was investigated from 6 to 96 h exposure at 25 °C (Shi et al. 2015). In their gills, HSP60 transcript level was significantly raised at 12 h after cooper contamination, but decreased at 48 h and it went up again at 96 h. By this time, HSP10 reached 7.8-fold also in the gills. Zinc provoked the maximum expression of HSP10 and HSP60 after 12 h in the gills as well (4.8- and 7.2-fold respectively). While Cu^{2+} did not change hepatopancreas Hsp transcript expression, HSP10 and HSP60 were at their strongest levels after 24 h with Zn^{2+}. For cadmium exposure, HSP10 mRNA maximum expression was at 12 h in the gills. Being phylogenetically close to *P. monodon* (Shen et al. 2007), chinese shrimp *Fenneropenaeus chinensis* juveniles (Decapoda, Penaeidae) responded to 3.17 mg.L^{-1} (Cu^{2+}) and 0.281 mg.L^{-1} (Cd^{2+}) contamination with a progressive HSP70 gene upregulation at 25 °C in a short-term exposure (72 h) (Luan et al. 2010). At 12 h, HSP70 gene was up to its highest expression level, but it went sharply down at 72 h. At 24 h exposure, HCS70 gene (constitutive) was slightly induced, but went back to lowest levels at 72 h. No overexpression of HSP70 gene occurred in cadmium media and HCS70 was greatly inhibited. Environmental heat stress can be extremely deleterious to post-larvae of the shrimp *Penaeus vannamei* (Decapoda, Penaeidae), but it also protects them from metal toxicity (Sung et al. 2017). Thereupon shrimps pre-heated at 38 °C had HSP70 expression increased and a better survival rate (5 to 15%) after exposures to Cu^{2+} (LC_{50} = 7.0 mg.L^{-1}) and Zn^{2+} (LC_{50} = 3.5 mg.L^{-1}) up to 24 h.

In the case of euryhaline crab *Portunus trituberculatus* (Decapoda, Portunidae), HSP60 seems to be dynamically expressed to face osmotic pressure changes. HSP60 mRNA expression time oscillates from 12 to 48 h in gills of *P.trituberculatus* under different salinity challenges (Xu and Qin 2012). High HSP60 mRNA expression can decrease from 12 h to 24 h and be upregulated at 48 h and reduced again after

120 h in high salinities. During low salinity stress, mRNA level went up at 12 h, and decreased to normal levels at 48 h but it arose again at 120 h. Under 58 mg.L^{-1}/Cu^{2+} 1 h-treatment *P. trituberculatus* genes HSP90–1 and HSP90–2, which are 78.4% and 86.7% identical with one another, operate in a contrasting way at 20 °C (Zhang et al. 2009). With HSP90-type 1 transcripts being mainly expressed in gills, HSP90-type 2 mRNAs were significantly upregulated in hepatopancreas and gills.

11.7.4 Rotifers (Monogononta)

Rotifers make up a crucial part of the zooplankton in coastal areas where pollution has being often ubiquitous and still a major concern along coasts of developing countries. As for their freshwater counterparts, marine rotifers are less studied as models to evaluate metal toxicity and Hsp expression. So far *Brachionus* spp. has been used to determine transcriptome responses to trace metals. It has been shown that HSP70 in *Brachionus calyciflorus* (Monogononta) is decreasing with aging (Yang et al. 2014), but the difference between HSP70 mRNA expression between young and adult rotifers following metal exposure is still less understood. Noteworthy, H_2O_2 (0.1 or 0.2 mM) induced expression of HSP40, HSP60, HSP70 and HSP90 (~30- to 90-fold) within 1 h in *B. calyciflorus*, but Hsp levels slightly dropped after 2 h (5- to 10-fold). Dietary restriction also improves the mRNA expression of HSP40, HSP60, HSP70 and HSP90 genes (Yang et al. 2014). Unfed rotifers *Brachionus plicatilis* (Monogononta) reached 50% of mortality in 24 h when treated with 80.9 μg.L^{-1}/Cu^{2+}. Fed siblings died over a 133.4 μg.L^{-1} dose in the exposure media (Arnold et al. 2011). Up to 48 h, these concentrations were ~13.4 μg.L^{-1} for starving individuals and ~20.8 μg.L^{-1} for the fed ones. Different levels of transcriptional expression of HSP20, HSP70 and HSP90α-1 appeared in neonates of *Brachionus koreanus* (Monogononta) in a 48 h-exposure with 0.125–1 mg.L^{-1}/Cu^{2+} and 31–500 mg.L^{-1}/Cd^{2+} at 25 °C (Jung and Lee 2012). To cope with Cu^{2+} effects, HSP20, HSP70 and HSP90α-1 mRNA were enhanced in a concentration-dependent manner at 24 h. Under middle- or high-doses of Cu^{2+}, HSP90α-1 m-RNA level was increased by 40-fold compared to controls while HSP27 and HSP90β mRNA levels were only upregulated at 48 h. With cadmium, HSP70 and HSP90α-1 mRNA expression was higher at 24 h and HSP90β mRNA was even modulated 24 h earlier in cadmium treatments.

11.7.5 Cnidarians (Scleractinia and Hexacorallia)

In cnidarians, as many as 5 to 6 major small Hsp isoforms have been identified (Downs et al. 2000). Small Hsp (αB-crystallin, HSP22, HSP23, HSP26 and HSP28) can be induced after metal contamination in some cnidarians studied. Adults of sea anemone *Anemonia viridis* (Hexacorallia, Actiniidae) showed increasing expression of HSP28.6 gene after cadmium and lead 12 h-treatments from low to mid level doses (2, 10 and 50 μg.L^{-1}) at 18 °C (Nicosia et al. 2014). Transcript was up-regulated by all doses of Pb^{2+}, but with a highest expression reached 5.4-fold at 50 μg.L^{-1}. Cadmium elevated HSP28.6 gene expression by 8-fold at 50 μg.L^{-1}. Another sea anemone species, *Nematostella vectensis* (Hexacorallia, Edwardsiidae), seemed to have a metal-specific regulation once challenged with 0.3–0.7 μM of Hg^{2+}, Cu^{2+}, Cd^{2+} and Zn^{2+} in single treatments up to 24 h at ~18 °C (Elran et al. 2014). About 80% of the up-regulated transcripts and 60% of the down-transcripts were similarly regulated in both cooper and mercury test-media. Cadmium and zinc shared less regulated genes though. The most highly regulated genes were those from the HSP20 group after contamination by Hg^{+2}. There was also an upregulation of transcription factors such as Egr1, AP1 and NF-kB just one hour after contamination showing a fast response to metal toxicity. Differently from solitary anemones, corals have a sessile life habit and might be definitely more vulnerable to environmental pollution carried by sediment transportation in estuaries. The 70 kDa chaperone has an important role in protecting coral colonies explants of *Montastraea franksi* (Scleractinia, Merulinidae) exposed to Cu^{2+} at 27 °C. Polyps had HSP70 gene expression significantly upregulated by 100 μg.L^{-1} in 4 h-exposure, and by 30 and 100 μg.L^{-1} in 8 h-exposure (Venn et al. 2009).

11.7.6 Sponges (Poecilosclerida, Chondrosiida and Hadromerida)

Compared to their freshwater siblings, resilience of marine sponges against metal toxicity appears to be a little more understood. A 5 day-exposure of sponge *Crambe crambe* (Poecilosclerida, Crambeidae) to Cu^{2+} (30 and 100 μg.L^{-1}) in laboratory showed that HSP52 and HSP72 were expressed at 17 °C (Agell et al. 2001). Yet significant differences were only found with HSP72 at 30 μg.L^{-1} dose, both were inhibited by 100 μg.L^{-1}/Cu^{2+}. While HSP72 was mobilized in short-term, HSP52 was accumulated in a long-term. In the field measurements, HSP54 is even found highly expressed in native sponges living in Cu^{2+}-contaminated environment and in individuals recently transplanted into it. Correspondingly, some tolerance to cooper seemed to emerge even though sponges accumulated 120 μg.g^{-1}/Cu^{2+} in field and laboratory groups. Comparatively *Chondrosia reniformis* (Chondrosiida, Condrosiidae) did not show any HSP70 mobilization after two treatments with Cu^{2+} (20 and 100 μg.L^{-1}) lasting 5 days at 17 °C (Cebrian et al. 2006). Healthy sponges experimentally transplanted to copper-polluted harbor site had even a lower survival rate 4 months after and no HSP70 expression. The species *Suberites domuncula* (Hadromerida, Suberitidae) incubated with increasing concentrations of Cd^{2+} (0.562 to 562 mg.L^{-1}) up to 6 days had a increase in the number of DNA single strand breaks at 16 °C (Schröder et al. 1999). In this work, a maximum induction of HSP70 was detected with 1 mg.L^{-1}/Cd^{2+} after 12 h, the same time with the maximum of DNA damage observed.

11.8 Conclusions

Our knowledge about Hsp modulation and induction in aquatic invertebrates facing metal contamination has greatly advanced in the last two decades. Species inhabiting different environments such as freshwater lakes, streams, estuaries and coastal ecosystems exhibit their own mechanisms to dynamically mobilize the inducible heat shock proteins. Overlapping expression of Hsp (proteins, mRNA and genes) and the compensatory mechanisms found with anti-oxidant system and metallothionein expression provided a more complete view of cytoprotective mechanisms especially in bivalves. As such, environmental parameters like temperature and salinity combined with metal toxicity have brought more data about the metabolic limits for Hsp to efficiently cope with pollution effects. Of particular interest is the use of parasites as proxy to assess the effects of metal toxicity in both host and parasites. As stated by Orsini et al. (2016), illuminating the link between genes and environment perturbations is an exciting yet challenging goal. In this context, understanding metabolism and transcription of stress-inducible Hsp at low nominal concentrations in chronic exposures have the potential to show a more accurate cell response in aquatic invertebrates. As emphasized throughout this review, exposing animal models to very high dosages does not bring useful data to unravel the mechanisms truly related to real aquatic systems, either targeted by intermittent imputs or already impacted by pollution, and where trace metals are found in much lower concentrations. It should be also mentioned that as metal overdose provokes extreme intoxication and lethality effects, other mechanisms allowing survival will probably be rather more important than Hsp response itself. As another further step, we should mention the need for studies regarding earth rare elements toxicity and the transcription of Hsp, especially for freshwater invertebrates. Up to the present moment, more work should equally reach invertebrate clades whose data remains unavailable.

Acknowledgements This study was partly supported by the Natural Sciences and Engineering Research Council of Canada (NSERC) - Discovery grant program (E.P.)

References

Agell G et al (2001) Does stress protein induction by copper modify natural toxicity in sponges? Environ Toxicol Chem 20:2588–2593

Agell G et al (2004) Molecular and organism biomarkers of copper pollution in the ascidian *Pseudodistoma crucigaster*. Mar Pollut Bull 48:759–767

Alejandrino A et al (2011) Convergent and parallel evolution in life habit of the scallops (Bivalvia: Pectinidae). BMC Evol Biol 11(1):164

Anestis A et al (2007) Behavioral, metabolic, and molecular stress responses of marine bivalve *Mytilus galloprovincialis* during long-term acclimation at increasing ambient temperature. Am J Phys Regul Integr Comp Phys 293(2):R911–R921

Arnold WR et al (2011) Acute and chronic toxicity of copper to the euryhaline rotifer, *Brachionus plicatilis* ("L" Strain). Arch Environ Contam Toxicol 60:250–260. https://doi.org/10.1007/s00244-010-9556-8

Bagarinao T (1992) Sulfide as an environmental factor and toxicant: tolerance and adaptations in aquatic organisms. Aquat Toxicol 24(1–2):21–62

Bell RA, Kramer JR (1999) Structural chemistry and geochemistry of silver-sulfur compounds: critical review. Environ Toxicol Chem 18(1):9–22

Beyer J et al (2017) Blue mussels (Mytilus edulis) as sentinel organisms in coastal pollution monitoring: a review. Mar Environ Res 130:338–365

Binelli A et al (2015) Does Zebra mussel (*Dreissena polymorpha*) represent the freshwater counterpart of *Mytilus* in ecotoxicological studies? A critical review. Environ Pollut 196:386–403

Bosch TCG, Praetzel G (1991) The heat shock response in *Hydra*: immunological relationship of hsp60, the major heat shock protein of *Hydra vulgaris*, to the ubiquitous hsp70 family. Hydrobiologia 216/217:513–517

Bosch TCG et al (1988) Thermotolerance and synthesis of heat-shock proteins—these responses are present in *Hydra attenuata* but absent in *Hydra oligactis*. Proc Natl Acad Sci U S A 85:7927–7931

Boscolo-Papo M et al (2014) Histopathology and stress biomarkers in the clam *Venerupis philippinarum* from the Venice lagoon (Italy). Fish Shellfish Immunol 39(1):42–50

Bremner I (1998) Manifestations of copper excess. Am J Clin Nutr 67(5):1069S–1073S

Browne CL et al (2007) Extracellular heat shock protein 70 has novel functional effects on sea urchin eggs and coelomocytes. J Exp Biol 210(Pt7):1275–1287

Buckley BA et al (2001) Adjusting the thermostat : the threshold induction temperature for the heat-shock response in intertidal mussels (genus *Mytilus*) changes as a function of thermal history. J Exp Biol 3579:3571–3579

Campbell PGC, Couillard Y (2004) Prise en charge et détoxification des métaux chez les organismes aquatiques. In: Écotoxicologie moléculaire. Principes fondamentaux et perspectives de développement, Presses de l'Université du Québec, p 462

Canadian Environmental Quality Guidelines (CEQG) (1999) *Canadian sediment quality guidelines for the protection of aquatic life*. Cooper. No. 1299; ISBN 1–896997–34-1

Canadian Environmental Quality Guidelines (CEQG) (2003) *Canadian water quality guidelines for the protection of aquatic life. Mercury (Inorganic mercury and methyl mercury)*. No. 1299; ISBN 1–896997–34-1

Canadian Environmental Quality Guidelines (CEQG) (2014) *Canadian water quality guidelines for the protection of aquatic life. Cadmium*. No. 1299; ISBN 1–896997–34-1

Cebrian E et al (2006) Response of the mediterranean sponge *Chondrosia reniformis nardo* to copper pollution. Environ Pollut 141(3):452–458

De Chaffoy D et al (1978) On the permeability and formation of the embryonic cuticle during development in vivo and in vitro of *Artemia salina* embryos. Differentiation 12:99–109

Chen CY et al (1999) Molecular and demographic measures of arsenic stress in *Daphnia pulex*. Hydrobiology 401:229–238

Cheng F et al (2015) Identification and expression analysis of a heat-shock protein 70 gene in *Polycelis* sp. Cell Stress Chaperones 20(6):907–915. https://doi.org/10.1007/s12192-015-0608-x

Choi Y et al (2008) Cadmium affects the expression of heat shock protein 90 and metallothionein mRNA in the Pacific oyster, Crassostrea gigas. Comp Bioch Physiol C Toxicol Pharmacol 147:286–292

Clark MS, Peck SL (2009) HSP70 heat shock proteins and environmental stress in Antarctic marine organisms: a mini-review. Mar Genomics 2(1):11–18

Clayton ME et al (2000) Different expression patterns of heat shock proteins Hsp60 and Hsp70 in zebra mussels (*Dreissena polymorpha*) exposed to copper and tributyltin. Aquat Toxicol 47:213–226

Conte M et al (2011) Expression of hsp90 mediates cytoprotective effects in the gastrodermis of planarians. Cell Stress Chaperones 16:33–39

Coyle P et al (2002) Metallothionein: the multipurpose protein. Cell Mol Life Sci 59:627–647

De Coninck DIM et al (2014) Genome-wide transcription profiles reveal genotype-dependent responses of biological pathways and gene-families in *Daphnia* exposed to single and mixed stressors. Environ Sci Technol 48:3513–3522

De Pomerai D (1996) Heat-shock proteins as biomarkers of pollution. Hum Exp Toxicol 15:279–285

Degger N et al (2009) Innovative fluorescence detection technique for metals in cestode egg-shells. Afr Zool 44(2):204–207

Dimauro I et al (2016) Exercise-induced ROS in heat shock proteins response. Free Radic Biol Med 98:46–55

Downs CA et al (2000) A molecular biomarker system for assessing the health of coral (*Montastraea faveolata*) during heat stress. Marine Biotechnol 2:533–544

Eeman M, Deleu M (2010) From biological membranes to biomimetic model membranes. Biotechnol Agron Soc Environ 14(4):719–736

Efremova SM et al (2002) Heat shock protein Hsp70 expression and DNA damage in Baikalian sponges exposed to model pollutants and wastewater from Baikalsk pulp and paper plant. Aquat Toxicol 57:267–280

Elran RON et al (2014) Early and late response of *Nematostella vectensis* transcriptome to heavy metals. Mol Ecol 23(19):4722–4736

Eze MO (1991) Phase transitions in phospholipid bilayers: Lateral phase separations play vital roles in biomembranes. Biochem Educ 19(4):204–208

Fabbri E et al (2008) HSP expression in bivalves. Invertebr Surviv J 5(2):135–161

Franco JL et al (2006) Antioxidant status and stress proteins in the gills of the brown mussel *Perna perna* exposed to zinc. Chem Biol Interact 160:232–240

Franzellitti S, Fabbri E (2005) Differential HSP70 gene expression in the Mediterranean mussel exposed to various stressors. Biochem Biophys Res Commun 336:1157–1163

Franzellitti S, Fabbri E (2006) Cytoprotective responses in the Mediterranean mussel exposed to Hg^{2+} and CH_3Hg^+. Biochem Biophys Res Commun 351:719–725

Gaetke LM, Chow CK (2003) Copper toxicity, oxidative stress, and antioxidant nutrients. Toxicology 189(1–2):147–163

Gao Q et al (2007) cDNA cloning and mRNA expression of heat shock protein 90 gene in the haemocytes of zhikong scallop *Chlamys farreri*. Comp Biochem Physiol B: Biochem Mol Biol 147:704–715

Garrido C et al (2012) The small heat shock proteins family: the long forgotten chaperones. Int J Biochem Cell Biol 44:1588–1592

Gheorghe S et al (2017) Metals toxic effects in aquatic ecosystems: modulators metals toxic. In: Water quality. https://doi.org/10.5772/65744

Gilbert BM, Avenant-Oldewage A (2017) Trace element and metal sequestration in vitellaria and sclerites, and reactive oxygen intermediates in a freshwater monogenean, *Paradiplozoon ichthyoxanthon*. PLoS One 12(5):e0177558. https://doi.org/10.1371/journal.pone.0177558

Giudice G et al (1999) Studies on heat shock proteins in sea urchin development. Dev Growth Differ 41(4):375–380

Gomes T et al (2013) Differential protein expression in mussels *Mytilus galloprovincialis* exposed to nano and ionic ag. Aquat Toxicol 136–137:79–90

Gomes T et al (2014) Proteomic response of mussels *Mytilus galloprovincialis* exposed to CuONPs and Cu^{2+}: an exploratory biomarker discovery. Aquat Toxicol 155:327–336

González-Fernández C et al (2017) Interactive effects of nutrition, reproductive state and pollution on molecular stress responses of mussels, *Mytilus galloprovincialis* Lamarck, 1819. Mar Environ Res 131:103–115

Gu XH et al (2012) Overexpression of heat shock protein 70 and its relationship to intestine under acute heat stress in broilers: 2. Intestinal oxidative stress. Poult Sci 91(4):790–799

Gupta SC et al (2010) Heat shock proteins in toxicology : how close and how far ? Life Sci 86(11–12):377–384

Haap T et al (2016) Metallothionein and Hsp70 trade-off against one another in *Daphnia magna* cross-tolerance to cadmium and heat stress. Aquat Toxicol 170:112–119
Haap T, Köhler H (2009) Cadmium tolerance in seven *Daphnia magna* clones is associated with reduced hsp70 baseline levels and induction. Aquat Toxicol 94(2):131–137. https://doi.org/10.1016/j.aquatox.2009.06.006
Halliwell B, Gutteridge JMC (1999) Free radicals in biology and medicine, 3rd edn. Oxford University Press, Oxford
Hochachka PW, Somero GN (2002) Biochemical adaptation: mechanism and process in physiological evolution. Oxford University Press, New York, 466 p
Hofmann G, Somero G (1995) Evidence for protein damage at environmental temperatures: seasonal changes in levels of ubiquitin conjugates and hsp70 in the intertidal mussel *Mytilus trossulus*. J Exp Biol 198(Pt 7):1509–1518
Holmstrup M et al (2010) Interactions between effects of environmental chemicals and natural stressors: a review. Sci Total Environ 408:3746–3762
Howlett S et al (1983) Induction of heat shock proteins in early embryos of *Arbacia punctulata*. Biol Bull 165:500
Ivanina A et al (2008) Effects of cadmium on cellular protein and glutathione synthesis and expression of stress proteins in eastern oysters, *Crassostrea virginica* (Gmelin). J Exp Biol 211(Pt4):577–586
Ivanina A et al (2009) Effects of elevated temperature and cadmium exposure on stress protein response in eastern oysters *Crassostrea virginica* (Gmelin). Aquat Toxicol 91:245–254
Jaishankar M et al (2014) Toxicity, mechanism and health effects of some heavy metals. Interdiscip Toxicol 7(2):60–72
Jomova K et al (2012) Redox active metal-induced oxidative stress in biological systems. Transit Met Chem 37(2):127–134
Joyner-Matos J et al (2006) Increased expression of stress proteins in the surf clam *Donax variabilis* following hydrogen sulfide exposure. Comp Biochem Physiol A Mol Integr Physiol 145:245–257
Jung MY, Lee YM (2012) Expression profiles of heat shock protein gene families in the monogonont rotifer *Brachionus koreanus* - exposed to copper and cadmium. Toxicol Environ Heal Sci 4(4):235–242
Kalmar B, Greensmith L (2009) Induction of heat shock proteins for protection against oxidative stress. Adv Drug Deliv Rev 61(4):310–318
Kim EK, Choi EJ (2010) Pathological roles of MAPK signaling pathways in human diseases. Biochim Biophys Acta (BBA) - Mol Basis Dis 1802(4):396–405
Kim B-M et al (2014a) UV-B Radiation-induced oxidative stress and p38 signaling pathway involvement in the benthic copepod *Tigriopus japonicus*, vol 167. Comparative Biochemistry and Physiology, Part C, pp 15–23. https://doi.org/10.1016/j.cbpc.2014.08.003
Kim B-M et al (2014b) Heavy metals induce oxidative stress and trigger oxidative stress-mediated heat shock protein (Hsp) modulation in the intertidal copepod *Tigriopus japonicus*. Comp Biochem Physiol, Part C 166:65–74. https://doi.org/10.1016/j.cbpc.2014.07.005
Klumpen E et al (2017) Reactive oxygen species (ROS) and the heat stress response of *Daphnia pulex*: ROS-mediated activation of hypoxia-inducible factor 1 (HIF-1) and heat shock factor 1 (HSF-1) and the clustered expression of stress genes. Biol Cell 109:39–64
Kourtidis A et al (2006) Identification of several cytoplasmic HSP70 genes from the Mediterranean mussel (*Mytilus galloprovincialis*) and their long-term evolution in Mollusca and Metazoa. J Mol Evol 62:446–459
Lannig G et al (2006) Temperature-dependent effects of cadmium on mitochondrial and whole-organism bioenergetics of oysters (*Crassostrea virginica*). Mar Environ Res 62(Suppl):S79–S82
Lapresta-Fernández A et al (2012) Nanoecotoxicity effects of engineered silver and gold nanoparticles in aquatic organisms. Trends Anal Chem 32(797):40–59

Lee S et al (2006) Expression of heat shock protein and hemoglobin genes in *Chironomus tentans* (Diptera, Chironomidae) larvae exposed to various environmental pollutants : a potential biomarker of freshwater monitoring. Chemosphere 65:1074–1081
Lee JY et al (2016) Transport of pyruvate into mitochondrial is involved in methylmercury toxicity. Sci Rep 6:21528. https://doi.org/10.1038/srep21528
Leung KMY, Furness RW (2001) Survival, growth, metallothionein and glycogen levels of *Nucella lapillus* (L.) exposed to sub-chronic cadmium stress : the influence of nutritional state and prey type. Mar Environ Res 52(2):173–194
Li C et al (2010) Identification of two small heat shock proteins with different response profile to cadmium and pathogen stresses in *Venerupis philippinarum*. Cell Stress Chaperones 15(6):897–904
Li J et al (2012) The Hsp90 chaperone machinery: conformational dynamics and regulation by co-chaperones. Biochim Biophys Acta (BBA) – Mol Cell Res 1823(3):624–635
Li H et al (2013) Identification of a small HSP gene from hard clam *Meretrix meretrix* and its potential as an environmental stress biomarker. Aquat Biol 18:243–252
Li S et al (2014) The trans-generation effect during pulsed cadmium exposure: tolerance and induction of hsp70. Ecotoxicol Environ Saf 107C:300–305. https://doi.org/10.1016/j.ecoenv.2014.06.003
Liping YOU et al (2014) Expression profiles of two small heat shock proteins and antioxidant enzyme activity in *Mytilus galloprovincialis* exposed to cadmium at environmentally relevant concentrations. Chin J Oceanol Limnol 32(2):334–343
Lister KN et al (2010) Sea ice protects the embryos of the Antarctic Sea urchin *Sterechinus neumayeri* from oxidative damage due to naturally enhanced levels of UV-B radiation. J Exp Biol 213(11):1967–1975
Liu D et al (2011) Mitochondrial pathway of apoptosis in the hepatopancreas of the freshwater crab *Sinopotamon yangtsekiense* exposed to cadmium. Aquat Toxicol 105(3–4):394–402
Liu T et al (2012) Comprehensive review on the HSC70 functions, interactions with related molecules and involvement in clinical diseases and therapeutic potential. Pharmacol Ther 136:354–374
Liu D et al (2013) Cd-induced apoptosis through the mitochondrial pathway in the hepatopancreas of the freshwater crab *Sinopotamon henanense*. PLoS One 8(7):1–10
Liu H et al (2017) Sunlight promotes fast release of hazardous cadmium from widely-used commercial cadmium pigment. Environ Sci Technol 51(12):6877–6886. https://doi.org/10.1021/acs.est.7b00654
Luan W et al (2010) Identification of a novel inducible cytosolic Hsp70 gene in Chinese shrimp *Fenneropenaeus chinensis* and comparison of its expression with the cognate Hsc70 under different stresses. Cell Stress Chaperones 15(1):83–93
Luo L et al (2014) Metal accumulation and differentially expressed proteins in gill of oyster (*Crassostrea hongkongensis*) exposed to long-term heavy metal-contaminated estuary. Fish Shellfish Immunol 38(2):318–329
Ma KX et al (2014) Molecular characterization of the glucose-regulated protein 78 (GRP78) gene in planarian *Dugesia japonica*. Comp Biochem Physiol Part B 171:12–17
MacRae T (2015) Stress tolerance during diapause and quiescence of the brine shrimp, *Artemia*. Cell Stress Chaperones 21(1):9–18
Magesky A et al (2017) Silver nanoparticles and dissolved silver activate contrasting immune responses and stress-induced heat shock protein expression in sea urchin. Environ Toxicol Chem 36:1872–1886
Maglott DR (1982) Heat shock response in the Atlantic sea urchin, *Arbacia punctulala*. Experientia 39:268–270
Marigómez I et al (2002) Cellular and subcellular distribution of metals in molluscs. Microsc Res Tech 56(5):358–392
Martínez-Paz P et al (2013) Characterization of the small heat shock protein Hsp27 gene in *Chironomus riparius* (Diptera) and its expression profile in response to temperature

changes and xenobiotic exposures. Cell Stress Chaperones 19:529. https://doi.org/10.1007/s12192-013-0479-y

Martínez-Paz PP et al (2016) Cadmium in vivo exposure alters stress response and endocrine-related genes in the freshwater snail *Physa acuta*. New biomarker genes in a new model organism. Environ Pollut:1–10

Martín-Folgar R, Martínez-Guitarte JL (2017) Cadmium alters the expression of small heat shock protein genes in the aquatic midge *Chironomus riparius*. Chemosphere 169:485–492

Matranga V et al (2012) Long-term environmental exposure to metals (cu, cd, Pb, Zn) activates the immune cell stress response in the common European sea star (*Asterias rubens*). Mar Environ Res 76:122–127

Meng J et al (2017) Cadmium effects on DNA and protein metabolism in oyster (*Crassostrea gigas*) revealed by proteomic analyses. Mar Environ Res 131:103–115

Migliaccio O et al (2014) Stress response to cadmium and manganese in *Paracentrotus lividus* developing embryos is mediated by nitric oxide. Aquat Toxicol 156:125–134

Moore MN (1985) Cellular responses to pollutants. Mar Pollut Bull 16(4):134–139

Moraga D et al (2005) Stress response in Cu^{2+} and Cd^{2+} exposed oysters (*Crassostrea gigas*): an Immunohistochemical approach. Comp Biochem Physiol C Toxicol Pharmacol 141:151–156

Morimoto RI (1998) Regulation of the heat shock transcriptional response: cross talk between a family of heat shock factors, molecular chaperones, and negative regulators. Genes Dev 12:3788–3379

Müller WEG et al (2007) Cold stress defense in the freshwater sponge *Lubomirskia baicalensis* role of okadaic acid produced by symbiotic dinoflagellates. FEBS J 274:23–36

Nagpal NK (2004) Water quality guidelines for cobalt/Ambient water quality guidelines for cobalt: overview. Ministry of water, land and air protection: Water protection section, British Columbia; ISBN 0-7726-5229-5

Negri A et al (2013) Transcriptional response of the mussel *Mytilus galloprovincialis* (Lam.) following exposure to heat stress and copper. PLoS One 8(6):e66802

Nemmiche S (2017) Oxidative signaling response to cadmium exposure. Toxicol Sci 156(1):4–10. https://doi.org/10.1093/toxsci/kfw222

Nicosia A et al (2014) Characterization of small HSPs from *Anemonia viridis* reveals insights into molecular evolution of alpha crystallin genes among cnidarians. PLoS One 9(9):e105908

Nicosia A et al (2015) Transcriptional and biochemical effects of cadmium and manganese on the defense system of *Octopus vulgaris* paralarvae. Biomed Res Int 437328. https://doi.org/10.1155/2015/437328

Nishizawa J et al (1999) Reactive oxygen species play an important role in the activation of heat shock factor in ischemic-reperfused heat. Circulation 99:934–941

Orsini L et al (2016) *Daphnia magna* Transcriptome by RNA-Seq across 12 environmental stressors. Sci Data 3(160030):1–15

Oweson C et al (2008) Manganese effects on haematopoietic cells and circulating coelomocytes of *Asterias rubens* (Linnaeus). Aquat Toxicol 89(2):75–81

Pari L, Murugavel P (2007) Diallyl tetrasulfide improves cadmium induced alterations of acetylcholinesterase, ATPases and oxidative stress in brain of rats. Toxicology 234(1–2):44–50

Patra RC et al (2011) Oxidative stress in lead and cadmium toxicity and its amelioration. Vet Med Int 457327

Peng C et al (2015) Toxic effects of chronic sub-lethal Cu^{2+}, $Pb^{2+}$and Cd^{2+} on antioxidant enzyme activities in various tissues of the blood cockle, *Anadara granosa*. J Residuals Sci Tech 12(3):125–131

Pestana JLT et al (2016) Non-lethal heat shock increases tolerance to metal exposure in brine shrimp. Environ Res 151:663–670. https://doi.org/10.1016/j.envres.2016.08.037

Piano A et al (2004) Expression of cytoprotective proteins, heat shock protein 70 and metallothioneins, in tissues of *Ostrea edulis* exposed to heat and heavy metals. Cell Stress Chaperones 9(2):134–142

Pinsino A et al (2010) Sea urchin embryos as an in vivo model for the assessment of manganese toxicity: developmental and stress response effects. Ecotoxicology 19(3):555–562
Pinsino A et al (2015) Titanium dioxide nanoparticles stimulate sea urchin immune cell phagocytic activity involving TLR/p38 MAPK-mediated signaling pathway. Sci Rep 5(August):14492. http://www.nature.com/articles/srep14492
Planelló R et al (2010) Effect of acute exposure to cadmium on the expression of heat-shock and hormone-nuclear receptor genes in the aquatic midge *Chironomus riparius*. Sci Total Environ 408(7):1598–1603
Plusquin M et al (2012) Physiological and molecular characterization of cadmium stress in *Schmidtea mediterranea*. Int J Dev Biol 56:183–191
Porte PFLA (2005) *Mytilus trossulus* hsp70 as a biomarker for arsenic exposure in the marine environment : laboratory and real-world results. Biomarkers 10(December):417–428
Powell SR (2000) The antioxidant properties of zinc. J Nutr 130(5S):1447S–1454S
Qian Y, Tiffany-Castiglioni E (2003) Lead-induced endoplasmic reticulum (ER) stress responses in the nervous system. Neurochem Res 28(1):153–162
Radlowska M, Pempkowiak J (2002) Stress-70 as indicator of heavy metals accumulation in blue mussel *Mytilus edulis*. Environ Int 27(8):605–608
Reeders HH, Bij de Vaate A (1990) Zebra mussels (*Dreissena polymorpha*): a new perspective for water quality management. Hydrobiologia 2001201:437–450
Reimer PS (1999) Environmental effects of manganese and proposed freshwater guidelines to protect aquatic life in British Columbia (Master's degree thesis). University of British Columbia
Rhee JS et al (2009) Heat shock protein (Hsp) gene responses of the intertidal copepod *Tigriopus japonicus* to environmental toxicants. Comp Biochem Physiol C: Toxicol Pharmacol 149(1):104–112. https://doi.org/10.1016/j.cbpc.2008.07.009
Rios-Arana J et al (2005) Heat shock protein 60 (HSP60) response of *Plationus patulus* (Rotifera: Monogononta) to combined exposures of arsenic and heavy metals. Hydrobiologia 60:577–585
Roccheri MC et al (1982) Heat-shock proteins in sea urchin embryos. Transcriptional and post-transcriptional regulation. Differentiation 22(3):175–178
Roccheri MC et al (2004) Cadmium induces the expression of specific stress proteins in sea urchin embryos. Biochem Biophys Res Commun 13(1):80–87
Rocha CT, Souza MM (2012) The influence of lead on different proteins in gill cells from the freshwater bivalve, *Corbicula fluminea*, from defense to repair biomarkers. Arch Environ Contam Toxicol 62:56–67
Romano G et al (2011) Nitric oxide mediates the stress response induced by diatom aldehydes in the sea urchin *Paracentrotus lividus*. PLoS One 6(10):2–10
Sanders BM (1993) Stress proteins in aquatic organisms : an environmental perspective. Crit Rev Toxicol 23:49–75
Sanders BM et al (1991) Relationships between accumulation of a 60kDa stress protein and scope-for-growth in *Mytilus edulis* exposed to a range of copper concentrations. Hydrobiologia 31:81–97
Satoh N et al (2014) Chordate evolution and the three-phylum system. Proc R Soc B 281(1794). https://doi.org/10.1098/rspb.2014.1729
Schieber M, Chandel NS (2014) ROS function in redox signaling and oxidative stress. Curr Biol 24(10):R453–R462 https://doi.org/10.1016/j.cub.2014.03.034
Schill RO et al (2006) Quiescent gemmules of the freshwater sponge, *Spongilla lacustris* (Linnaeus, 1759), contain remarkably high levels of Hsp70 Stress Protein and hsp70 Stress gene mRNA. J Exp Zool 305A:449–457
Schröder HC et al (1999) Induction of DNA strand breaks and expression of HSP70 and GRP78 homolog by cadmium in the marine sponge *Suberites domuncula*. Arch Environ Contam Toxicol 36(1):47–55
Schumpert C et al (2014) Relationship between heat shock protein 70 expression and life span in *Daphnia*. Mech Ageing Dev 139:1–10. https://doi.org/10.1016/j.mad.2014.04.001

Shen X et al (2007) The complete mitochondrial genomes of two common shrimps (*Litopenaeus vannamei* and *Fenneropenaeus chinensis*) and their phylogenomic considerations. Gene 403:98–109

Shi J et al (2015) Characterization and function analysis of Hsp60 and Hsp10 under different acute stresses in black tiger shrimp, *Penaeus monodon*. Cell Stress Chaperones 21(2):295–312

Sinclair JA et al (2015) Ambient water quality guidelines for cadmium – technical report. Water protection and sustainability branch: environmental sustainability and strategic policy division. Ministry of Environment, British Columbia, p 114

Singer C et al (2005) Induction of heat shock proteins (hsp70) in the zebra mussel (*Dreissena polymorpha*) following exposure to platinum group metals (platinum, palladium and rhodium): comparison with lead and cadmium exposures. Aquat Toxicol 75:65–75

Snell TW, Janssen CR (1995) Rotifers in ecotoxicology: a review. Hydrobiologia 313:231. https://doi.org/10.1007/BF00025956

Sokolova IM et al (2012) Energy homeostasis as an integrative tool for assessing limits of environmental stress tolerance in aquatic invertebrates. Mar Environ Res 79:1–15. http://www.ncbi.nlm.nih.gov/pubmed/22622075

Song L et al (2006) The cDNA cloning and mRNA expression of heat shock protein 70 gene in the haemocytes of bay scallop (*Argopecten irradians*, Lamarck 1819) responding to bacteria challenge and naphthalin stress. Fish Shellfish Immunol 21(4):335–345

Su HL et al (2009) The disruption of bacterial membrane integrity through ROS generation induced by nanohybrids of silver and clay. Biomaterials 30(30):5979–5987. https://doi.org/10.1016/j.biomaterials.2009.07.030

Sung YY et al (2017) Non-lethal heat shock induces Hsp70 synthesis and promotes tolerance against heat, ammonia and metals in post-larvae of the white leg shrimp *Penaeus vannamei* (Boone, 1931). Aquaculture 483:21–16. https://doi.org/10.1016/j.aquaculture.2017.09.034

Sures B, Radszuweit H (2015) Pollution-induced heat shock protein expression in the amphipod *Gammarus roeseli* is affected by larvae of *Polymorphus minutus* (Acanthocephala). J Helminthol 81(2):191–197

Tang S et al (2015) Chemosphere distinct expression profiles of stress defense and DNA repair genes in *Daphnia pulex* exposed to cadmium, zinc, and quantum dots. Chemosphere 120:92–99. https://doi.org/10.1016/j.chemosphere.2014.06.011

Tedengren M et al (1999) Heavy metal uptake, physiological response and survival of the blue mussel (*Mytilus edulis*) from marine and brackish waters in relation to the induction of heat-shock protein. Hydrobiology 70:261–269

Temenouga G et al (2003) Stress protein response and catalase activity in freshwater planarian *Dugesia* (*Girardia*) *schubarti* exposed to copper. Ecotoxicol Environ Saf 56:351–357

Thompson E et al (2015) Proteomic analysis of Sydney rock oysters (*Saccostrea glomerata*) exposed to metal contamination in the field. Environ Pollut 170:102–112

Toh HS et al (2014) Chemical interactions between silver nanoparticles and thiols: a comparison of mercaptohexanol against cysteine. Sci China Chem 57:1199

Valko M et al (2016) Redox and non-redox metal induced formation of free radicals and their role in human disease. Arch Toxicol 90(1):1–37

Varotto L et al (2013) DNA damage and transcriptional changes in the gills of *Mytilus galloprovincialis* exposed to nanomolar doses of combined metal salts (cd, cu, hg). PLoS One 8(1):e54602

Venn A et al (2009) P-glycoprotein (multi-xenobiotic resistance) and heat shock protein gene expression in the reef coral *Montastraea franksi* in response to environmental toxicants. Aquat Toxicol 93:188–195

Wadige C et al (2014) Effects of lead-spiked sediments on freshwater bivalve, *Hyridella australis* : linking organism metal exposure-dose-response. Aquat Toxicol 149:83–93. https://doi.org/10.1016/j.aquatox.2014.01.017

Wang J et al (2012) Cadmium induces hydrogen peroxide production and initiates hydrogen peroxidedependent apoptosis in the gill of freshwater crab, *Sinopotamon henanense*. Comp Biochem Physiol Part C 156:195–201

Wang L et al (2013) The molluscan HSP70s and their expression in hemocytes. Invertebr Surviv J 10(1):77–83
Wang M et al (2017a) Different transcriptomic responses of two marine copepods, *Tigriopus japonicus* and *Pseudodiaptomus annandalei*, to a low dose of mercury chloride ($HgCl_2$). Aquat Toxicol 187:124. https://doi.org/10.1016/j.aquatox.2017.03.018
Wang Q et al (2017b) Molecular cloning, sequencing, and expression profiles of heat shock protein 90 (HSP90) in *Hyriopsis cumingii* exposed to different stressors: temperature, cadmium and *Aeromonas hydrophila*. Aquac Fish 90:4–11
Wu H et al (2013) Proteomic and metabolomic responses of clam *Ruditapes philippinarum* to arsenic exposure under different salinities. Aquat Toxicol 15:136–137. https://doi.org/10.1016/j.aquatox.2013.03.020
Xu Q, Qin Y (2012) Molecular cloning of heat shock protein 60 (Pt HSP60) from *Portunus trituberculatus* and its expression response to salinity stress. Cell Stress Chaperones 60:589–601
Xu D et al (2017) The novel transcriptional factor HP1BP3 negatively regulates Hsp70 transcription in *Crassostrea hongkongensis*. Sci Rep (March) 7:1–10. https://doi.org/10.1038/s41598-017-01573-y
Yang J et al (2013) Effects of cadmium on lipid storage and metabolism in the freshwater crab *Sinopotamon henanense*. PLoS One 8(10):e77569
Yang J et al (2014) Changes in the expression of four heat shock proteins during the aging process in *Brachionus calyciflorus* (Rotifera). Cell Stress Chaperones 19:33–52
Yannick G et al (2017) Response of the pearl oyster *Pinctada margaritifera* to cadmium and chromium : identification of molecular biomarkers. Mar Pollut Bull. https://doi.org/10.1016/j.marpolbul.2017.03.012 118:420
Yiin SJ, Lin TH (1995) Lead-catalyzed peroxidation of essential unsaturated fatty acid. Biol Trace Elem Res 50(2):167–172
Yoshimi T et al (2009) Induction profile of HSP70-cognate genes by environmental pollutants in Chironomidae. Environ Toxicol Pharmacol 28:294–301
Zahoor Z et al (2010) Larval excretory-secretory products from the parasite *Schistosoma mansoni* modulate HSP70 protein expression in defense cells of its snail host, *Biomphalaria glabrata*. Cell Stress Chaperones 15:639–650
Zeeshan M et al (2017) Ecotoxicological assessment of cobalt using Hydra model : ROS, oxidative stress, DNA damage, cell cycle arrest, and apoptosis as mechanisms of toxicity. Environ Pollut. https://doi.org/10.1016/j.envpol.2016.12.042 224:54
Zhang G et al (2012) The oyster genome reveals stress adaptation and complexity of shell formation. Nature 490(7418):49–54. https://doi.org/10.1038/nature11413
Zhang L et al (2010a) The responsive expression of heat shock protein 22 gene in zhikong scallop *Chlamys farreri* against a bacterial challenge. Aquac Res 41(2):257–266
Zhang L et al (2010b) The involvement of HSP22 from Bay scallop *Argopecten irradians* in response to heavy metal stress. Mol Biol Rep 37(4):1763–1771. https://doi.org/10.1007/s11033-009-9603-6
Zhang A et al (2013) A small heat shock protein (sHSP) from *Sinonovacula constricta* against heavy metals stresses. Fish Shell Immunol Short Commun 34:1605–1610
Zhang X et al (2014) Induction of hsp70, hsp90, and catalase activity in planarian *Dugesia japonica* exposed to cadmium. Toxicology and industrial Health 1-8. https://doi.org/10.1177/0748233714561488
Zhang X-Y et al (2009) Identification of two hsp90 genes from the marine crab, *Portunus trituberculatus* and their specific expression profiles under different environmental conditions. Comp Biochem Physiol, Part C 150:465–473

Chapter 12
Heat Shock-Induced Transcriptional and Translational Arrest in Mammalian Cells

Anshika Goenka, Rashmi Parihar, and Subramaniam Ganesh

Abstract The heat shock response is a highly conserved cellular stress response pathway that protects cells against stress-induced damages. The heat shock response is characterized by the activation of the heat shock factor, a transcription factor, which in turn regulate the expression of heat shock proteins which help the cells to recover from the stress-induced damages. While the mechanism behind the activation of heat shock factor and the expression of heat shock proteins are well studied, our understanding on the other aspects of the heat shock response pathway – mainly the global suppression of transcriptional and the translational processes – has been very limited. Thus, this chapter would focus more on the recent studies elucidating the mechanism behind the heat shock-induced transcriptional and post-transcriptional repression of gene expression. Given the importance of the stress response pathways in cell survival, the chapter would focus on the discoveries made in the mammalian systems and their impact on human health and disease.

Keywords Chromatin dynamics · Heat shock · Heat shock factor 1 · Non-coding RNA · Stress · Stress granules · Transcription · Translation

---

A. Goenka
Department of Biological Sciences and Bioengineering, Indian Institute of Technology, Kanpur, India

Syngene International Ltd., Biocon Park, Bengaluru, India

R. Parihar · S. Ganesh (✉)
Department of Biological Sciences and Bioengineering, Indian Institute of Technology, Kanpur, India
e-mail: sganesh@iitk.ac.in

A. A. A. Asea, P. Kaur (eds.), *Heat Shock Proteins and Stress*,
Heat Shock Proteins 15, https://doi.org/10.1007/978-3-319-90725-3_12

Abbreviations

CREBBP	CREB-binding protein
eIF2	Eukaryotic initiation factor 2
hnRNPs	Heterogeneous nuclear ribonucleoproteins
HSF1	Heat shock factor 1
HSP	Heat shock proteins
HSR	Heat shock response
IR	Intron retention
NMD	Non-sense mediated degradation
RBPs	mRNA-binding proteins
RNA pol II	RNA polymerase II
Sat3	Satellite III transcripts
SGs	Stress granules
SINEs	Short interspersed elements
SRSF1	Serine/arginine-rich splicing factor 1
TSS	Transcription start site

12.1 Introduction

Notwithstanding the existence of homeostatic mechanisms, cells forming the organ systems in the complex multicellular organisms such as mammals at times face sudden changes in the cellular environment that are harmful to the cellular physiology. For example, fever or hyperthermia lead to an increased load of misfolded proteins within the cell, which in turn activates a cascade of cellular response pathways which are collectively called as the heat shock response (HSR) (Morimoto 1998). Besides the heat stress, several other stimuli, such as the free radicals, heavy metals, ischemia, and infection, also induce HSR (Morimoto 1998). The heat shock factor – a conserved transcription factor – is thought to activate the HSR, resulting in a global change in the transcriptional, translational and post-translation processes. The HSR in a multicellular organism was first discovered in *Drosophila*, as a "puffing" pattern of the polytene chromosome in response to a heat shock (Lindquist 1986; Lakhotia 1989). Further metabolic labeling studies in flies revealed that while the expression of a majority of proteins was down-regulated, the expression of a set of proteins, known now as heat shock proteins (HSP), are up-regulated upon exposure to a heat shock (Lindquist 1986). HSP, which are molecular chaperones, help the synthesis and folding of other proteins (Lindquist 1986). During a heat shock, the heat shock-induced HSP help to prevent the misfolding of proteins (Powers et al. 2009). Devastating human neurodegenerative disorders, like the Huntington disease, are thought to result from chronic misfolding of proteins, thereby causing an imbalance in the protein homeostasis (Powers et al. 2009). Therefore, activating the heat shock response has been a long sought therapeutic intervention for these diseases (Maheshwari et al. 2014). More than 100 human diseases with protein conformation abnormalities are known in the literature (Westerheide and Morimoto

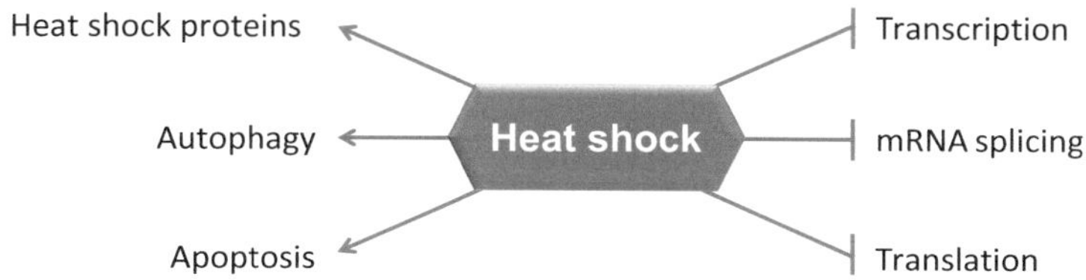

Fig. 12.1 Schematic diagram showing the positive and negative regulatory roles of heat shock response pathway in cellular physiology. Heat shock induces the expression of heat shock proteins and the autophagic process while a lethal thermal stress can also trigger apoptosis. On the other hand, heat shock response brings about global suppression of transcription, mRNA splicing and translational processes

2005) and the perturbations in the HSR could be playing a potential role in their pathophysiology.

The regulation of the HSR is a well-studied phenomenon and is known to be orchestrated by the transcription factor Heat Shock Factor 1 (HSF1). HSF1 is normally sequestered in the cell in the form of inactive monomers and released when the misfolded protein load increases in the cell for them to trimerize and acquire a site-specific DNA binding activity (Sarge et al. 1993; Westwood and Wu 1993). It then induces the transcription of HSP and other cytoprotective proteins (Sarge et al. 1991). Activation of HSF1 during the heat stress, in turn, is regulated by a ribonucleoprotein complex containing translation elongation factor eEF1A and a non-coding RNA HSR1 (heat shock RNA-1) (Shamovsky et al. 2006). One of the immediate effects of heat shock is translational shut-down and breakdown of the cytoskeleton complex. As a result, there is a free pool of eEF1A which then forms a complex with HSR1 non-coding RNA. Further, the HSR1–eEF1A complexes assist in the HSF1 trimerization and in its subsequent DNA binding activity (Shamovsky et al. 2006).

Since the discovery of genes coding for HSP in *Drosophila* (Lindquist 1986), the mechanism of upregulation of these proteins and their functional significance had been the major interest of study for a long time. However, the impact of heat shock on the cellular physiological processes such as DNA replication and repair, transcription, splicing, and translation had been thoroughly investigated (Velichko et al. 2013; Velichko et al. 2012) (see Fig. 12.1). Some of the early studies which suggested a global transcription repression to occur during heat shock had mostly been performed in *Drosophila* (Lindquist 1986). Interestingly it had been shown that the expression of "high molecular weight" RNA species, such as rRNA and mRNA precursors, were repressed during a heat shock while the expression of the "low molecular weight" RNA species i.e. those produced by RNA Pol III, remain unaffected (Kantidze et al. 2015). Moreover, gene expression data from mammalian cells using oligonucleotide hybridization chips showed several hundred genes to be repressed by more than two-fold during the heat shock (Kantidze et al. 2015). The extent of transcriptional repression is thought to be dependent upon the temperature or the duration of stress and also the cell type exposed to the stress (Kantidze et al. 2015). The possible mechanism of regulation of such repression has been uncovered only recently. Thus, this chapter would focus more on the recent studies elucidating the mechanism behind the heat shock-induced transcriptional and post-transcriptional repression of gene expression.

12.1.1 Heat Shock Response and Chromatin Dynamics

The DNA is packed into nucleosomes which in turn are packed into high-order structures. This order of nucleosome packaging in chromatin is crucial to gene expression. During a heat shock the chromatin structure decondenses at the promoter elements of the genes coding for the HSP. The coordinated function of various proteins and protein complexes such as histone variants, histone-modifying enzymes, histone chaperones and chromatin remodeling complexes bring about a change the chromatin structure by allowing the displacement of nucleosomes upon exposure to a heat shock (Li et al. 2007). Such heat shock-induced changes in gene expression are preceded by changes in the promoter chromatin dynamics thereby altering the kinetics of RNA polymerase II-promoter binding (Teves and Henikoff 2013). Binding of the Heat Shock factor 1 (HSF1) to the HSP gene promoters initiates the release of RNA Pol II into the elongation phase of transcription which was initially paused about 30bp downstream of Transcription Start Site (TSS) (Rougvie and Lis 1988). Regulation of gene expression by RNA Pol II pausing is known to occur in up to 30% of the genes in the genome (Levine 2011), including the heat shock genes (Levine 2011). Thus, the RNA Pol II is bound to the gene promoter even when the gene is not being transcribed. The paused Pol II binds to factors that induce pausing (Missra and Gilmour 2010; Wu et al. 2003) including components of the RNAi pathway (Cernilogar et al. 2011). During a heat shock exposure, the active HSF1 binds to the HSP gene promoters and induces the elongation factor P-TEFb kinase to bind to the paused RNA Pol II and relieve the inhibitory factors, thereby facilitating the transcription elongation by RNA Pol II (Peterlin and Price 2006).

Concurrent with the expression of HSP, the HSR is marked by the reduction in the transcriptional, translation and splicing events in the cell which collectively bring about global changes in the gene expression patterns, which eventually help the cell under heat stress by reducing the load of misfolded protein (Shalgi et al. 2014). This is ensued by loss of Pol II from the promoter region of active genes as reported in the salivary glands of *Drosophila* (Jamrich et al. 1977) and in the mammalian systems wherein non-coding RNAs induced upon heat shock inhibit the contacts between Pol II and the promoter DNA (Yakovchuk et al. 2009). Such repression does not operate on the HSP gene promoters since selective RNAse complexes remove the non-coding RNA repressors and facilitate the binding of RNA Pol II (Shamovsky et al. 2006)

12.1.2 Heat Shock Response and Transcriptional Regulation

As discussed above, the heat shock response is known to be associated with global transcription repression (Lindquist 1986), however, the mechanism behind such a global change was not understood well for a very long time. Recent studies indicate critical roles for long non-coding RNAs in this process. For example, a class of human ncRNAs called *Alu* RNA, and its mouse homologue *B2* RNA transcribed from short interspersed elements (SINEs), were shown to regulate different aspects of the heat-induced transcription repression (Allen et al. 2004). *Alu* and *B2* RNAs are known to interact directly with RNA Pol II and occupy the promoter of repressed genes in a complex with RNA Pol II (Yakovchuk et al. 2009). Further, the *Alu* ncRNA prevents pol II from engaging the promoter DNA by preventing appropriate contacts between the polymerase and the promoter, both upstream and downstream of the TATA box during closed complex formation by RNA Pol II in initiating transcription (Mariner et al. 2008). As a result, an inert transcriptional complex is formed at the promoter with the polymerase being trapped in the complex thereby resulting in gene repression. During the heat shock response, transcription of human *Alu* and mouse *B2* RNA increase several folds (Li et al. 1999; Liu et al. 1995). The increased RNA acts as a *trans*-acting repressor of transcription during cellular heat shock response by the above mechanism. Interesting knockdown of the non-coding RNA molecules during heat shock relieves the transcript repression significantly thus establishing the direct role of the RNA in transcription repression (Mariner et al. 2008). A very similar role for another non-coding RNA, known as Satellite III transcripts (Sat3), in HSR was shown in the human cell lines. The Sat3 transcripts which are human specific are synthesized upon heat shock and remain associated with their locus of transcription (Jolly et al. 2004). Intriguingly, these transcripts co-localize with transcription factors such as HSF1 and CREB-binding protein (CREBBP; also known as CBP), RNA polymerase II and RNA-binding proteins such as SRSF1 (serine/arginine-rich splicing factor 1; also known as SF2 or ASF), KHDRBS1 (also known as Sam68), and several heterogeneous nuclear ribonucleoproteins (hnRNPs) (Weighardt et al. 1999; Denegri et al. 2001; Metz et al. 2004; Chiodi et al. 2004; Jolly et al. 2004). A recent study demonstrates that Sat3 transcripts are involved in transcription repression by sequestering the transcription co-activator CREBBP to its locus, thus rendering it less/unavailable for the transcription of its target genes during heat shock (Goenka et al. 2016). Interestingly, knockdown of either the Sat3 RNA or the SRSF1 redistributes CREBBP from the nSBs into the nucleus in the heat-shocked cells (Goenka et al. 2016), reflecting the hierarchy in the cascade from HSF1 to Sat3 to SRSF1 to CREBBP which essentially functions as a co-activator and has more than 400 potential sites in the genome (Zhang et al. 2005); its modulation thus could have cascading effects on secondary and tertiary targets. Thus, the Sat3 RNA mediates the transcriptional repression observed during heat shock suggests a role analogous to the *hsr-omega* transcripts in *Drosophila* (Jolly and Lakhotia 2006).

12.1.3 Heat Shock Response and RNA Splicing

One of the key consequences of the meticulously regulated heat shock response is the altered splicing events or inhibition of splicing as observed during heat stress. A recent study reports a significant inhibition in splicing, particularly by the mechanism of Intron Retention (IR), during a heat stress (Hussong et al. 2017). Notably, the introns which are perfectly spliced in normal cells were retained in the pre-mRNA during heat shock. As a result, there is a significant reduction in the transcript abundance due to the introduction of premature termination codon and the mis-spliced transcript being degraded through the non-sense mediated degradation (NMD) pathway (Hussong et al. 2017). One of the critical factors that regulate the alternative splicing is an acetylated histone binding protein named BRD4 (bromodomain protein 4). During a heat shock, BRD4 gets recruited to the Sat3-positive nuclear stress bodies in an HSF1-dependent manner, depleting its concentration from gene cassettes thereby resulting in delayed or inhibition of splicing events, leading to the degradation of abnormally processed mRNAs (Hussong et al. 2017).

12.1.4 Heat Shock Response and Translational Control

Translational regulation has a vital role in maintaining cellular homeostasis; for example, in mammals, high temperatures cause an immediate halt of global protein synthesis, which recovers when cells are returned to normal temperature (Banerji et al. 1984; De Benedetti and Baglioni 1984). Heat shock induced translation regulation was first described in *Drosophila* by Storti et al. (Storti et al. 1980) and later in other organisms, cells, and tissues (Spriggs et al. 2010; Richter et al. 2010) and in mammalian cells translation regulation during heat shock was first described in chicken reticulocytes by Banerji et al. (Banerji et al. 1984). One of the most commonly used mechanisms for inhibiting global translation during heat shock in mammalian cells is by phosphorylation of Eukaryotic Initiation Factor 2 (eIF2) (see Fig. 12.2), a critical factor involved in the initiation of translation (Trachsel et al. 1978; Rowlands et al. 1988; De De Benedetti and Baglioni 1986; Dubois et al. 1989). Phosphorylation of eIF2α is regulated by various stress-activated eIF2α kinases (Holcik and Sonenberg 2005). HSP act as a chaperone for eIF2 α -specific kinases and regulates their activation (Matts et al. 1992, 1993). Intriguingly, in response to heat shock, phosphorylation of eIF2a can induce the formation of stress granules (SGs) (see Fig. 12.2). SGs are the dynamic non-membranous cytosolic foci which range in size 0.1 μm to 2 μm in size (Anderson and Kedersha 2006). SGs are primarily composed of un-translated mRNAs, translation initiation components, and other proteins which effects mRNA function like the TDP43 and G3BP1 (Anderson and Kedersha 2006, 2008; Buchan and Parker 2009). SGs are known to recruit various RNA-binding proteins (such as HuR, TIA1, SMN, TDP-43, FUS, hnRNPA1, ATXN2, VCP) (Anderson and Kedersha 2008; Buchan and Parker 2009; Wolozin 2012; McDonald et al. 2011; Bosco et al. 2010), transcription factors (SRC3, Rpb4p) (Yu et al. 2007; Lotan et al. 2005), signaling molecules (dsRNA dependent protein kinase (PKR)) (Reineke and Lloyd 2015), RNA helicases (Dhh1/RCK, Rhau, P54 RNA helicase) (Sheth and Parker 2003; Swisher and Parker 2010; Chalupníková et al. 2008; Zampedri et al. 2016), and nucleases (TSN) (Shao et al. 2017; Weissbach and Scadden 2012). SGs are known to harbor microRNAs,

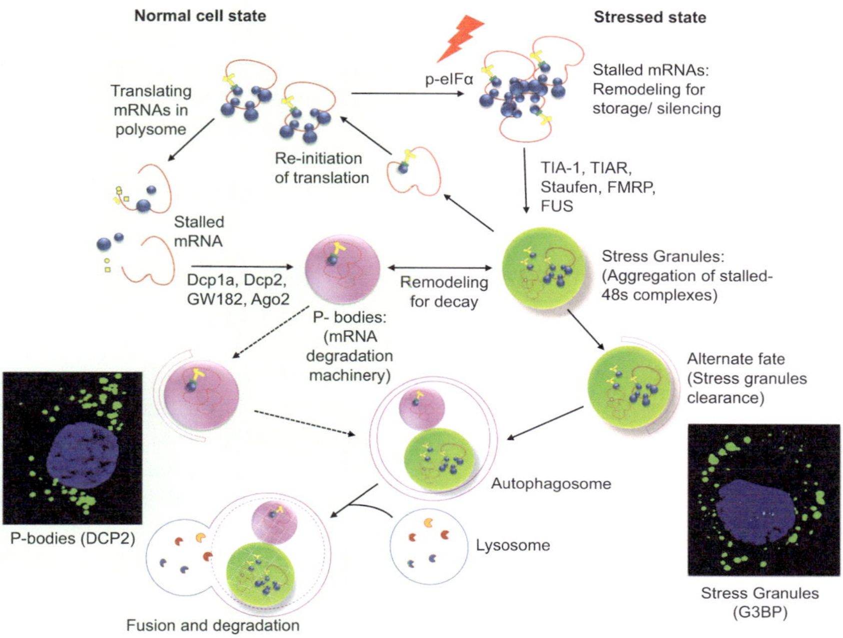

Fig. 12.2 Schematic diagram showing the fate the mRNA during a normal cell state (*left side*) or during a stressed state (*right side*), as indicated. During the normal state, the translated mRNA might get recruited to the processing bodies (P-bodies). The P-bodies represent a lose assembly of mRNA and P-body specific proteins (such as the Dcp1a, Dcp2, GW182, and Ago2) and are involved in mRNA storage and degradation. The fate P-bodies in the cell are not clear; one line of evidences indicates that they might get degraded via the autophagic process. When the cells are exposed to a stress, such as heat shock, the heat shock response pathway is activated which in turn arrests the translation process. This is done via the formation of a specialized structure called stress granules which recruit the mRNA-ribosome complex. The stress granule-specific proteins, such as TIA-1, TIAR, Staufen, FMRP, and FUS, are thought to arrest the translational process. The mRNA-ribosome complex may resume translation during the recovery phase or the damaged complex may be degraded via the P-bodies or the autophagic process. The fluorescence images shown at the bottom of the schematic visualize the P-bodies (stained for Dcp2; green color) (*left side*) in a normal state or the stress granule (stained for G3BP; green color) (*right side*) upon exposure to a heat shock. The nuclei were visualized by staining with DAPI (blue color)

argonaute proteins, mRNA-editing enzymes, and proteins required for transposon activity (Leung and Sharp 2013). Once formed, the SGs act as the site of mRNA triage to sequester mRNAs for stability or degradation (Anderson and Kedersha 2008). Further, the transcripts are facilitated from stress granules to processing bodies (PBs) for degradation (Anderson and Kedersha 2006; Anderson and Kedersha 2008; also, see below). Processing bodies are the cytoplasmic domains for the RNA decay machinery (Eulalio et al. 2007a). PBs represent a loose assembly of mRNA decay factors (Dcp1/Dcp2, Hedls), and the factors involved in the NMD and decapping machinery (Pérez-Ortín et al. 2013; Eulalio et al. 2007b). The cellular recovery process from a heat stress is accompanied by the disassembly of SGs. The assembly and disassembly of SGs are dependent on various mRNA-binding proteins (RBPs) (Emde and Hornstein 2014), and thus represent essential components of the cellular stress response mechanism. Abnormal or aberrant functions of SGs are associated with several diseases in humans, including cancer and neurodegenerative disorders (Mahboubi and Stochaj 2017).

12.1.5 Heat Shock Response and Proteolytic Pathways

Besides altering the transcription and translational processes, heat shock also affects the half-life of proteins. Proteins are five times more stable and 900 times more abundant to their corresponding mRNAs that encode them in the mammalian cells (Schwanhäusser et al. 2011). Secondly, since proteins are the tertiary products, perturbations in protein stability and function would have an immediate impact on the cell as compared to the effect of heat shock on RNAs. Heat shock is known to induce unfolding of protein and their aggregations (Parag et al. 1987), thus these aberrant protein forms need to be selectively degraded (Goldberg and St. John 1976; Hershko et al. 1983). Ubiquitin proteasome system is well-known degradation machinery and also thought to be clear heat-denatured proteins in mammalian cells (Hershko et al. 1983; Ciechanover et al. 1984; Bond and Schlesinger 1985; Finley et al. 2004). To reduce the overload of abnormal proteins the heat shock response is also known to activate proteolytic processes (Dokladny et al. 2015). Intriguingly, HSF1, a master of heat shock response is known to regulate autophagy – the bulk degradative pathway in the cell via transcription of an autophagy gene, *Atg*-7 and induces autophagy (Liu et al. 2010). Indeed, a regulatory cooperation between the heat shock response pathway and the autophagy pathway has been proposed and demonstrated (Dokladny et al. 2013; Jain et al. 2017). HSF1 was shown regulate the LC3 lipidation – a critical step in the activation of autophagy (Dokladny et al. 2013), suggesting a well-coordinated interaction between the two homeostatic pathways. Besides the degradation of aberrant proteins, the heat shock-induced autophagy is also known to clear SGs and processing bodies via autophagy (Buchan et al. 2013; Walters et al. 2015; Mateju et al. 2017) (see Fig. 12.2).

12.1.6 Heat Shock Response in Infection

Heat shock response is not limited to exposure to external stress; it can be induced to stimulate the host immune response during an infection including fever (Hasday and Singh 2000). HSF1 serves as an important factor for immunoregulatory signaling and the expression of several pro-inflammatory factors (reviewed in Hasday and Singh 2000). Fever acts as a protective effect in response to the infection in the body either by directly inducing heat shock response in the host or by the activation of pathogen heat shock proteins (Kluger 1978). A study by Kluger et al showed a significant increase of survival in ectothermic vertebrates during infection with a gram-negative pathogen. Similar findings were observed in goldfish, sockeye salmon, crickets, grasshoppers, rainbow trout followed by experiments in mice model (Kluger 2005). A 2°C increase of the core body temperature of mice infected with a lethal form of herpes simplex virus leads to up to 85% increase in the survival of infected animals (Schmidt and Rasmussen 1960; Bell and Moore 1974;, Jiang et al. 2000). Conversely, blocking fever in animals infected with microbes lead to the poor survival of the host (Bernheim and Kluger 1976; Vaughn et al. 1980; Esposito 1984; Kurosawa et al. 1987). The HSR is thought to protect cells against the fever by preventing the heat-induced protein denaturation and by activating proteolytic processes. A more direct role for HSR in the pro-inflammatory response has also been documented. For example, the HSF1 induced heat shock proteins are known to modulate the activity of NF-κB at different levels (Leite et al. 2016). In general, HSR is a primary defense mechanism that protects cells from apoptosis, by clearing the abnormally misfolded proteins and inhibiting the pro-apoptotic proteins (Stefani and Dobson 2003; Stankiewicz et al. 2005; Lanneau et al. 2008; Ali et al. 2010; Upadhyay et al. 2016). The importance of the proteolytic processes in cell survival has come from the studies on neurodegenerative disorders (Mittal and Ganesh 2010). A majority of these disorders are associated with the accumulation of abnormally folded proteins in the neurons and activation of HSP was shown to protect the neurons from the toxicity of these aggregates (Dong et al. 2005; Kumar et al. 2007; Nagel et al. 2008; Jung et al. 2008; Kalia et al. 2010), indirectly by inhibiting the pro-inflammatory and pro-apoptotic signaling pathways. Thus, the HSR response found in various neurodegenerative disorders, ageing, dementia, and cancer possibly represent a failed attempt by the cellular machinery.

12.2 Conclusions

The discovery of heat shock response pathway leads to the thorough characterization of HSF1, its targets and the functional significance of heat shock proteins coded by them. Studies have shown that the activation of HSR is not just limited to heat shock, that it is a conserved pro-survival pathway and is activated by exposure to the variety of stressor, including infection. The HSR is not just limited to the activation of a set of genes coding for HSP. As discussed above, the HSR is accompanied by global changes in the transcriptional, translational and post-translational changes in the cell exposed to a stress. While much advancement has been made in understanding the role of HSF1 in activating the genes coding for the HSP, our understanding of the mechanism behind the HSR in modulating the global changes in the transcription, translation, and post-translational processes have been very limited. Secondly, most of the studies on HSR have restricted the analyses to cellular processes and thus HSR was considered as a cell-autonomous process. Perhaps this is due to the model systems used for the studies – such as cell lines and the yeast model. Thus there is a greater need for studies on HSR as a systemic response to thermal stress in multicellular organisms. Indeed, the possible involvement of neuronal signaling in cell-non-autonomous mode of heat shock response has been shown in worms (Prahlad et al. 2008), that this may involve serotonergic signaling (Tatum et al. 2015) and that such mechanism could operate for the protein homeostasis processes (Sala et al. 2017). Thus, appropriate genetic screens are required to understand cell-non-autonomous HSRs in mammalian species not only to understand the HSR *per se* but also to relate HSR in disease conditions.

Acknowledgements Subramaniam Ganesh is a Tata Innovation Fellow and P.K. Kelkar Endowed Chair Professor at IIT Kanpur. Research work on cellular stress response in his laboratory is supported by the Department of Biotechnology, Govt. of India (Grant Number: BT/HRD/35/01/01/2017). Rashmi Parihar is a SERB Young Scientists and is supported by a research grant from the Science and Engineering Research Board, Govt. of India (Grant Number: YSS/2015/001818).

References

Ali YO, Kitay BM, Zhai RG (2010) Dealing with misfolded proteins: examining the neuroprotective role of molecular chaperones in neurodegeneration. Molecules 15:6859–6887

Allen TA, Von KS, Goodrich JA, Kugel JF (2004) The SINE-encoded mouse B2 RNA represses mRNA transcription in response to heat shock. Nat Struct Mol Biol 11:816–821

Anderson P, Kedersha N (2006) RNA granules. J Cell Biol 172:803–808

Anderson P, Kedersha N (2008) Stress granules: the Tao of RNA triage. Trends Biochem Sci 33:141–150

Banerji SS, Theodorakis NG, Morimoto RI (1984) Heat shock-induced translational control of HSP70 and globin synthesis in chicken reticulocytes. Mol Cell Biol 4:2437–2448

Bell JF, Moore GJ (1974) Effects of high ambient temperature on various stages of rabies virus infection in mice. Infect Immun 10:510–515

Bernheim HA, Kluger MJ (1976) Fever: effect of drug-induced antipyresis on survival. Science 193:237–239

Bond U, Schlesinger MJ (1985) Ubiquitin is a heat shock protein in chicken embryo fibroblasts. Mol Cell Biol 5:949–956

Bosco DA, Lemay N, Ko HK, Zhou H et al (2010) Mutant FUS proteins that cause amyotrophic lateral sclerosis incorporate into stress granules. Hum Mol Genet 19:4160–4175

Buchan JR, Parker R (2009) Eukaryotic stress granules: the ins and outs of Translation. Mol Cell 36:932–941

Buchan JR, Kolaitis RM, Taylor JP, Parker R (2013) Eukaryotic stress granules are cleared by autophagy and Cdc48/VCP function. Cell 153:1461–1474

Cernilogar FM, Onorati MC, Kothe GO et al (2011) Chromatin-associated RNA interference components contribute to transcriptional regulation in Drosophila. Nature 480:391–395

Chalupníková K, Lattmann S, Selak N et al (2008) Recruitment of the RNA helicase RHAU to stress granules via a unique RNA-binding domain. J Biol Chem 283:35186–35198

Chiodi I, Corioni M, Giordano M et al (2004) RNA recognition motif 2 directs the recruitment of SF2/ASF to nuclear stress bodies. Nucleic Acids Res 32:4127–4136

Ciechanover A, Finley D, Varshavsky A (1984) Ubiquitin dependence of selective protein degradation demonstrated in the mammalian cell cycle mutant ts85. Cell 37:57–66

De Benedetti A, Baglioni C (1984) Inhibition of mRNA binding to ribosomes by localized activation of dsRNA-dependent protein kinase. Nature 311:79–81

De Benedetti A, Baglioni C (1986) Activation of hemin-regulated initiation factor-2 kinase in heat-shocked HeLa cells. J Biol Chem 261:338–342
Denegri M, Chiodi I, Corioni M, Cobianchi F, Riva S, Biamonti G (2001) Stress-induced nuclear bodies are sites of accumulation of pre-mRNA processing factors. Mol Biol Cell 12:3502–3514
Dokladny K, Zuhl MN, Mandell M (2013) Regulatory coordination between two major intracellular homeostatic systems: heat shock response and autophagy. J Biol Chem 288:14959–14972
Dokladny K, Myers OB, Moseley PL (2015) Heat shock response and autophagy-cooperation and control. Autophagy 11:200–213
Dong Z, Wolfer DP, Lipp HP, Büeler H (2005) Hsp70 gene transfer by adeno-associated virus inhibits MPTP-induced nigrostriatal degeneration in the mouse model of Parkinson disease. Mol Ther 11:80–88
Dubois MF, Galabru J, Lebon P, Safer B, Hovanessian AG (1989) Reduced activity of the interferon-induced double-stranded RNA-dependent protein kinase during a heat shock stress. J Biol Chem 264:12165–12171
Emde A, Hornstein E (2014) miRNAs at the interface of cellular stress and disease. EMBO J 33:1428–1437
Esposito AL (1984) Aspirin impairs antibacterial mechanisms in experimental pneumococcal pneumonia. Am Rev Respir Dis 130:857–862
Eulalio A, Behm-Ansmant I, Izaurralde E (2007a) P bodies: at the crossroads of post-transcriptional pathways. Nat Rev Mol Cell Biol 8:9–22
Eulalio A, Behm-Ansmant I, Schweizer D, Izaurralde E (2007b) P-body formation is a consequence, not the cause, of RNA-mediated gene silencing. Mol Cell Biol 27:3970–3981
Finley D, Ciechanover A, Varshavsky A (2004) Ubiquitin as a central cellular regulator. Cell. 23 116(2 Suppl):S29–S32
Goenka A, Sengupta S, Pandey R et al (2016) Human satellite-III non-coding RNAs modulate heat-shock-induced transcriptional repression. J Cell Sci 129:3541–3552
Goldberg AL, St. John AC (1976) Intracellular protein degradation in mammalian and bacterial cells: Part 2. Annu Rev Biochem 45:747–803
Hasday JD, Singh IS (2000) Fever and the heat shock response: distinct, partially overlapping processes. Cell Stress Chaperones 5:471–480
Hershko A, Heller H, Elias S, Ciechanover A (1983) Components of ubiquitin-protein ligase system. Resolution, affinity purification, and role in protein breakdown. J Biol Chem 258:8206–8214
Holcik M, Sonenberg N (2005) Translational control in stress and apoptosis. Nat Rev Mol Cell Biol 6:318–327
Hussong M, Kaehler C, Kerick M et al (2017) The bromodomain protein BRD4 regulates splicing during heat shock. Nucleic Acids Res 45:382–394
Jain N, Rai A, Mishra R, Ganesh S (2017) Loss of malin, but not laforin, results in compromised autophagic flux and proteasomal dysfunction in cells exposed to heat shock. Cell Stress Chaperones 22:307–315
Jamrich M, Greenleaf AL, Bautz EK (1977) Localization of RNA polymerase in polytene chromosomes of Drosophila melanogaster. Proc Natl Acad Sci U S A 74:2079–2083
Jiang Q, Cross AS, Singh IS et al (2000) Febrile core temperature is essential for optimal host defense in bacterial peritonitis. Infect Immun 68:1265–1270
Jolly C, Lakhotia SC (2006) Human sat III and Drosophila hsr omega transcripts: a common paradigm for regulation of nuclear RNA processing in stressed cells. Nucleic Acids Res 34:5508–5514
Jolly C, Metz A, Govin J et al (2004) Stress-induced transcription of satellite III repeats. J Cell Biol 164:25–33
Jung AE, Fitzsimons HL, Bland RJ, During MJ, Young D (2008) HSP70 and constitutively active HSF1 mediate protection against CDCrel-1-mediated toxicity. Mol Ther 16:1048–1055
Kalia SK, Kalia LV, McLean PJ (2010) Molecular chaperones as rational drug targets for Parkinson's disease therapeutics. CNS Neurol Disord Drug Targets 9:741–753

Kantidze OL, Velichko AK, Razin SV (2015) Heat Stress-Induced Transcriptional Repression. Biochemistry (Mosc.) 80:990–993
Kluger MJ (1978) The evolution and adaptive value of fever. Am Sci 66:38–43
Kluger MJ (2005) The adaptive value of fever. In: Mackowiak PA (ed) Fever: Basic Mechanisms and Management. Raven Press, New York, pp 105–124
Kumar P, Ambasta RK, Veereshwarayya V et al (2007) CHIP and HSPs interact with beta-APP in a proteasome-dependent manner and influence Abeta metabolism. Hum Mol Genet 16:848–864
Kurosawa S, Kobune F, Okuyama K, Sugiura A (1987) Effects of antipyretics in rinderpest virus infection in rabbits. J Infect Dis 155:991–997
Lakhotia SC (1989) The 93D heat shock locus of Drosophila melanogaster: modulation by genetic and developmental factors. Genome 31:677–683
Lanneau D, Brunet M, Frisan E, Solary E, Fontenay M, Garrido C (2008) Heat shock proteins: essential proteins for apoptosis regulation. J Cell Mol Med 12:743–761
Leite JSM, Cruzat VF, Krause M et al (2016) Physiological regulation of the heat shock response by glutamine: implications for chronic low-grade inflammatory diseases in age-related conditions. Nutrire 41:17
Leung AK, Sharp PA (2013) Quantifying Argonaute proteins in and out of GW/P-bodies: implications in microRNA activities. Adv Exp Med Biol 768:165–182
Levine M (2011) Paused RNA polymerase II as a developmental checkpoint. Cell 145:502–511
Li T, Spearow J, Rubin CM, Schmid CW (1999) Physiological stresses increase mouse short interspersed element (SINE) RNA expression in vivo. Gene 239:367–372
Li B, Carey M, Workman JL (2007) The role of chromatin during transcription. Cell 128:707–719
Lindquist S (1986) The heat-shock response. Annu Rev Biochem 55:1151–1191
Liu WM, Chu WM, Choudary PV, Schmid CW (1995) Cell stress and translational inhibitors transiently increase the abundance of mammalian SINE transcripts. Nucleic Acids Res 23:1758–1765
Liu TT, Hu CH, Tsai CD, Li CW, Lin YF, Wang JY (2010) Heat stroke induces autophagy as a protection mechanism against neurodegeneration in the brain. Shock 34:643–648
Lotan R, Bar-On VG, Harel-Sharvit L, Duek L, Melamed D, Choder M (2005) The RNA polymerase II subunit Rpb4p mediates decay of a specific class of mRNAs. Genes Dev 19:3004–3016
Mahboubi H, Stochaj U (2017) Cytoplasmic stress granules: Dynamic modulators of cell signaling and disease. Biochim Biophys Acta 1863:884–895
Maheshwari M, Bhutani S, Das A et al (2014) Dexamethasone induces heat shock response and slows down disease progression in mouse and fly models of Huntington's disease. Hum Mol Genet 23:2737–2751
Mariner PD, Walters RD, Espinoza CA et al (2008) Human Alu RNA is a modular transacting repressor of mRNA transcription during heat shock. Mol Cell 29:499–509
Mateju D, Franzmann TM, Patel A et al (2017) An aberrant phase transition of stress granules triggered by misfolded protein and prevented by chaperone function. EMBO J 36:1669–1687
Matts RL, Xu Z, Pal JK, Chen JJ (1992) Interactions of the heme-regulated eIF-2 alpha kinase with heat shock proteins in rabbit reticulocyte lysates. J Biol Chem 267:18160–18167
Matts RL, Hurst R, Xu Z (1993) Denatured proteins inhibit translation in hemin-supplemented rabbit reticulocyte lysate by inducing the activation of the heme-regulated eIF-2 alpha kinase. Biochemistry 32:7323–7328
McDonald KK, Aulas A, Destroismaisons L et al (2011) TAR DNA-binding protein 43 (TDP-43) regulates stress granule dynamics via differential regulation of G3BP and TIA-1. Hum Mol Genet 20:1400–1410
Metz A, Soret J, Vourc'h C, Tazi J, Jolly C (2004) A key role for stress-induced satellite III transcripts in the relocalization of splicing factors into nuclear stress granules. J Cell Sci 117:4551–4558
Missra A, Gilmour DS (2010) Interactions between DSIF (DRB sensitivity inducing factor), NELF (negative elongation factor), and the Drosophila RNA polymerase II transcription elongation complex. Proc Natl Acad Sci U S A 107:11301–11306

Mittal S, Ganesh S (2010) Protein quality control mechanisms and neurodegenerative disorders: Checks, balances and deadlocks. Neurosci Res 68:159–166

Morimoto RI (1998) Regulation of the heat shock transcriptional response: cross talk between a family of heat shock factors, molecular chaperones, and negative regulators. Genes Dev 12:3788–3796

Nagel F, Falkenburger BH, Tönges L et al (2008) Tat-Hsp70 protects dopaminergic neurons in midbrain cultures and in the substantia nigra in models of Parkinson's disease. J Neurochem 105:853–864

Parag HA, Raboy B, Kulka RG (1987) Effect of heat shock on protein degradation in mammalian cells: involvement of the ubiquitin system. EMBO J 6:55–61

Pérez-Ortín JE, Alepuz P, Chávez S, Choder M (2013) Eukaryotic mRNA decay: methodologies, pathways, and links to other stages of gene expression. J Mol Biol 425:3750–3775

Peterlin BM, Price DH (2006) Controlling the elongation phase of transcription with P-TEFb. Mol Cell 23:297–305

Powers ET, Morimoto RI, Dillin A, Kelly JW, Balch WE (2009) Biological and chemical approaches to diseases of proteostasis deficiency. Annu Rev Biochem 78:959–991

Prahlad V, Cornelius T, Morimoto RI (2008) Regulation of the cellular heat shock response in Caenorhabditis elegans by thermosensory neurons. Science 320:811–814

Reineke LC, Lloyd RE (2015) The stress granule protein G3BP1 recruits protein kinase R to promote multiple innate immune antiviral responses. J Virol 89:2575–2589

Richter K, Haslbeck M, Buchner J (2010) The heat shock response: life on the verge of death. Mol Cell 40:253–266

Rougvie AE, Lis JT (1988) The RNA polymerase II molecule at the 5' end of the uninduced hsp70 gene of D. melanogaster is transcriptionally engaged. Cell 54:795–804

Rowlands AG, Montine KS, Henshaw EC, Panniers R (1988) Physiological stresses inhibit guanine-nucleotide-exchange factor in Ehrlich cells. Eur J Biochem 175:93–99

Sala AJ, Bott LC, Morimoto RI (2017) Shaping proteostasis at the cellular, tissue, and organismal level. J Cell Biol 216:1231–1241

Sarge KD, Zimarino V, Holm K, Wu C, Morimoto RI (1991) Cloning and characterization of two mouse heat shock factors with distinct inducible and constitutive DNA-binding ability. Genes Dev 5:1902–1911

Sarge KD, Murphy SP, Morimoto RI (1993) Activation of heat shock gene transcription by heat shock factor 1 involves oligomerization, acquisition of DNA-binding activity, and nuclear localization and can occur in the absence of stress. Mol Cell Biol 13:1392–1407

Schmidt J, Rasmussen AJ (1960) The influence of environmental temperature on the course of experimental herpes simplex infection. J Infect Dis 107:356–360

Schwanhäusser B, Busse D, Li N, Dittmar G et al (2011) Global quantification of mammalian gene expression control. Nature 473:337–342

Shalgi R, Hurt JA, Lindquist S, Burge CB (2014) Widespread inhibition of posttranscriptional splicing shapes the cellular transcriptome following heat shock. Cell Rep 7:1362–1370

Shamovsky I, Ivannikov M, Kandel ES, Gershon D, Nudler E (2006) RNA-mediated response to heat shock in mammalian cells. Nature 440:556–560

Shao J, Gao F, Zhang B et al (2017) Aggregation of SND1 in stress granules is associated with the microtubule cytoskeleton during heat shock stimulus. Anat Rec (Hoboken). https://doi.org/10.1002/ar.23642

Sheth U, Parker R (2003) Decapping and decay of messenger RNA occur in cytoplasmic processing bodies. Science 300:805–808

Spriggs KA, Bushell M, Willis AE (2010) Translational regulation of gene expression during conditions of cell stress. Mol Cell 40:228–237

Stankiewicz AR, Lachapelle G, Foo CP, Radicioni SM, Mosser DD (2005) Hsp70 inhibits heat-induced apoptosis upstream of mitochondria by preventing Bax translocation. J Biol Chem 280:38729–38739

Stefani M, Dobson CM (2003) Protein aggregation and aggregate toxicity: new insights into protein folding, misfolding diseases and biological evolution. J Mol Med (Berl) 81:678–699

Storti RV, Scott MP, Rich A, Pardue ML (1980) Translational control of protein synthesis in response to heat shock in D. melanogaster cells. Cell 22:825–834
Swisher KD, Parker R (2010) Localization to, and effects of Pbp1, Pbp4, Lsm12, Dhh1, and Pab1 on stress granules in Saccharomyces cerevisiae. PLoS One 5:e10006
Tatum MC, Ooi FK, Chikka MR et al (2015) Neuronal serotonin release triggers the heat shock response in C. elegans in the absence of temperature increase. Curr Biol 25:163–174
Teves SS, Henikoff S (2013) The heat shock response: A case study of chromatin dynamics in gene regulation. Biochem Cell Biol 91:42–48
Trachsel H, Ranu RS, London IM (1978) Regulation of protein synthesis in rabbit reticulocyte lysates: purification and characterization of heme-reversible translational inhibitor. Proc Natl Acad Sci U S A 75:3654–3658
Upadhyay M, Bhadauriya P, Ganesh S (2016) Heat shock modulates the subcellular localization, stability, and activity of HIPK2. Biochem Biophys Res Commun 472:580–584
Vaughn LK, Veale WL, Cooper KE (1980) Antipyresis: its effect on mortality rate of bacterially infected rabbits. Brain Res Bull 5:69–73
Velichko AK, Petrova NV, Kantidze OL, Razin SV (2012) Dual effect of heat shock on DNA replication and genome integrity. Mol Biol Cell 23:3450–3460
Velichko AK, Markova EN, Petrova NV, Razin SV, Kantidze OL (2013) Mechanisms of heat shock response in mammals. Cell Mol Life Sci 70:4229–4241
Walters RW, Muhlrad D, Garcia J, Parker R (2015) Differential effects of Ydj1 and Sis1 on Hsp70-mediated clearance of stress granules in Saccharomyces cerevisiae. RNA 21:1660–1671
Weighardt F, Cobianchi F, Cartegni L et al (1999) A novel hnRNP protein (HAP/SAF-B) enters a subset of hnRNP complexes and relocates in nuclear granules in response to heat shock. J Cell Sci 112:1465–1476
Weissbach R, Scadden AD (2012) Tudor-SN and ADAR1 are components of cytoplasmic stress granules. RNA 18:462–471
Westerheide SD, Morimoto RI (2005) Heat shock response modulators as therapeutic tools for diseases of protein conformation. J Biol Chem 280:33097–33100
Westwood JT, Wu C (1993) Activation of Drosophila heat shock factor: conformational change associated wqith a monomer-to-trimer transition. Mol Cell Biol 13:3481–3486
Wolozin B (2012) Regulated protein aggregation: stress granules and neurodegeneration. Mol Neurodegener 7:56
Wu CH, Yamaguchi Y, Benjamin LR et al (2003) NELF and DSIF cause promoter proximal pausing on the hsp70 promoter in Drosophila. Genes Dev 17:1402–1414
Yakovchuk P, Goodrich JA, Kugel JF (2009) B2 RNA and Alu RNA repress transcription by disrupting contacts between RNA polymerase II and promoter DNA within assembled complexes. Proc Natl Acad Sci U S A 106:5569–5574
Yu C, York B, Wang S, Feng Q, Xu J, O'Malley BW (2007) An essential function of the SRC-3 coactivator in suppression of cytokine mRNA translation and inflammatory response. Mol Cell 25:765–778
Zampedri C, Tinoco-Cuellar M, Carrillo-Rosas S et al (2016) Zebrafish P54 RNA helicases are cytoplasmic granule residents that are required for development and stress resilience. Biol Open 5:1473–1484
Zhang X, Odom DT, Koo S-H et al (2005) Genome-wide analysis of cAMP-response element binding protein occupancy, phosphorylation, and target gene activation in human tissues. Proc Natl Acad Sci USA 102:4459–4464

Chapter 13
HSP70 Mediated Stress Modulation in Plants

Rashid Mehmood Rana, Azhar Iqbal, Fahad Masoud Wattoo, Muhammad Azam Khan, and Hongsheng Zhang

Abstract Stresses induced by abiotic and biotic factors badly effect sessile natured plants. Several innate mechanism has been established by plants during evolution. Expression of specialized proteins under harsh conditions to mitigate the effects is one of the conserved mechanisms. Heat shock proteins (hsp) are one of those conserved proteins. Among HSP, Hsp70 holds critical role in stress tolerance as well as development. HSP70 are structurally conserved across the genomes and reported to mitigate stress at cellular level. At the onset of stress, HSP70 releases HSF to translocate and transcribe HSP resulting in increased tolerance. While, abundant HSP70 bind to HSF to down-regulate the expression of HSP. Hsp70 associated with mitochondria and endoplasmic reticulum respond to accumulation of unfolded proteins (unfolded protein response) and repair them. Moreover, they are also associated to control programmed cell death in plants by interacting with BAG proteins.

Keywords Heat shock · Heat shock proteins · HSP70 · Misfolding · UPR

Abbreviations

ABA	Abscisic acid
BAG	Bcl-2 associated athanogenes
BiPs	Binding immunoglobulin proteins
DnaJ	Bacterial homologue of Hsp40
DnaK	Bacterial homologue of Hsp70

R. M. Rana (✉) · A. Iqbal · F. M. Wattoo
Department of Plant Breeding and Genetics, PMAS-Arid Agriculture University, Rawalpindi, Pakistan

M. A. Khan
Department of Horticulture, PMAS-Arid Agriculture University, Rawalpindi, Pakistan

H. Zhang
State Key Laboratory of Crop Genetics and Germplasm Enhancement, Nanjing Agricultural University, Nanjing, China

A. A. A. Asea, P. Kaur (eds.), *Heat Shock Proteins and Stress*,
Heat Shock Proteins 15, https://doi.org/10.1007/978-3-319-90725-3_13

ER	Endoplasmic reticulum
HR	Hypersensitive response
HS	Heat shock
HSC	Heat shock cognate protein
HSF	Heat shock transcription factor
HSP	Heat shock protein
Hsp-BP	Hsp binding proteins
HSR	Heat shock response
NBD	Nucleotide binding domain
NEF	Nucleotide exchange factor
ROS	Reactive oxygen species
SBD	Substrate binding domain
sHSP	Small heat shock proteins
UPR	Unfolded protein response

13.1 Introduction

Ever increasing atmospheric temperature is becoming an emerging threat to life on earth including sessile natured plants. The secondary effects of temperature could be increase in the spells of drought. Along with climatic elements, plants also face various biotic stress including pathogens (viruses, bacteria, fungi etc.) (Shao et al. 2007). Studies on evolution depicted that plants developed various innate stress mitigation systems to address their adoption to dry and hot environment (Feder and Hofmann 1999). A consortium of heat shock proteins (Hsp) could be regarded as one of the most important and established system in this context. The proteins degraded as a result of heat stress are repaired by HSP. A vast variety of these HSP have been reported in plants and have been reported to play vital role in a vast range of stresses. These heat shock proteins have been divided in various subclasses including Hsp100, Hsp90, Hsp70, Hsp60 and small HSP (sHSP) (Saidi et al. 2009). Among these, Hsp70 (DnaK-like) superfamily is involved in protein folding, degradation, transport as well as modulation of stress response processes throughout the cell (Wang et al. 2004). The members of Hsp70 superfamily characterized by the presence of N-terminal adenosine triphosphatase (ATPase) domain, substrate-binding domain, and C-terminal domain. Plants possess four types of Hsp70 based on their subcellular localization viz. cytoplasm, mitochondrion, chloroplast and endoplasmic reticulum (ER) (Karlin and Brocchieri 1998). Being ubiquitous, Hsp70 proteins perform chaperone and housekeeping role in plants (Gupta and Golding 1993). They have been characterized to play important role in transportation of proteins across the organelles as well as folding of newly synthesized proteins (Mayer and Bukau 2005). Heat stress triggers the production certain HSP70 where they participate in repair of misfolded/denatured proteins (Hartl and Hayer-Hartl 2002). Here, we summarized the types, structure and molecular mechanisms of plant Hsp70 proteins. A detail about some important interactions of HSP70 proteins

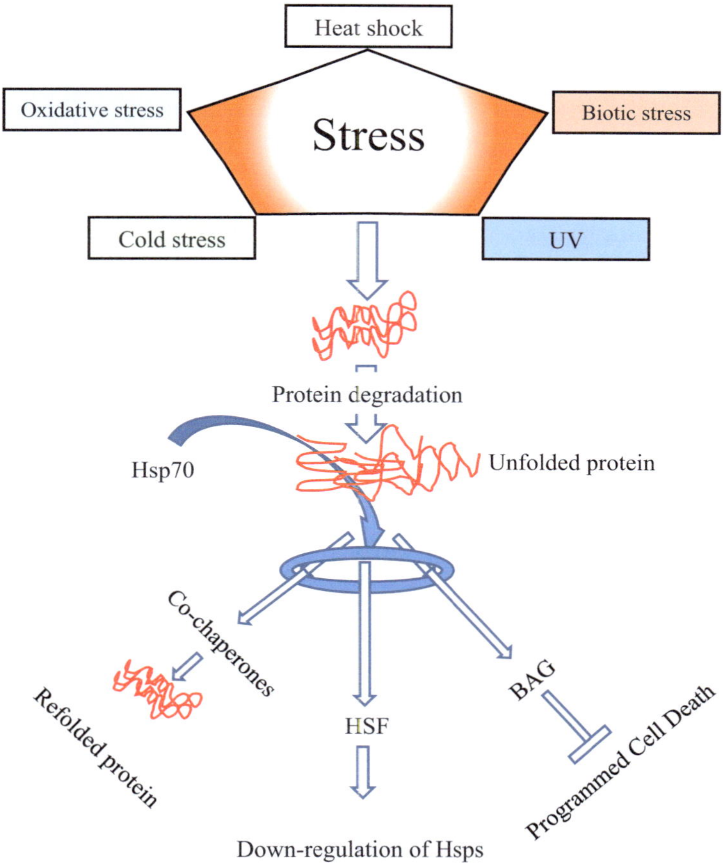

Fig. 13.1 Role of HSP70 in stress modulation. Unfolded protein accumulation activates HSP70 where it works along with co-chaperone to refold the unfolded proteins. Free molecules of HSP70 suppress HSF and down regulate the expression of Hsps. Some members of HSP70 family may interact with BAG (Bcl-2 associated athanogenes) to prevent programed cell death

with other proteins to regulate the mechanism of stress tolerance is also given, that will provide an understanding of these fundamental pathways (Fig. 13.1).

13.1.1 HSP70 Family

The Hsp70 superfamily contains Hsp70 as well as Hsp110 (Hsp70 homoloues) protein families. It has been reported in various plant species (Ray et al. 2016). The Arabidopsis genome has at least 18 members of Hsp70 superfamily containing 14 HSP70 while 4 belong to Hsp110/SSE. Five members of Hsp70 in Arabidopsis resides in cytosol; three found in endoplasmic reticulum (lumen), two are present in mitochondria, three resides in chloroplast and location of last one is not known (Sung et al. 2001a, b). In spinach genome, 12 Hsp70 family members has been reported (Li et al. 1999). The expression profiling of both Arabidopsis and spinach Hsp70 genes depicted that chaperone activity of this family is directly enhanced in response to adverse environmental conditions like cold, heat, salinity and drought as well as chemical stresses and ultimately resulted in improved stress tolerance in plants. Functional analysis of rice Hsp70 genes revealed 32 representative members including 24 Hsp70 while 8 Hsp110 (Sarkar et al. 2013).

13.1.2 Structure of HSP70

Hsp70 chaperones consist of highly conserved N terminal ATPase domain of ~44-kD (also known as nucleotide binding domain, NBD), a substrate binding domain of ~18-kD (SBD) and a highly variable C domain of ~10-kD (lid). C-terminal domain is complex in structure and consisted of five elongated α-helices responsible for structural changed during attaching and releasing of substrate (Sung et al. 2001a, b). The C terminal lid is responsible for keeping the substrate at SBD, which itself is associated to hydrophobic region expressed in non-native substrate. ATPase domain executes regulatory functions (Dragovic et al. 2006). The SBD binds transiently to hydrophobic region of binding protein, and is controlled by NBD. For the functioning of Hsp70 protein, it involves two co-chaperones in form of J-domain and nucleotide exchange factor (NEFs). Binding of J-domain stimulates hydrolysis of ATP that in turn traps the exposed hydrophobic region of the substrate and blocks conformation of SBD. NEFs take part in conversion of ATP-ADP by catalysis, the release of ADP from Hsp70 leads to convert the open conformation Hsp70 by releasing the substrate. Protein families like Mgel in yeast mytochondria, GrpE in E. coli, Hsp binding proteins (Hsp-BP) and Hsp 110 acts as NEFs (Storozhenko et al. 1996; Zhang et al. 2010).

13.1.3 HSP70 Motifs and their Functions

The motif for the mitochondrial group is PEAEYEEAKK, for the cytosolic group is EEVD and for the chloroplast group is PEGDVIDADFTDSK (Guy and Li 1998). The C terminal cytosolic motif EEVD of Hsp70 proteins interacts with co-chaperones containing numerous degenerate 34 amino acid repeats known as tetratricopeptide repeats. Hsp70 performs various biological functions with co-chaperones or in association with other chaperones. While DnaK system effectively solubilizes and refolds small protein aggregates. The aggregates of larger proteins are solubilized with Hsp100/ClpB (Diamant et al. 2003).

13.1.4 Subcellular Localization of HSP70

Hsp70 proteins perform specific functions with respect to the site of their presence in the subcellular components. Plant HSP70 has been reported in nucleus, cytosol, mitochondria, chloroplast and endoplasmic reticulum (Karlin and Brocchieri 1998). Nuclear HSP70 protect DNA from damages, cytosolic HSP70 prevent protein aggregation, assists in folding of de novo proteins (Cho and Choi 2009). HSP70 related to endoplasmic reticulum (ER) are usually known as luminal binding HSP70 or BiPs (Binding immunoglobulin Proteins) precursor and associated in unfolded protein response (UPR). Mitochondrial HSP70 function in protein import/translocation as well as mitochondrial biogenesis. Chloroplastic HSP70 protects against photo-inhibition during heat stress (Mayer and Bukau 2005).

13.1.5 Expression of HSP70

Cytosolic HSP70 may be expressed constitutively or only under stress conditions. The constitutively expressed HSP70 are called as cognate HSP70 or Hsc70s. The location of HSP70 expression and their interaction with co-chaperones determines the exact function. The expression of HSP70 has been reported under abiotic stress like cold, heat, salt, and drought as well as biotic stresses (viral, bacterial or fungal infections). Arabidopsis Hsp70 homolog AtHsp70–4 was responsive to most of the biotic and abiotic stresses (Sung et al. 2001a, b). Several rice HSP70 respond variably to stresses like salt, drought, heat and light. Sometimes mRNA is highly upregulated by some stresses while remain unaffected or even down-regulated by another type of stresses (Sarkar et al. 2013).

13.1.6 Function of HSP70

13.1.6.1 Nuclear HSP70

Nuclear HSP70 have been reported to play critically important functions. They have been found associated with synthesis of ribosomal RNA and translocation of substrates across nuclear membrane (Nollen et al. 2001). Nuclear HSP70 have also been reported in repair of nuclear damage caused by heat stress and protect nuclear components against heat stress. A nuclear Hsp70 (NtHsp70–1) in Nicotiana tabacum was found useful in preventing nuclear DNA from degradation and fragmentation due to heat stress (Cho and Choi 2009).

13.1.6.2 Cytosolic HSP70

The members of cytosolic HSP70 could be divided in to two groups based on their functions. The group responsible for intracellular targeting and protein folding are known as heat shock cognate 70 (Hsc70) protein. While, those involved in mitigating the effects of different stresses by refolding of denatured and aggregated proteins are called as HSP70 (Kominek *et al.* 2013). Cytosolic Hsc70s are reported to interact with Hip - a tetrameric protein where it stabilizes the ADP status of Hsc70 and increase the substrate holding activity of Hsp70. The Hsp70 works in collaboration with small heat shock proteins (sHSP) and acts as molecular chaperone during heat stress (Rousch et al. 2004). Hsp70 and sHsp17.6 has been reported during temperature stress (heat or cold) pretreatment in grape plants (Zhang et al. 2008). Cytosolic Hsp70 controls the abscisic acid (ABA) induced antioxidant response in maize. Abiotic stresses enhance the production of H_2O_2 that in turn increase antioxidant enzymatic activity by promoting the accumulation of Hsp70 and ABA level. The cytosolic Hsp70 in plants plays vital role in overcoming the stresses caused by salinity, heat, drought, ROS and ABA (Park and Seo 2015). Moreover, HSP70 also demonstrated their function in microbial pathogenesis like viral, or fungal infections. A cytoplasmic Hsp70 (NbHsp70c-1) from *Nicotiana benthamiana* effect viral infection by assisting virus in assembling proteins (Chen et al. 2008). Moreover, this protein was also found essential for INF1-mediated hypersensitive response (HR) against Phytophthora infestans (Kanzaki et al. 2003). Cytoplasmic HSP70 also regulates the activity of heat shock factor (HSF) and modulate the expression of heat shock response (HSR) machinery (Rana et al. 2016). Besides the general chaperonin activity, HSP70 assist in expression of other stress associated genes. Hsp70 binds with HSF (heat shock factor) and keep it monomerized. This prevents the activation of HSF and which ultimately cannot translocate in to the nucleus. It therefore downregulates the expression of HSP (Rana et al. 2016). Hsp70 also functions as modulator of signals transduction like protein kinase A, Protein kinase C and protein phosphatase. Hsp70 chaperones may broadly participate in modulating the expression of several down-stream genes in signal transduction pathway under both normal and stressed conditions (Wang et al. 2004).

13.1.6.3 Endoplasmic Reticulum HSP70

The main cause of endoplasmic stress is the accumulation of miss-folded or unfolded proteins in endoplasmic reticulum (ER) lumen. Stress activates the protective unfolded protein response (UPR) signaling pathway (Kleizen and Braakman 2004). The UPR includes ER resident molecular chaperone Hsp70 known as Binding immunoglobulin protein (BiPs). The chaperone recognizes miss-folded or unfolded proteins and expels them out of endoplasmic reticulum for degradation. Overexpression of BiP in transgenic tobacco (Nicoriana tabacum L. cv. Havan) resulted in drought tolerance and prevented dehydration of cell (Alvim et al. 2001).

Similarly BiP prevents drought induced leaf senescence in soybeans as demonstrated in transgenic line overexpressing BiP (Valente et al. 2008).

13.1.6.4 Mitochondrial Hsp70

Mostly synthesis of mitochondrial protein precursors takes place in cytosol and then transported into mitochondrial matrix. Mitochondrial Hsp70 (mtHsp70) plays an important role in mitochondrial biogenesis. As the protein precursor arrives in mitochondrial matrix, mtHsp70 binds with them and initiate refolding off substrate (Shi and Theg 2010). The mitochondrial Hsp70 has been reported to play significant role in development of mitochondrial matrix as well as protein import in pea (*pisum sativum* L.) (Dudley et al. 1997). Mitochondrial Hsp70 has also been characterized as chaperone and play important role abiotic stresses like salinity, high or low temperature, drought, and UV radiation (Liu et al. 2014).

13.1.6.5 Chloroplast HSP70

The heat shock protein Hsp70 is not merely associated with heat stress but also show expression under high light intensity, oxidative stress and pathological stresses (Yu et al. 2015). Chlamysomonas Hsp70 protects the cell from photo-inhibition while a chloroplast homolog TaHsp70 in bread wheat (*Triticum aestivum* L.) plays critical role in hypersensitive response (HR) provoked by stripe rust through jasmonic acid signaling pathway (Duan et al. 2011). Arabidopsis cpHsp70–1 is responsible for root growth under heat stress conditions during seed germination. It is reported that cpHsp70–1 and cpHsp70–2 are important for plant development and have diverse roles (Su and Li 2008). Rice (Orzya sativa) OsHspCP1 plays vital role during heat stress in development of chloroplast (Kim and An 2013).

13.1.7 HSP70 as Chaperone

Heat stress triggers the expression of HSP70 (70-kilo Dalton heat shock protein) chaperone, these chaperones along with other co-chaperones i.e., Hsp40, DnaJ, GrpE etc. produces a cellular machinery that helps in folding of de novo protein (Goloubinoff 2017). They also play an important role in refolding of non-native proteins and inhibit aggregation of protein under normal as well as stressed conditions (Kotak et al. 2007). HSP70 are also responsible for transportation of proteins and helps in proteolysis of denatured aggregated proteins through proteasomes or by lysosomes. The Hsp70 family members concerned with protein folding and translocation of precursors are expressed constitutively while other members are expressed only in stress conditions, as they enhance the capability of refolding and proteolysis rate of native protein (Hartl et al. 2011).

13.1.8 HSP70 and Unfolded Protein Response (UPR)

The mechanism of unfolded protein response (UPR) in animals and yeast are well understood where Hsp70/BiP proteins keep the aggregated proteins unfolded specially in ER to reduce accumulation of protein in stress situations (Hartl et al. 2011). Overexpression of BiP in tobacco reduces the induction of UPR and results in increased tolerance against tunicamycin (a UPR pathway activator) (Alvim et al. 2001).

13.1.9 HSP70 as Transporter

It has been reported that several members of Hsp70 molecular chaperone are associated with import of protein and its translocation to mitochondria and chloroplast (Haynes and Ron 2010). They are also involved in intercellular movement of protein and viruses from plasmodesmata (Aoki et al. 2002). The cytosolic Hsc70, with assistance of other chaperone e.g. 14–3-3 protein, a novel molecular chaperone protein, interact with protein precursor of mitochondria or chloroplast to keep them in unfolded state (Wang et al. 2004). Hsc70 may act as motor for transportation of protein precursor through membranes. The protein once reaches in stroma or matrix, the chloroplactic or mitochondrial Hsp70 proteins (cpHsp70 and mtHsp70 respectively), along with their co chaperones, react with protein precursors and cleaves the signal peptide or leader with the help of peptidase(Ueki and Citovsky 2011). The structural motif present on Hsc70 chaperone protein is compulsory for intercellular transport and shows interaction with plasmodesmatal translocation pathway (Aoki et al. 2002).

13.2 Conclusions

Conclusively, Hsp70 modulate the expression of HSP by binding to HSFs, while controls programmed cell death by interacting with BAG proteins during stresses. Moreover, HSP70 repair misfolded proteins and transport organelle proteins into their respective compartments.

Acknowledgements The authors would like to thank Dr. M Ishaq A Rehmani (Ghazi University, D.G. Khan, Pakistan) for useful discussion.

References

Alvim FC, Carolino SM, Cascardo JC, Nunes CC, Martinez CA, Otoni WC, Fontes EP (2001) Enhanced accumulation of bip in transgenic plants confers tolerance to water stress. Plant Physiol 126:1042–1054

Aoki K, Kragler F, Xoconostle-Cázares B, Lucas WJ (2002) A subclass of plant heat shock cognate 70 chaperones carries a motif that facilitates trafficking through plasmodesmata. Proc Natl Acad Sci 99:16342–16347

Chen Z, Zhou T, Wu X, Hong Y, Fan Z, Li H (2008) Influence of cytoplasmic heat shock protein 70 on viral infection of nicotiana benthamiana. Mol Plant Pathol 9:809–817

Cho EK, Choi YJ (2009) A nuclear-localized hsp70 confers thermoprotective activity and drought-stress tolerance on plants. Biotechnol Lett 31:597–606

Diamant S, Rosenthal D, Azem A, Eliahu N, Ben-Zvi AP, Goloubinoff P (2003) Dicarboxylic amino acids and glycine-betaine regulate chaperone-mediated protein-disaggregation under stress. Mol Microbiol 49:401–410

Dragovic Z, Broadley SA, Shomura Y, Bracher A, Hartl FU (2006) Molecular chaperones of the hsp110 family act as nucleotide exchange factors of hsp70s. EMBO J 25:2519–2528

Duan Y-H, Guo J, Ding K, Wang S-J, Zhang H, Dai X-W, Chen Y-Y, Govers F, Huang L-L, Kang Z-S (2011) Characterization of a wheat hsp70 gene and its expression in response to stripe rust infection and abiotic stresses. Mol Biol Rep 38:301–307

Dudley P, Wood C, Pratt J, Moore A (1997) Developmental regulation of the plant mitochondrial matrix located hsp70 chaperone and its role in protein import. FEBS Lett 417:321–324

Feder ME, Hofmann GE (1999) Heat-shock proteins, molecular chaperones, and the stress response: evolutionary and ecological physiology. Annu Rev Physiol 61:243–282

Goloubinoff P (2017) The hsp70 molecular chaperone machines. Front Mol Biosci 4

Gupta RS, Golding GB (1993) Evolution of hsp70 gene and its implications regarding relationships between archaebacteria, eubacteria, and eukaryotes. J Mol Evol 37:573–582

Guy CL, Li Q-B (1998) The organization and evolution of the spinach stress 70 molecular chaperone gene family. Plant Cell 10:539–556

Hartl FU, Hayer-Hartl M (2002) Molecular chaperones in the cytosol: from nascent chain to folded protein. Science 295:1852–1858

Hartl FU, Bracher A, Hayer-Hartl M (2011) Molecular chaperones in protein folding and proteostasis. Nature 475:324

Haynes CM, Ron D (2010) The mitochondrial upr–protecting organelle protein homeostasis. J Cell Sci 123:3849–3855

Kanzaki H, Saitoh H, Ito A, Fujisawa S, Kamoun S, Katou S, Yoshioka H, Terauchi R (2003) Cytosolic hsp90 and hsp70 are essential components of inf1-mediated hypersensitive response and non-host resistance to pseudomonas cichorii in nicotiana benthamiana. Mol Plant Pathol 4:383–391

Karlin S, Brocchieri L (1998) Heat shock protein 70 family: multiple sequence comparisons, function, and evolution. J Mol Evol 47:565–577

Kim S-R, An G (2013) Rice chloroplast-localized heat shock protein 70, oshsp70cp1, is essential for chloroplast development under high-temperature conditions. J Plant Physiol 170:854–863

Kleizen B, Braakman I (2004) Protein folding and quality control in the endoplasmic reticulum. Curr Opin Cell Biol 16:343–349

Kominek J, Marszalek J, Neuveglise C, Craig EA, Williams BL (2013) The complex evolutionary dynamics of Hsp70s: a genomic and functional perspective. Genome Biol Evol 5(12):2460–2477

Kotak S, Larkindale J, Lee U, von Koskull-Döring P, Vierling E, Scharf K-D (2007) Complexity of the heat stress response in plants. Curr Opin Plant Biol 10:310–316

Li Q-B, Haskell DW, Guy CL (1999) Coordinate and non-coordinate expression of the stress 70 family and other molecular chaperones at high and low temperature in spinach and tomato. Plant Mol Biol 39:21–34

Liu S, Wang J, Cong B, Huang X, Chen K, Zhang P (2014) Characterization and expression analysis of a mitochondrial heat-shock protein 70 gene from the antarctic moss pohlia nutans. Polar Biol 37:1145–1155

Mayer M, Bukau B (2005) Hsp70 chaperones: cellular functions and molecular mechanism. Cell Mol Life Sci 62:670

Nollen EA, Kabakov AE, Brunsting JF, Kanon B, Höhfeld J, Kampinga HH (2001) Modulation of in vivo HSP70 chaperone activity by Hip and Bag-1. J Biol Chem 276(7):4677–4682
Park C-J, Seo Y-S (2015) Heat shock proteins: a review of the molecular chaperones for plant immunity. Plant Pathol J 31:323
Rana RM, Khan MA, Shah MK, Ali Z, Zhang H (2016) Insights into the mechanism of heat shock mitigation through protein repair, recycling and degradation. In: Heat shock proteins and plants. Springer International Publishing, pp 103–119
Ray D, Ghosh A, Mustafi SB, Raha S (2016) Plant stress response: Hsp70 in the spotlight. In: Heat shock proteins and plants. Springer, pp 123–147
Rousch JM, Bingham SE, Sommerfeld MR (2004) Protein expression during heat stress in thermo-intolerant and thermo-tolerant diatoms. J Exp Mar Biol Ecol 306:231–243
Saidi Y, Finka A, Muriset M, Bromberg Z, Weiss YG, Maathuis FJ, Goloubinoff P (2009) The heat shock response in moss plants is regulated by specific calcium-permeable channels in the plasma membrane. Plant Cell 21:2829–2843
Sarkar NK, Kundnani P, Grover A (2013) Functional analysis of hsp70 superfamily proteins of rice (oryza sativa). Cell Stress Chaperones 18:427–437
Shao H-B, Guo Q-J, Chu L-Y, Zhao X-N, Su Z-L, Hu Y-C, Cheng J-F (2007) Understanding molecular mechanism of higher plant plasticity under abiotic stress. Colloids Surf B: Biointerfaces 54:37–45
Shi L-X, Theg SM (2010) A stromal heat shock protein 70 system functions in protein import into chloroplasts in the moss physcomitrella patens. Plant Cell 22:205–220
Storozhenko S, De Pauw P, Kushnir S, Van Montagu M, Inzé D (1996) Identification of an arabidopsis thaliana cdna encoding a hsp70-related protein belonging to the hsp110/sse1 subfamily. FEBS Lett 390:113–118
Su P-H, Li H-m (2008) Arabidopsis stromal 70-kd heat shock proteins are essential for plant development and important for thermotolerance of germinating seeds. Plant Physiol 146:1231–1241
Sung DY, Kaplan F, Guy CL (2001a) Plant hsp70 molecular chaperones: protein structure, gene family, expression and function. Physiol Plant 113:443–451
Sung DY, Vierling E, Guy CL (2001b) Comprehensive expression profile analysis of the arabidopsis hsp70 gene family. Plant Physiol 126:789–800
Ueki S, Citovsky V (2011) To gate, or not to gate: regulatory mechanisms for intercellular protein transport and virus movement in plants. Mol Plant 4:782–793
Valente MAS, Faria JA, Soares-Ramos JR, Reis PA, Pinheiro GL, Piovesan ND, Morais AT, Menezes CC, Cano MA, Fietto LG (2008) The er luminal binding protein (bip) mediates an increase in drought tolerance in soybean and delays drought-induced leaf senescence in soybean and tobacco. J Exp Bot 60:533–546
Wang W, Vinocur B, Shoseyov O, Altman A (2004) Role of plant heat-shock proteins and molecular chaperones in the abiotic stress response. Trends Plant Sci 9:244–252
Yu A, Li P, Tang T, Wang J, Chen Y, Liu L (2015) Roles of hsp70s in stress responses of microorganisms, plants, and animals. Biomed Res Int 2015:1
Zhang J-H, Wang L-J, Pan Q-H, Wang Y-Z, Zhan J-C, Huang W-D (2008) Accumulation and subcellular localization of heat shock proteins in young grape leaves during cross-adaptation to temperature stresses. Sci Hortic 117:231–240
Zhang JX, Wang C, Yang CY, Wang JY, Chen L, Bao XM, Zhao YX, Zhang H, Liu J (2010) The role of arabidopsis atfes1a in cytosolic hsp70 stability and abiotic stress tolerance. Plant J 62:539–548

Chapter 14
Small Heat Shock Proteins in Stress Response of Higher Eukaryotes

Annika Strauch and Martin Haslbeck

Abstract Small heat shock proteins (sHsps) are promiscuous molecular chaperones which form dynamic, spherical oligomers. Besides their main function in balancing proteostasis upon stress, sHsps are implicated in a variety of regulatory cascades especially in eukaryotic cells. The chaperone activity of sHsps is regulated by specific triggers as heat/cold stress, starvation, pH changes, chemical, biomodulators or posttranslational modifications. In this context, switching between activation and inactivation is achieved by modulating the composition of the ensemble of sHsp oligomers. This scheme of regulation allows a tremendous variability in contact sites and, thus, high flexibility in the specific, mechanistically properties of single sHsps. Regarding the evolutionary diversification and high sequence variability of sHsps it is not surprising that their cellular functions seem to be manifold. Here we summarize the current knowledge on the common structural and functional properties of sHsps. Focusing on *Drosophila melanogaster* as an exemplary model, we especially revisit the expression of sHsps at different stress conditions as well as changes in their expression throughout the development and aging of higher eukaryotes.

Keywords α-crystallin · Aging · Molecular chaperones · Regulation of sHsps · Stress response · Structure of sHsps

Abbreviations

Aα	Amyloid beta
ACD	α-crystallin domain
ATP	Adenosine triphosphate
BAG-3	BCL2 associated athanogene 3
CAMA	Chaperone assisted macroautophagy
CTR	C-terminal region

A. Strauch · M. Haslbeck (✉)
Department Chemie, Technische Universität München, Garching, Germany
e-mail: martin.haslbeck@tum.de

A. A. A. Asea, P. Kaur (eds.), *Heat Shock Proteins and Stress*,
Heat Shock Proteins 15, https://doi.org/10.1007/978-3-319-90725-3_14

Da	Dalton
Hsp	Heat-shock protein
Hsps	Heat-shock proteins
MAPK	Mitogen-activated protein kinase
mRNA	Messenger ribonucleic acid
NTR	N-terminal region
NMR	Nuclear magnetic resonance
pH	Potential of hydrogen
PQC	Protein quality control
PTM	Post-translation modifications
sHsp	Small heat-shock protein
sHsps	Small heat-shock proteins

14.1 Introduction

sHsps were among the first proteins detected as part of the cellular stress response. Early experiments studying gene expression after heat stress induced appearance of expanded chromosomal puffs in *Drosophila* (Jamrich et al. 1977; Ritossa 1962; Tissieres et al. 1974). In the following years, the corresponding stress genes were analyzed and sHsps were identified as part of the heat stress response (Ashburner and Bonner 1979; Craig and McCarthy 1980; Wadsworth et al. 1980). sHsps were initially named according to their low molecular mass of the respective monomers (Jakob et al. 1993), and besides Hsp100, Hsp90, Hsp70/Hsp40 and Hsp60 (Chaperonins) they present one of the five main families of molecular chaperones (Lindquist and Craig 1988). However, they present the least conserved family of molecular chaperones (Basha et al. 2012; Kriehuber et al. 2010; Waters et al. 2008).

sHsps have been found in all three domains of life and are the most omnipresent molecular chaperones besides proteins of the Hsp60 family (Kriehuber et al. 2010). Interestingly, sHsps have also been found in cyanophages (Bourrelle-Langlois et al. 2016; Maaroufi and Tanguay 2013) but it remains elusive what function they fulfil in the replication cycle of the phage. Another striking feature is the high variation of the number of sHsps genes in different organisms (Kriehuber et al. 2010). The general trend is an enrichment of sHsps genes throughout evolution. E.g. man encodes 10 sHsps (Fontaine et al. 2003; Kappe et al. 2003), *Drosophila melanogaster* encodes 12 (Morrow and Robert 2015) and *Caenorhabditis elegans* 16 (Candido 2002) while bacteria and archaea usually encode only 1-2 sHsps (Kriehuber et al. 2010). Nevertheless, it should be noted that some pathogen microorganisms like *Helicobacter pylori*, *Borrelia burgdorferi* and *Neisseria meningitidis* (Kappe et al. 2002; Narberhaus 2002) do not encode a single sHsp gene (Fig. 14.1). The highest number of sHsps genes is found in plants which can encode up to 37 sHsps (Basha et al. 2012; Waters 2013; Waters et al. 2008; Zhang et al. 2015). The potential reason for the high number of sHsps in plants is thought to be the fixed location of plants and the inability to avoid stress situations by simply moving away. As the main

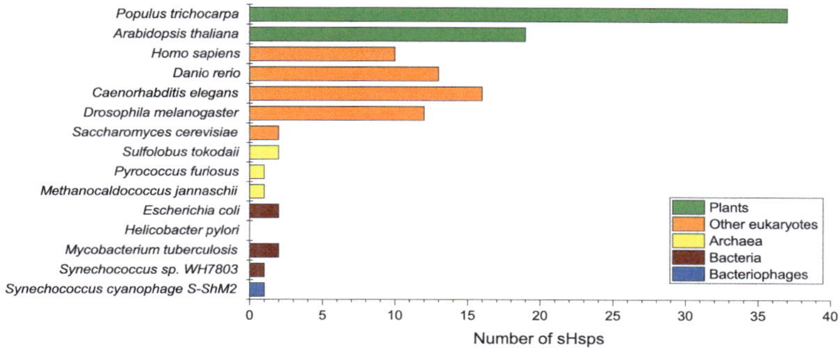

Fig. 14.1 Number of sHsps in different organisms. The bar graph represents the number of sHsps in several organisms which were further classified in bacteriophages, bacteria, archaea, plants and other eukaryotes. A clear trend concerning the number of sHsps is obvious: archaea, bacteria and bacteriophages mostly encode one or two sHsps while the number of sHsps in eukaryotes is notably higher

function of sHsps is to act as a first line of defense in the stress response where they promiscuously stabilize other cellular proteins.

However, this interpretation does not essentially explain why e.g. a motile worm like *C. elegans* would need 16 (Candido 2002) or a fly like *D. melanogaster* 12 (Morrow and Robert 2015) different genes for sHsps while e.g. yeast as a stationary eukaryote only encodes for two different sHsps (Petko and Lindquist 1986; Wotton et al. 1996). The general characteristics of sHsps can be summarized in terms of the following hallmarks (Basha et al. 2012; Haslbeck et al. 2005; Haslbeck and Vierling 2015):

(i) The presence of a conserved α-crystallin domain of 80-100 amino acids in length, which represents the signature motif for sHsps.
(ii) A molecular weight between ~ 12 and 43 kDa of the monomer.
(iii) The ability to form dynamic, oligomeric assemblies commonly consisting of 12 to 32 subunits or at least dimers.
(iv) ATP-independent chaperone activity in terms of aggregation suppression of diverse substrate proteins.

Within the proteostasis network of the cell, sHsps act as an ATP-independent buffer system. Able to recognize and bind unfolding proteins in various situations, sHsps protect them from irreversible aggregation. sHsps keep their substrates in a folding competent state and beyond that, enable the release and reactivate of the substrates by the ATP-dependent, Hsp70 and/or Hsp100 chaperone systems (Basha et al. 2012; Haslbeck et al. 2005). Besides this stress protection sHsps seem to be also involved in several regulatory processes throughout the life cycle of especially eukaryotic cells.

In eukaryotes and especially in human, the two most prominent members of the family of sHsps are αA- and αB-crystallin, which make up to 30 % of the total protein amount of the eye lens (Horwitz 2003; Slingsby et al. 2013). In eye lens the protein concentration is up to 450 g/mL in order to reach the required refraction index for clear vision. The two sHsps help to maintain the visual properties of the lens by enhancing the solubility of aggregation-prone lenticular proteins in this crowded environment (Fagerholm et al. 1981). αA-crystallin is predominantly expressed in the eye and only detected in small amounts in a few other tissues e.g. liver, kidney and pancreas (Deng et al. 2010; Kato et al. 1991; Uhlen et al. 2015; Wilhelm et al. 2014). αB-crystallin is additional expressed in many other tissues highlighting its more general importance (Arrigo 2013; Carra et al. 2013; Kannan et al. 2012; Mymrikov and Haslbeck 2015; Wilhelm et al. 2014). Together with Hsp27, another omnipresent sHsp in the cytosol of human cells, αB-crystallin is implicated in a variety of diseases which are typically also characterized by the deregulation of the physiological expression level of the human sHsps (HSPBs) (Bakthisaran et al. 2015; Mymrikov and Haslbeck 2015).

The affected diseases range from different myopathies, ischemia, diabetes, multiples sclerosis to neurological disorders such as Alzheimer's, Parkinson, Creutzfeld-Jacob as well as diverse cancers and inflammatory disorders (Arrigo and Gibert 2014; Bakthisaran et al. 2015; Kannan et al. 2012; Sun and MacRae 2005). The common trait of the available data indicates that elevated levels of mammalian sHsps protect against the proteotoxic damage induced throughout disease progression (Bakthisaran et al. 2015; Kampinga and Garrido 2012). However, enhanced levels of sHsps are not always favorable for the organism e.g. by supporting the growth of tumor cells and reducing the susceptibility for cancer therapy (Arrigo and Gibert 2014; Arrigo et al. 2007; Bakthisaran et al. 2015; Ciocca and Calderwood 2005; Kang et al. 2008). On the other hand, reduced or insufficient levels of sHsps can also lead to diseases (Andley et al. 2001; Ojha et al. 2011). Thus, for most diseases, sHsps are presumably not causative and do not directly regulate disease progression. The observed effects and implications of sHsps are thought to be secondary and correlated to the general fate of the cell (Arrigo 2013; Kampinga and Garrido 2012). However, exceptions are represented by diseases which are caused by mutations in a sHsp like cataract which is caused by the R120G mutation of αB-crystallin or neuropathies caused by the K141N mutation in HSPB8 (Irobi et al. 2004; Perng et al. 1999; Sun and MacRae 2005).

14.1.1 Architecture of sHSPs

sHsps share a conserved organization of their structure. In this context the two most prominent features are the conserved α-crystallin domain (ACD) which represents a dimerization motive and the ability of sHsps to form dynamic, oligomeric assemblies (Fig. 14.2) (Delbecq and Klevit 2013; Haslbeck et al. 2005). In the primary sequence the conserved ACD is flanked by highly variable N-terminal (NTR) and C-terminal regions (CTR) (Fig. 14.2a) (Caspers et al. 1995; de Jong et al. 1998; Kriehuber et al. 2010). This overall high sequence variability is especially remarkable in comparison to other molecular chaperone families which are highly conserved like the Hsp70 family (Brocchieri et al. 2008; Karlin and Brocchieri 1998)

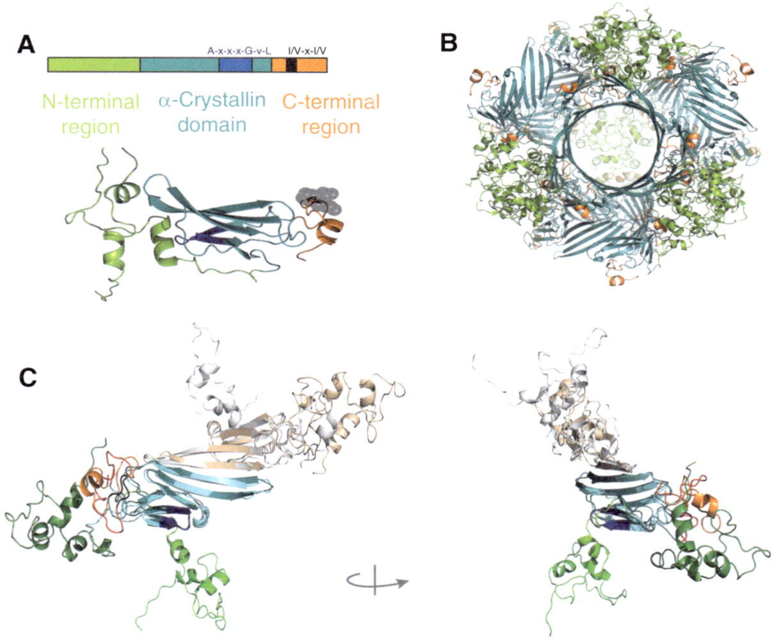

Fig. 14.2 Structure of sHsps. The domain organization of sHsps is divided into a N-terminal region (light green), α-crystallin domain (turquoise) and the C-terminal region (orange). The A-x-x-x-G-v-L motif (blue) and the I/V-x-I/V- motif (black) are further highlighted. Moreover one αB-crystallin monomer (extracted from PDB 2YGD) is colored according to the domain organization. Oligomeric structure of αB-crystallin (PDB: 2YGD). The N-terminal region (light green), α-crystallin domain (turquoise), C-terminal region (orange) and the I/V-x-I/V (black) are highlighted. Two dimeric building blocks were extracted from the αB-crystallin structure (PDB: 2YGD) and aligned, thereby it should be highlighted that the N- and C-terminal regions show different structural arrangements within the oligomer. The upper monomers are colored silver and bronze. The lower monomers were colored according to the domain organization in (A) with light variations in the color intensity

and Hsp90 family (Gupta 1995). The conserved ACD (or Hsp20 domain) is around 94 amino acids long (Kriehuber et al. 2010) and represents the signature motif of sHsps. It shows the highest sequence identity within the typical primary sequence of sHsps and in several sequence alignments it has been shown that especially the A-x-x-x-n-G-v-L sequence motif, located at the end of the ACD, represents the most conserved part (Caspers et al. 1995; de Jong et al. 1998; Narberhaus 2002). Nevertheless, it should be noted that the overall conservation of the ACD is quite low when compared to other protein domains. While the ACDs of the human eye lens sHsps, αA-crystallin and αB-crystallin show a sequence identity of around 50 %, the comparison of the ACD of human αB-crystallin and the ACD of yeast Hsp26 shows a sequence identity of only 20 %.

The α-crystallin domain forms a compact immunoglobulin-like β-sandwich which is composed of seven antiparallel strands β3 to β9. (Fig. 14.2 c) These seven strands are arranged in two β-sheets. One sheet is made up of three β-strands (β3-β9-β8) and the other one is composed of four β-strands (β4-β5-β6+β7). The two sheets are connected via a short inter-domain loop (Delbecq and Klevit 2013; Jehle et al. 2011; Van Montfort et al. 2001a; van Montfort et al. 2001b). Recombinant expression and structural analysis of the several isolated ACDs of different mammalian sHsps revealed the formation of dimers via the interaction of the β6+β7 interface (Bagneris et al. 2009; Baranova et al. 2011; Laganowsky et al. 2010). Interestingly, such a "β7-interface" dimer is also present in Sip-1, sHsps from the nematode *C. elegans* (Fleckenstein et al. 2015). TSP36 from the parasitic flatworm *Taenia saginata* on the other hand exhibits an alternative assembly theme for the dimer (Stamler et al. 2005). In this so called "β6-swapped" dimer the two monomers are assembled by a reciprocal swap of the β6-strand. Thereby the β6-strand is interacting with the β2-strand of the neighboring monomer. This type of dimer is also common in plants, yeast, archaea and bacterial sHsps (Bepperling et al. 2012; Hanazono et al. 2013; Kim et al. 1998; van Montfort et al. 2001b).

Besides a high variability in sequences, the NTR also shows a variation in length (Kriehuber et al. 2010). On average the NTR is 56 amino acids long (Kriehuber et al. 2010), but the length varies between 25 or even less amino acids in the special case of the Hsp12 sub-family of sHsps from *C.* elegans (Candido 2002) and up to 240 amino acids in yeast Hsp42 (Haslbeck et al. 2004a; Specht et al. 2011; Wotton et al. 1996). In contrast to the ACD, structural information on the NTR are limited. So far, only the crystal structure of Hsp14.1 of *Sulfolobus solfataricus* is available were the NTR is fully resolved. Here the NTR forms a long α-helix which interacts with the respective helix of the neighboring dimer in higher oligomers of Hsp14.1. Additionally, the pseudo atomic structure of αB-crystallin indicates the formation of helical elements in the NTR (Braun et al. 2011; White et al. 2006). Nevertheless, according to investigations by NMR, mass spectrometry and proteolysis, the structure of the NTR seems to be highly flexible (Jehle et al. 2011). Recombinant expressed NTRs from αA- and αB-crystallin and other sHsps were described to be insoluble. The low solubility of the NTR is commonly ascribed to its highly hydrophobic nature which bases on typically high content of phenylalanines and other hydrophobic amino acids (Asomugha et al. 2011; Merck et al. 1992).

Another typical feature of the NTR of eukaryotic sHsps is the presence of phosphorylation sites which are commonly used to regulate and modulate the assembly of higher oligomers as described in more detail below. Recently a previously uncharacterized, alternative splice variant (isoform 2) of human αA-crystallin with an exchanged NTR was described (Preis et al. 2017). In contrast to the canonical αA-crystallin this variant predominantly forms a monomer–dimer equilibrium and displays only low chaperone activity. However, the variant with the alternative NTR is able to integrate into oligomers of canonical αA- and αB-crystallin as well as their hetero-oligomers, enhancing their chaperone activity. Thus, alternative mRNA splicing of αA-crystallin introduces a new NTR and modulates the structural and functional properties of oligomeric ensemble (Preis et al. 2017). The CTR is the shortest of the three parts of the primary sequence of sHsps with an average length of 10 amino acids (Kriehuber et al. 2010). The CTR can be divided into the highly conserved I-x-I/V motif, a C-terminal tail which describes the sequence from the ACD to the I-x-I/V and the so-called C-terminal extension which represents the sequence part between the I-x-I/V to the C-terminus (Fig 14.2c) (Haslbeck et al. 2016; Hilton et al. 2013; Treweek et al. 2015). The I-x-I motif binds into the hydrophobic groove formed by the β4- and β8-strands of the ACD of a neighboring monomer (Kim et al. 1998; van Montfort et al. 2001b).

14.1.2 Formation of Oligomeric Structures

The dimers formed by the ACD typically represent the basic building blocks which are assembled to higher oligomers hierarchically by contacts in the NTR and CTR. E.g. in the case of human AB-crystallin, a 24-mer is formed by three dimers assembling into a hexamer via contacts formed by the CTR and four such hexamers further associate into the 24-mer through contacts within the NTR. Interestingly, this hierarchical assembly principle seems to be conserved among sHsps indicating that the assembly type and the total number of subunits in the oligomers can be modulated by the respective variations in the N- and CTR (Fig. 14.2b) (Braun et al. 2011; Delbecq and Klevit 2013).

In context of oligomer assembly it is also remarkable that sHsps populate different oligomeric states at equilibrium (Fig. 14.3). The oligomers are commonly polydisperse and represent dynamic ensembles which constantly exchange subunits whereby the degree of heterogeneity varies for different members of the family (Ehrnsperger et al. 1999; Fleckenstein et al. 2015; Mymrikov et al. 2017; Peschek et al. 2009). This ability of sHsps to generate oligomers of different sizes from a single sequence is linked to the above described hierarchical and modular architecture and seems to be correlated directly with their chaperone activity (Basha et al. 2012; Delbecq and Klevit 2013; Fleckenstein et al. 2015; Garrido et al. 2012; Haslbeck et al. 1999; McHaourab et al. 2009).

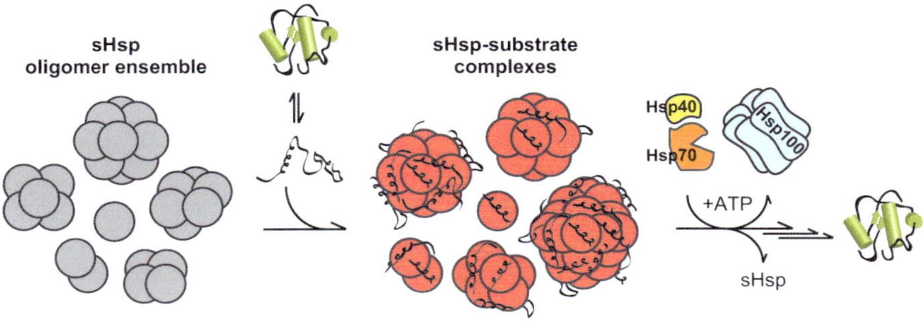

Fig. 14.3 Mechanism of the chaperone function of sHsps. Under stress conditions sHsps bind partially unfolded proteins in an energy-independent manner and keep them in a folding-competent state in sHsps/substrate complexes. The substrates are subsequently refolded by downstream ATP-dependent chaperone systems. (Figure and figure legend taken from Haslbeck et al. (2015))

E.g. as already mentioned, an α-helix in the NTR of *S. solfataricus* Hsp14.1 forms the contact to the neighboring dimer. Interestingly, crystallisation of a functionally inactive variant of the proteins (point mutation A102D) and the deletion of the C-terminal four amino acids resulted in bending of the helix in the NTR which correlates with changes in the oligomeric state and chaperone activity of Hsp14.1 (Liu et al. 2015a; Liu et al. 2015b). Additionally, in the two human α-crystallins, the SRLFDQFFG sequence motif is present in both proteins and locates to one of the proposed α-helixes within the first thirty amino acids of the NTR. Deletion of this motif led to a reduction in oligomeric size but also to an increase of chaperone activity (Pasta et al. 2003) similar to the changes induced by alternative splicing of the NTR (Preis et al. 2017). This further indicates that helixes formed in the NTR are important for higher order oligomerization. Similarly, the A truncated variant of Hsp26 from Saccharomyces cerevisiae, leaking the first 30 amino acids showed a lower chaperone activity and reduced stability of complexes with non-native protein compared to the wild type protein (Haslbeck et al. 2004b) indicating that the assembly hierarchy was evolutionary conserved.

14.1.3 Function of Small Heat Shock Proteins

Already in the early 1990s, first *in vitro* studies demonstrated that sHsps bind denatured proteins and prevent them from irreversible aggregation in an ATP-independent manner (Basha et al. 2004; Horwitz 1992; Jakob et al. 1993; Studer and Narberhaus 2000). Meanwhile, the general mechanistic concept views sHsps as the molecular chaperone system which acts fast and directly when cells and organisms encounter stress conditions (Fig. 14.3) (Bepperling et al. 2012; Haslbeck et al. 2004a; Haslbeck and Vierling 2015; McHaourab et al. 2009; Treweek et al. 2015). They stabilize aggregation-prone proteins usually by the formation of rather stable sHps-substrate complexes but cannot rescue already aggregated substrates. This is why sHsps are often called holdase in comparison to the ATP-dependent foldase like Hsp70 which are able to actively refold substrate proteins (Hartl et al. 2011; Richter et al. 2010). Similar to the sHsps itself, the formed sHsp-substrate complexes form a polydisperse ensemble of different species in dependence of the ratio of substrate to sHsp (Fig. 14.3) (Haslbeck et al. 2016; Lee et al. 1997). Nevertheless, non-native proteins trapped in sHsp-substrate complexes remain folding-competent and can be refolded by the ATP-dependent Hsp70/Hsp40 and or Hsp100 system (Ehrnsperger et al. 1997; Lee et al. 1997; Lee and Vierling 2000; Stege et al. 1994). Thus, within the protein homeostasis network of the cell, sHsps function as a buffer system protecting them from irreversible aggregation until other ATP-dependent chaperone systems are capable of restoring protein homeostasis (Fig. 14.3) (Haslbeck and Vierling 2015).

However, besides this tightly binding and complexing of substrate proteins, many sHsps additionally or alternatively are able to transiently bind substrate proteins (Lambert et al. 2013; Lindner et al. 2001). The substrates then are released spontaneously from the sHsps and refold on their own or with the help of ATP-dependent chaperones. Some sHsps, e.g. the dimeric Hsp17.7 from *Deinococcus radiodurans* seems to interact only transiently with its substrate and is unable to from stable sHsp/substrate complexes (Bepperling et al. 2012). Besides the involvement of sHsps in protecting the proteome during stress conditions, they also are incorporated at the cellular decision points between protein refolding and degradation of substrates (Fernandez-Fernandez et al. 2017). In this context the cytosolic Hsp70 proteins seem to act as master regulators and sHsps are especially involved when chaperone assisted macroautophagy (CAMA) is triggered (Arndt et al. 2010; Behl 2016; Chiang et al. 1989; Fernandez-Fernandez et al. 2017; Young 2010). In macroautophagy a specific protein complex presents ubiquinated proteins to the forming phagophore for further degradation. In the human cytosol the basic component of this presentation complex is Hsp70 which is assisted by BAG-3, a cochaperone acting as a molecular link to HspB8 (Ganassi et al. 2016). BAG-3 and the human sHsp HspB8 form a stable complex which is essential for CAMA (Carra et al. 2008; Fuchs et al. 2009). However, it remains elusive if substrate proteins attached to HspB8 are also transferred to the phagophore or if only ubiquinated proteins are transferred which are linked via CHIP, an additional Hsp70-cochaperone which is also integrated in the complex.

Besides this involvement in the regulation of protein degradation, sHsps seem to serve to monitor and regulate cell fate more directly. Whenever the constitutively present amount of sHsps in the cell is needed to complex unfolding proteins during stress defense, their availability for specific client binding decreases. Such a regulatory role of a sHsp in proteostasis is represented by the stabilization of procaspase-3 by human Hsp27 and αB-crystallin. This interaction is independent of a general stress condition (Acunzo et al. 2012; Pandey et al. 2000; Voss et al. 2007). To generate caspase-3, an essential activator of apoptosis, the N-terminal prodomain has to be cleaved off. The sHsps binding to the prodomain and inhibiting its cleavage and thus act as negative regulators of apoptosis (Acunzo et al. 2012; Pandey et al. 2000). The binding of procaspase-3 by Hsp27 is further controlled by the phosphorylation state of Hsp27. Specific phosphorylation of Hsp27 by MAPK-activated protein kinase 2/3 (Arrigo 2011; Ehrnsperger et al. 1999; Landry et al. 1992) which is also enhanced upon stress, triggers the release of procaspase-3 and induction of apoptosis (Gonzalez-Mejia et al. 2010). Additionally, Hsp27 binds to several other proteins of the apoptotic cascade (Acunzo et al. 2012; Bakthisaran et al. 2015; Voss et al. 2007) rendering it to a sensory and regulatory tool for the fate of the cell (Acunzo et al. 2012). Further examples for such regulatory implication of sHsps in proteostasis would be diapause prolongation in *Artemia* (MacRae 2016) and modulation of the intermediate filament network of cells (Elliott et al. 2013; Kayser et al. 2013; Quinlan and Ellis 2013).

14.1.4 Activation of Small Heat Shock Proteins Upon Stress Conditions

On the basis of the efficient and promiscuous binding of sHsps to non-native proteins, one might imagine that the constitutive presence of sHsps might lead to dominant negative effects in cellular proteostasis. In this context, compelling evidence indicates that the promiscuous substrate binding activity of sHsps is tightly regulated by external triggers which induce transition into a state of increased chaperone activity (Basha et al. 2004; Haslbeck and Vierling 2015; Haslbeck et al. 1999; Stromer et al. 2003; Treweek et al. 2015). The activation is achieved by changes in the distribution of species in the ensemble of the sHsp oligomers. Commonly a higher content of smaller species, often including dimers (Basha et al. 2010; Delbecq and Klevit 2013) leads to higher activity. Up to now at least the following triggers have been described to enable activation of sHsps (Bakthisaran et al. 2016; Bakthisaran et al. 2015; Benesch et al. 2008; Gaestel 2002; Haslbeck et al. 2015; Studer and Narberhaus 2000):

(i) the presence of unfolded or partially folded substrates,
(ii) the stress situation (e.g. changes in the environmental temperature and pH),
(iii) post-translational modifications (PTMs, in particular phosphorylation) and
(iv) hetero-oligomerization

The dynamic nature of sHsp oligomers always allows potential substrate binding sites, which are sequestered in the oligomers, to become exposed to a certain extent (Giese and Vierling 2002; Lindner et al. 2001; McHaourab et al. 2009; Shashidharamurthy et al. 2005; Yang et al. 1999). The presence of unfolding substrate proteins that interact with these binding sites leads to a shift in the equilibrium of the sHsp ensemble towards the more active species. In the cellular environment these sHsp ensemble dynamics act as a sensing mechanism that monitors the presence of non-native proteins and triggers activation (Benesch et al. 2008; Shi et al. 2013).

That heat stress temperature (or more generally, a stressor itself) is an activating trigger, was shown for several sHsps. The shift from physiological to heat stress temperatures, which usually requires only an increase of a few degrees leading to enhanced dissociation and, thus, activation of the sHsps. Similar to the presence of substrate, the temperature stimulus represents a very effective and rapid trigger for the activation of sHsps which allows organisms to rapidly stabilize the proteome without the need for new protein synthesis (Basha et al. 2012; Haslbeck et al. 2015). Besides heat stress, cold stress is also inducing the expression of sHsps. This was mostly reported and analyzed for plant sHsps like carrot Hsp17.7 (Song and Ahn 2011) as well as *Camellia sinensis* Hsp17.7, Hsp18.1 and Hsp21.8 (Wang et al. 2017).

Additionally, changes in pH seem to be a common trigger to activate sHsps. The cellular pH of eukaryotes typically ranges between 4.7 and 8. The lowest pH values are observed in the endosomes and secretory granules while the highest are found in the mitochondria. The pH in the cytosol, endoplasmic reticulum and nucleus is usually 7.2 (Casey et al. 2010). The overall pH of normal tissues (e.g. cerebral tissue) is between 7.0 and 7.2 (McVicar et al. 2014). *C. elegans* Sip1 represents a sHsp which is exclusively expressed in oocytes and embryos (Fleckenstein et al. 2015). Interestingly, the chaperone activity of Sip1 is optimal at the acidic conditions (pH 6.3) as they are present in the nematode eggs. Similarily, the chaperone activity of human αB-crystallin seems to be mediated by a single conserved histidine which acts as a pH switch in the physiologically relevant pH range of 6.5-7.5 (Rajagopal et al. 2015). Additionally, human Hsp27 responds to pH via a pH-dependent protonation of the structurally analogous histidine (Clouser and Klevit 2017). Like for Sip1 and αB-crystallin, this protonation reduces the stability of the dimer interface of the ACD, changes oligomeric size and modestly increases chaperone activity. Together this suggests a conserved mechanism of pH-dependent structural regulation among the human sHsps.

14.1.4.1 Phosphorylations/ PTMs

Phosphorylation or generally, post-translational modifications, appear to be a more specific trigger to regulate sHsps in eukaryotes. All human sHsps are regulated by phosphorylation in response to stress (Derham and Harding 1999; Gaestel 2002; Kantorow and Piatigorsky 1998; Kostenko and Moens 2009). E.g. human Hsp27 is phosphorylated at three sites, S15, S78 and S82, by a MAP-kinase cascade (Arrigo 2011; Kostenko and Moens 2009). Again, sHsp phosphorylation shifts the distribution of the sHsp ensemble towards smaller species (Haslbeck and Vierling 2015; Jovcevski et al. 2015) indicating that phosphorylation primarily affects N-terminal contacts in the oligomer. The negative charges incorporated in the NTR upon phosphorylation are thought to destabilize the respective subunit interfaces. In the pseudo-atomic model of αB-crystallin, the three phosphorylation sites are found in special proximity which would allow destabilizing the oligomer in a titratable manner upon enhanced phosphorylation (Jehle et al. 2011; Peschek et al. 2013).

Also non-mammalian sHsps are phosphorylated, e.g. phosphorylated species of Hsp22 from maize mitochondria (Lund et al. 2001) and yeast Hsp26 have been described (Bentley et al. 1992; Bodenmiller et al. 2010; Ficarro et al. 2002). However, it remains unknown whether phosphorylation of non-mammalian sHsps has a similar regulatory influence on their activity. Other post-translational modifications including deamination, oxidation, glycation or the attachment of methylglyoxal, have also been described to have regulatory influence on the chaperone activity of sHsps (Chalova et al. 2014; Chen et al. 2001; Eaton et al. 2002; Oya-Ito et al. 2006; Satish Kumar et al. 2004). Similar to post translational modifications, hetero-oligomer formation as regulatory tool of sHsps activity is especially crucial to eukaryotic cells where many different sHsps are present in dependence of the respective cell and tissue type.

The most prominent sHsp hetero-oligomer is formed in the human eye lens by the αA- and αB-crystallin (Horwitz et al. 1999; Kannan et al. 2012) but hetero-oligomer formation was also shown for most other human sHsps (Arrigo 2013). In a human muscle cell, eight cytosolic sHsps are present in moderate to high amounts in parallel. Interestingly, not all of them are able to form hetero-oligomers. They can be divided roughly into two classes based on their expression and their potential to interact with each other. Most of the members of these two classes only form hetero-oligomers with sHsps from the same class, but some members inter-connect the two classes by their ability to interact with members of both classes (Arrigo 2013; Bakthisaran et al. 2015; Fontaine et al. 2005; Mymrikov and Haslbeck 2015; Mymrikov et al. 2012; Sun et al. 2004).

Intriguingly, not only the activity per se but also the substrate spectra of the human sHsps become modified when they are incorporated into hetero-oligomers (Arrigo 2013). It also should be noted that the phosphorylation of the human sHsps additionally influences their hetero-oligomer formation properties and, thus, the two stimuli in combination allow a highly complex and variable adjustment of the activity of the respective sHsps (Arrigo 2013). This suggests that the formation of hetero-oligomers is tightly regulated and the co-existence of homo-oligomeric and hetero-oligomeric ensembles is possible. Furthermore, the at least theoretically possible high number of different combinations of hetero-oligomers allows a tremendous variability in activity. Thus, this variability in activity might be the basis for the modulation of substrate recognition and define different cell type and situation dependent substrate spectra. Hereby, the expression of the respective sHsps in a specific cell type represents the first level of modulation of the potentially possible ensemble of homo- and hetero-oligomers.

14.1.5 Expression and Cellular Localization of sHSP

As already mentioned for the two α-crystallins the expression of sHsps is not limited to one tissue. In the human skeletal muscle eight different sHsps are present. In nerve cells (e.g. sciatic nerve) especially the levels of αB-crystallin and Hsp20 (HspB6) are enriched (Bakthisaran et al. 2015; Verschuure et al. 2003). Other human sHsps like αA-crystallin, which is predominantly expressed in lens and only in small amounts in other tissues, or HSPB9 and HSPB10 which are both highly specific to testis. (Bakthisaran et al. 2015; Fontaine et al. 2003; Kappe et al. 2001; Xun et al. 2015). But tissue or stage specific expression is a common feature also for sHsps of other eukaryotes. E.g. Sip1 from *C. elegans* represents a sHsp which is exclusively expressed in oocytes and embryos (Fleckenstein et al. 2015). Also, p26, a sHsp from the brine shrimp *Artemia*, is only found during embryonic diapause (King and MacRae 2012; MacRae 2016). Comparable results were also found for the Hsp22 from *Calanus finmarchicus*, a marine copepod (Aruda et al. 2011). Even in plants, expression and differentially regulation during developmental has been reported for several sHsps (Kobayashi et al. 1994; Volkov et al. 2005; Wehmeyer and Vierling 2000).

Similarly, *D. melanogaster* Hsp23, Hsp26 and Hsp27 are also expressed in high levels during embryonal stages (Arrigo and Ahmad-Zadeh 1981; Glaser and Lis 1990; Glaser et al. 1986; Marin et al. 1993; Pauli et al. 1989; Zimmerman et al. 1983) while for Hsp22 only low amounts of mRNA were detected (Pauli et al. 1989). But it should be noted that in contrast to Sip1 which is exclusively expressed in the oocytes and during embryonal development (Fleckenstein et al. 2015; Linder et al. 1996), the four *Drosophila* sHsps were expressed during several developmental stages including larval and pupal stages as well as in adult flies (Glaser et al. 1986; Marin and Tanguay 1996; Pauli et al. 1989), further highlighting the complexity of the sHsp-network. In this context the analysis of the Sip1 interactome showed that it has a specific substrate spectrum, including many proteins that are essential for embryonic development (Fleckenstein et al. 2015). Additionally, the substrate spectra of individual sHsps of one organism show substantial overlaps but are not identical what further highlights the specific need for a distinguished developmental stage (Fleckenstein et al. 2015; Haslbeck et al. 2004a; Mymrikov et al. 2017).

In addition to the developmental and tissue regulated expression, different cellular localizations have been reported for sHsps from diverse organisms. Again, plants show the highest diversity regarding the subcellular localization of sHsps. Here sHsps are found in the cytosol, chloroplasts, mitochondria, nucleus and peroxisomes (Ma et al. 2006; Sanmiya et al. 2004; Wang et al. 2004b; Waters 2013; Waters and Vierling 1999). For example, *Arabidopsis thaliana* encodes for two sHsps which are localized to the peroxisomal matrix, three which localize to mitochondria, one localizing to the endoplasmic reticulum and one found in chloroplasts (Ma et al. 2006; Siddique et al. 2008). However, the sub-cellular distribution of sHsps to several compartments and organelles is not limited to higher eukaryotes. In the parasitic unicellular alveolate *Toxoplasma gondii,* five sHsps are expressed. They are found in mitochondria, located to the outer leaflet of the inner membrane complex, membrane–associated or present in the cytosol (de Miguel et al. 2009; de Miguel et al. 2005).

But one of the most prominent examples is *D. melanogaster*. Here Hsp22 and Hsp27 together with Hsp23 and Hsp26 present the best analyzed sHsps concerning their localization. Hsp22 has been reported to localize to mitochondrial matrix (Morrow et al. 2000), Hsp27 to the nucleus (Beaulieu et al. 1989; Kallappagoudar et al. 2010; Michaud et al. 2008) while Hsp23 and Hsp26 are both located to the cytoplasm (Arrigo and Ahmad-Zadeh 1981; Marin et al. 1993). In addition, it was shown that the localization of Hsp27 is not restricted to the nucleus. During oogenesis the *D. melanogaster* sHsp is also found in the cytoplasm (Marin and Tanguay 1996) indicting that developmental regulation and sub-cellular localization are intercalated.

Changes of the cellular localization have also been reported in context of human sHsps, e.g. Hsp27 (HspB1) is also found in the nucleus upon heat stress (McClaren and Isseroff 1994; Vos et al. 2010). Likewise the pseudophosphorylated (S19D, S45D and S59D) mutant of αB-crystallin is imported into the nucleus (den Engelsman et al. 2013). Also baker´s yeast Hsp26 changes its localization from primary cytosolic to nucleolar upon heat stress (Rossi and Lindquist 1989). It should be kept in mind that besides the developmental and tissue specific expression at physiological conditions, the expression of sHsps can be dramatically enhanced upon several stress triggers. Besides heat stress, which might be the most prominent and intensive studied trigger for molecular chaperones, other stress conditions like cold, acidification, heavy metal ions or osmotic stress lead to enhanced expression levels of sHsps. E.g. in *D. melanogaster* the expression of all sHsps, albeit on a low level for CG14207, inducible by heat stress (Arrigo and Ahmad-Zadeh 1981; Ingolia and Craig 1982; Landis et al. 2012; Marin et al. 1993; Vos et al. 2016). Because of this stress protective effect of sHsps, their heterologous expression can lead to an increased stress resistance. E.g. the expression of the carrot Hsp17.7 in *S. cerevisiae* leads to an improvement of cell viability, protein solubility and an increase in growth at several stress conditions (Ko et al. 2017). But it should be noted that such a heterologous protection is not generally applicable.

In context of stress situations, aging represents one of the most challenging events for an organism regarding proteostasis (Ben-Zvi et al. 2009; Labbadia and Morimoto 2015; Taylor and Dillin 2011). Increasing age leads to higher genome instability and oxidative stress. This is further highlighted by a collapse of the proteostasis maintenance of cells and tissues which therefore affects the whole organism (Ben-Zvi et al. 2009; Fedarko 2011; Hoeijmakers 2009; Johnson et al. 1999; Taylor and Dillin 2011). Age induced changes in the general metabolism force the cell to spend a lot of energy on balancing cellular homoestasis. Thereby proteome stability and maintenance of the protein quality control (PQC) seem to be the most important processes which need to be stabilized (Powers et al. 2009). Overexpression of several sHsps is correlated to an extended life span which was reported for *D. melanogaste*r Hsp22, Hsp26 and Hsp27. In contrast, the depletion of the respective genes leads to a reduced life span (Morrow et al. 2004a; Morrow et al. 2004b; Wang et al. 2004a). Moreover, it was shown for Hsp22 that overexpression increases the resistance against oxidative stress (Morrow et al. 2004b). A similar effect was also reported for overexpression of the human Hsp27 in a neuronal cell line (Rogalla et al. 1999).

Studies analyzing the transcriptome of aged *D. melanogaster* flies found an induction of genes of the oxidative stress response (Landis et al. 2004; Zou et al. 2000) and the innate immune response (Pletcher et al. 2002) while the expression of gene for the mitochondrial metabolism and the electron transport chain are reduced (Landis et al. 2004; Pletcher et al. 2002). Comparable aging-dependent expression patterns were also found in *C. elegans* (McCarroll et al. 2004) and human tissues (Lu et al. 2004; Rodwell et al. 2004; Zahn et al. 2006). In this study, increased expression levels for certain sHsps seems to be a common feature (Walther et al. 2015). In *D. melanogaster* the expression of the mitochondrial Hsp22 is strongly induced already at an advanced age (Fig. 14.4) (King and Tower 1999; Landis et al. 2004; Morrow and Tanguay 2003) whereas the expression of the other three main sHsps remains unchanged (King and Tower 1999; Morrow and Tanguay 2003).

Interestingly, the changes in gene expression upon aging and oxidative stress are comparable (Landis et al. 2012). Moreover, the ability for the cell to respond to oxidative stress has been shown to be decreased during aging. Increasing

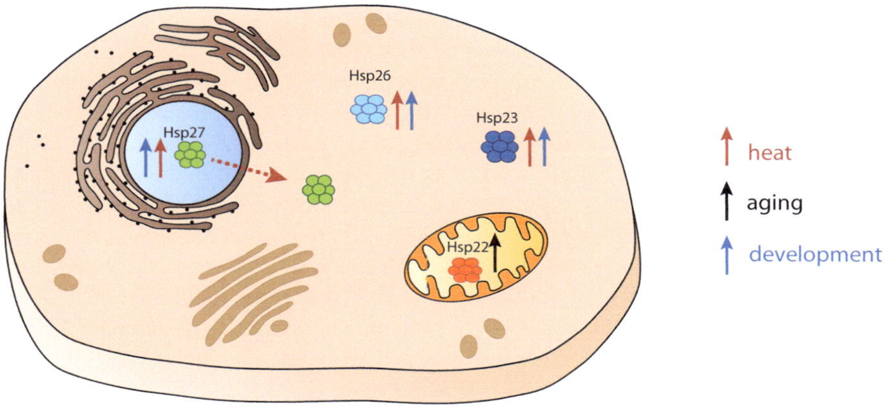

Fig. 14.4 Cellular distribution of *D. melanogaster* sHsps. The four main sHsps from *D. melanogaster* and their cellular localization. Hsp23 (dark blue) and Hsp26 (light blue) are located in the cytosol. Hsp22 (dark orange) is located in the mitochondria matrix and Hsp27 (green) is located in the nucleus. Vertical arrows indicate the enhanced expression levels of the four sHsps induced heat (red), aging (black) and during development (blue). Further under heat stress Hsp27 is also found in the cytosol which is indicated by the red dashed arrow. The cellular structure and organelles were simplified: nucleus (blue), endoplasmic reticulum (dark brown), ribosomes (black), Golgi apparatus and lysosomes (light brown)

malfunction of mitochondria and accumulation of altered mitochondria in several tissues during aging have been reported (Calleja et al. 1993; Sohal 1975; Tower 2011; Walker and Benzer 2004). Altogether, in the course of aging, sHsps possess a stabilizing function which is especially crucial in organisms at high age. Furthermore, at high age molecular chaperones are vitally important for the organism to prevent protein aggregation or to slow down formation of amyloid deposits, e.g. Aβ plaques in Alzheimer`s diseases (Ojha et al. 2011; Wilhelmus et al. 2006). Increased formation of deposits of aggregated protein and plaques are common in the aging organism (David et al. 2010; Reis-Rodrigues et al. 2012; Walther et al. 2015). Additionally, mutated αA-crystallin has been associated with an early-onset of cataract (Litt et al. 1998).

14.2 Conclusions

sHsps are virtually ubiquitous molecular chaperones. They share a conserved hierarchical organization of their oligomeric-structures which is based on the ACD, a conserved dimerization motif, and divergent NTRs and CTRs. The NTRs and CTRs dictate differences in the oligomer structure and its dynamics. Additionally, especially the NTRs are most likely responsible for substrate interaction. Currently we are far from understanding why in higher eukaryotes the number of sHsp genes is enriched. The expression of the different sHsps is strictly regulated and highly dependent on the cell or tissue type and the developmental stage of the organism. Additionally, the different sHsps show a characteristic cellular localization to organelles. Furthermore, the expression is enhanced and the distribution patterns are mediated in response to stress situations and throughout development. Initial results show that the substrate spectra of the respective sHsps diverge and are specifically adjusted in response to the specific needs of the respective cell. In the cell, the family of sHsps forms a highly regulated and interconnected network which is also tightly connected to the other chaperone systems. The detailed mechanisms of substrate recognition by sHsps still remain elusive and it is still enigmatic if and how such a seemingly tightly balanced network is involved in the regulation of the fate of a cell. As the sHsps network seems to be implicated in a whole diversity of different diseases as well as aging, they seem to represent interesting drug targets. However, a lot of additional work is needed to understand if and how the tuning of the function of specific sHsps can be used to influence disease propagation.

Acknowledgements We thank the Deutsche Forschungsgemeinschaft (SFB 1035) for financial support and Oliver Wenisch and Georgie Johnson for critical reading of the manuscript.

References

Acunzo J, Katsogiannou M, Rocchi P (2012) Small heat shock proteins HSP27 (HspB1), alphaB-crystallin (HspB5) and HSP22 (HspB8) as regulators of cell death. Int J Biochem Cell Biol 44:1622–1631

Andley UP, Song Z, Wawrousek EF, Brady JP, Bassnett S, Fleming TP (2001) Lens epithelial cells derived from alphaB-crystallin knockout mice demonstrate hyperproliferation and genomic instability. FASEB J 15:221–229

Arndt V, Dick N, Tawo R et al (2010) Chaperone-assisted selective autophagy is essential for muscle maintenance. Curr Biol 20:143–148

Arrigo AP (2011) Structure-functions of HspB1 (Hsp27). Methods Mol Biol 787:105–119

Arrigo AP (2013) Human small heat shock proteins: protein interactomes of homo- and hetero-oligomeric complexes: an update. FEBS Lett 587:1959–1969

Arrigo AP, Ahmad-Zadeh C (1981) Immunofluorescence localization of a small heat shock protein (hsp 23) in salivary gland cells of Drosophila melanogaster. Mol Gen Genet 184:73–79

Arrigo AP, Gibert B (2014) HspB1, HspB5 and HspB4 in human cancers: potent oncogenic role of some of their client proteins. Cancers (Basel) 6:333–365

Arrigo AP, Simon S, Gibert B et al (2007) Hsp27 (HspB1) and alphaB-crystallin (HspB5) as therapeutic targets. FEBS Lett 581:3665–3674

Aruda AM, Baumgartner MF, Reitzel AM, Tarrant AM (2011) Heat shock protein expression during stress and diapause in the marine copepod Calanus finmarchicus. J Insect Physiol 57:665–675

Ashburner M, Bonner JJ (1979) The induction of gene activity in drosophilia by heat shock. Cell 17:241–254

Asomugha CO, Gupta R, Srivastava OP (2011) Structural and functional properties of NH(2)-terminal domain, core domain, and COOH-terminal extension of alphaA- and alphaB-crystallins. Mol Vis 17:2356–2367

Bagneris C, Bateman OA, Naylor CE et al (2009) Crystal structures of alpha-crystallin domain dimers of alphaB-crystallin and Hsp20. J Mol Biol 392:1242–1252

Bakthisaran R, Tangirala R, Rao Ch M (2015) Small heat shock proteins: Role in cellular functions and pathology. Biochim Biophys Acta 1854:291–319
Bakthisaran R, Akula KK, Tangirala R, Rao Ch M (2016) Phosphorylation of alphaB-crystallin: Role in stress, aging and patho-physiological conditions. Biochim Biophys Acta 1860:167–182
Baranova EV, Weeks SD, Beelen S, Bukach OV, Gusev NB, Strelkov SV (2011) Three-dimensional structure of alpha-crystallin domain dimers of human small heat shock proteins HSPB1 and HSPB6. J Mol Biol 411:110–122
Basha E, Lee GJ, Demeler B, Vierling E (2004) Chaperone activity of cytosolic small heat shock proteins from wheat. Eur J Biochem 271:1426–1436
Basha E, Jones C, Wysocki V, Vierling E (2010) Mechanistic differences between two conserved classes of small heat shock proteins found in the plant cytosol. J Biol Chem 285:11489–11497
Basha E, O'Neill H, Vierling E (2012) Small heat shock proteins and alpha-crystallins: dynamic proteins with flexible functions. Trends Biochem Sci 37:106–117
Beaulieu JF, Arrigo AP, Tanguay RM (1989) Interaction of Drosophila 27,000 Mr heat-shock protein with the nucleus of heat-shocked and ecdysone-stimulated culture cells. J Cell Sci 92(Pt 1):29–36
Behl C (2016) Breaking BAG: the co-chaperone BAG3 in health and disease. Trends Pharmacol Sci 37:672–688
Benesch JL, Ayoub M, Robinson CV, Aquilina JA (2008) Small heat shock protein activity is regulated by variable oligomeric substructure. J Biol Chem 283:28513–28517
Bentley NJ, Fitch IT, Tuite MF (1992) The small heat-shock protein Hsp26 of Saccharomyces cerevisiae assembles into a high molecular weight aggregate. Yeast 8:95–106
Ben-Zvi A, Miller EA, Morimoto RI (2009) Collapse of proteostasis represents an early molecular event in Caenorhabditis elegans aging. Proc Natl Acad Sci U S A 106:14914–14919
Bepperling A, Alte F, Kriehuber T et al (2012) Alternative bacterial two-component small heat shock protein systems. Proc Natl Acad Sci U S A 109:20407–20412
Bodenmiller B, Wanka S, Kraft C et al (2010) Phosphoproteomic analysis reveals interconnected system-wide responses to perturbations of kinases and phosphatases in yeast. Sci Signal 3:rs4
Bourrelle-Langlois M, Morrow G, Finet S, Tanguay RM (2016) In vitro structural and functional characterization of the small heat shock proteins (sHSP) of the cyanophage S-ShM2 and its host, Synechococcus sp. WH7803. PLoS One 11:e0162233
Braun N, Zacharias M, Peschek J et al (2011) Multiple molecular architectures of the eye lens chaperone αB-crystallin elucidated by a triple hybrid approach. Proc Natl Acad Sci U S A 108:20491–20496
Brocchieri L, Conway de Macario E, Macario AJ (2008) hsp70 genes in the human genome: Conservation and differentiation patterns predict a wide array of overlapping and specialized functions. BMC Evol Biol 8:19
Calleja M, Pena P, Ugalde C, Ferreiro C, Marco R, Garesse R (1993) Mitochondrial DNA remains intact during Drosophila aging, but the levels of mitochondrial transcripts are significantly reduced. J Biol Chem 268:18891–18897
Candido EP (2002) The small heat shock proteins of the nematode Caenorhabditis elegans: structure, regulation and biology. Prog Mol Subcell Biol 28:61–78
Carra S, Seguin SJ, Landry J (2008) HspB8 and Bag3: a new chaperone complex targeting misfolded proteins to macroautophagy. Autophagy 4:237–239
Carra S, Rusmini P, Crippa V et al (2013) Different anti-aggregation and pro-degradative functions of the members of the mammalian sHSP family in neurological disorders. Philos Trans R Soc Lond B Biol Sci 368:20110409
Casey JR, Grinstein S, Orlowski J (2010) Sensors and regulators of intracellular pH. Nat Rev Mol Cell Biol 11:50–61
Caspers GJ, Leunissen JA, de Jong WW (1995) The expanding small heat-shock protein family, and structure predictions of the conserved "alpha-crystallin domain". J Mol Evol 40:238–248
Chalova AS, Sudnitsyna MV, Semenyuk PI, Orlov VN, Gusev NB (2014) Effect of disulfide crosslinking on thermal transitions and chaperone-like activity of human small heat shock protein HspB1. Cell Stress Chaperones 19:963–972

Chen SJ, Sun TX, Akhtar NJ, Liang JJ (2001) Oxidation of human lens recombinant alphaA-crystallin and cysteine-deficient mutants. J Mol Biol 305:969–976
Chiang HL, Terlecky SR, Plant CP, Dice JF (1989) A role for a 70-kilodalton heat shock protein in lysosomal degradation of intracellular proteins. Science 246:382–385
Ciocca DR, Calderwood SK (2005) Heat shock proteins in cancer: diagnostic, prognostic, predictive, and treatment implications. Cell Stress Chaperones 10:86–103
Clouser AF, Klevit RE (2017) pH-dependent structural modulation is conserved in the human small heat shock protein HSBP1. Cell Stress Chaperones 22:569–575
Craig EA, McCarthy BJ (1980) Four Drosophila heat shock genes at 67B: characterization of recombinant plasmids. Nucleic Acids Res 8:4441–4457
David DC, Ollikainen N, Trinidad JC, Cary MP, Burlingame AL, Kenyon C (2010) Widespread protein aggregation as an inherent part of aging in C. elegans. PLoS Biol 8:e1000450
de Jong WW, Caspers GJ, Leunissen JA (1998) Genealogy of the alpha-crystallin--small heat-shock protein superfamily. Int J Biol Macromol 22:151–162
de Miguel N, Echeverria PC, Angel SO (2005) Differential subcellular localization of members of the Toxoplasma gondii small heat shock protein family. Eukaryot Cell 4:1990–1997
de Miguel N, Braun N, Bepperling A et al (2009) Structural and functional diversity in the family of small heat shock proteins from the parasite Toxoplasma gondii. Biochim Biophys Acta 1793:1738–1748
Delbecq SP, Klevit RE (2013) One size does not fit all, the oligomeric states of alphaB crystallin. FEBS Lett 587:1073–1080
den Engelsman J, van de Schootbrugge C, Yong J, Pruijn GJ, Boelens WC (2013) Pseudophosphorylated alphaB-crystallin is a nuclear chaperone imported into the nucleus with help of the SMN complex. PLoS One 8:e73489
Deng M, Chen PC, Xie S et al (2010) The small heat shock protein alphaA-crystallin is expressed in pancreas and acts as a negative regulator of carcinogenesis. Biochim Biophys Acta 1802:621–631
Derham BK, Harding JJ (1999) Alpha-crystallin as a molecular chaperone. Prog Retin Eye Res 18:463–509
Eaton P, Fuller W, Shattock MJ (2002) S-thiolation of HSP27 regulates its multimeric aggregate size independently of phosphorylation. J Biol Chem 277:21189–21196
Ehrnsperger M, Graber S, Gaestel M, Buchner J (1997) Binding of non-native protein to Hsp25 during heat shock creates a reservoir of folding intermediates for reactivation. EMBO J 16:221–229
Ehrnsperger M, Lilie H, Gaestel M, Buchner J (1999) The dynamics of Hsp25 quaternary structure. Structure and function of different oligomeric species. J Biol Chem 274:14867–14874
Elliott JL, Der Perng M, Prescott AR, Jansen KA, Koenderink GH, Quinlan RA (2013) The specificity of the interaction between alphaB-crystallin and desmin filaments and its impact on filament aggregation and cell viability. Philos Trans R Soc Lond B Biol Sci 368:20120375
Fagerholm PP, Philipson BT, Lindstrom B (1981) Normal human lens - the distribution of protein. Exp Eye Res 33:615–620
Fedarko NS (2011) The biology of aging and frailty. Clin Geriatr Med 27:27–37
Fernandez-Fernandez MR, Gragera M, Ochoa-Ibarrola L, Quintana-Gallardo L, Valpuesta JM (2017) Hsp70 - a master regulator in protein degradation. FEBS Lett
Ficarro SB, McCleland ML, Stukenberg PT et al (2002) Phosphoproteome analysis by mass spectrometry and its application to Saccharomyces cerevisiae. Nat Biotechnol 20:301–305
Fleckenstein T, Kastenmuller A, Stein ML et al (2015) The Chaperone activity of the developmental small heat shock protein sip1 is regulated by pH-dependent conformational changes. Mol Cell 58:1067–1078
Fontaine JM, Rest JS, Welsh MJ, Benndorf R (2003) The sperm outer dense fiber protein is the 10th member of the superfamily of mammalian small stress proteins. Cell Stress Chaperones 8:62–69
Fontaine JM, Sun X, Benndorf R, Welsh MJ (2005) Interactions of HSP22 (HSPB8) with HSP20, alphaB-crystallin, and HSPB3. Biochem Biophys Res Commun 337:1006–1011

Fuchs M, Poirier DJ, Seguin SJ et al (2009) Identification of the key structural motifs involved in HspB8/HspB6-Bag3 interaction. Biochem J 425:245–255
Gaestel M (2002) sHsp-phosphorylation, enzymes, signaling pathways and functional implications. Prog Mol Subcell Biol 28:151–169
Ganassi M, Mateju D, Bigi I et al (2016) A surveillance function of the HSPB8-BAG3-HSP70 chaperone complex ensures stress granule integrity and dynamism. Mol Cell 63:796–810
Garrido C, Paul C, Seigneuric R, Kampinga HH (2012) The small heat shock proteins family: the long forgotten chaperones. Int J Biochem Cell Biol 44:1588–1592
Giese KC, Vierling E (2002) Changes in oligomerization are essential for the chaperone activity of a small heat shock protein in vivo and in vitro. J Biol Chem 277:46310–46318
Glaser RL, Lis JT (1990) Multiple, compensatory regulatory elements specify spermatocyte-specific expression of the Drosophila melanogaster hsp26 gene. Mol Cell Biol 10:131–137
Glaser RL, Wolfner MF, Lis JT (1986) Spatial and temporal pattern of hsp26 expression during normal development. EMBO J 5:747–754
Gonzalez-Mejia ME, Voss OH, Murnan EJ, Doseff AI (2010) Apigenin-induced apoptosis of leukemia cells is mediated by a bimodal and differentially regulated residue-specific phosphorylation of heat-shock protein-27. Cell Death Dis 1:e64
Gupta RS (1995) Phylogenetic analysis of the 90 kD heat shock family of protein sequences and an examination of the relationship among animals, plants, and fungi species. Mol Biol Evol 12:1063–1073
Hanazono Y, Takeda K, Oka T et al (2013) Nonequivalence observed for the 16-meric structure of a small heat shock protein, SpHsp16.0, from Schizosaccharomyces pombe. Structure 21:220–228
Hartl FU, Bracher A, Hayer-Hartl M (2011) Molecular chaperones in protein folding and proteostasis. Nature 475:324–332
Haslbeck M, Vierling E (2015) A first line of stress defense: small heat shock proteins and their function in protein homeostasis. J Mol Biol 427:1537–1548
Haslbeck M, Walke S, Stromer T et al (1999) Hsp26: a temperature-regulated chaperone. EMBO J 18:6744–6751
Haslbeck M, Braun N, Stromer T et al (2004a) Hsp42 is the general small heat shock protein in the cytosol of Saccharomyces cerevisiae. EMBO J 23:638–649
Haslbeck M, Ignatiou A, Saibil H et al (2004b) A domain in the N-terminal part of Hsp26 is essential for chaperone function and oligomerization. J Mol Biol 343:445–455
Haslbeck M, Franzmann T, Weinfurtner D, Buchner J (2005) Some like it hot: the structure and function of small heat-shock proteins. Nat Struct Mol Biol 12:842–846
Haslbeck M, Weinkauf S, Buchner J (2015) Regulation of the chaperone function of small Hsps. In: Tanguay, RMH LES (eds) The big book on small heat shock, vol 8. Springer International Publishing, Switzerland, pp 155–178
Haslbeck M, Peschek J, Buchner J, Weinkauf S (2016) Structure and function of α-crystallins: traversing from in vitro to in vivo. Biochim Biophys Acta 1860:149–166
Hilton GR, Lioe H, Stengel F, Baldwin AJ, Benesch JL (2013) Small heat-shock proteins: paramedics of the cell. Top Curr Chem 328:69–98
Hoeijmakers JH (2009) DNA damage, aging, and cancer. N Engl J Med 361:1475–1485
Horwitz J (1992) Alpha-crystallin can function as a molecular chaperone. Proc Natl Acad Sci U S A 89:10449–10453
Horwitz J (2003) Alpha-crystallin. Exp Eye Res 76:145–153
Horwitz J, Bova MP, Ding LL, Haley DA, Stewart PL (1999) Lens alpha-crystallin: function and structure. Eye (Lond) 13(Pt 3b):403–408
Ingolia TD, Craig EA (1982) Drosophila gene related to the major heat shock-induced gene is transcribed at normal temperatures and not induced by heat shock. Proc Natl Acad Sci U S A 79:525–529
Irobi J, Van Impe K, Seeman P et al (2004) Hot-spot residue in small heat-shock protein 22 causes distal motor neuropathy. Nat Genet 36:597–601

Jakob U, Gaestel M, Engel K, Buchner J (1993) Small heat shock proteins are molecular chaperones. J Biol Chem 268:1517–1520
Jamrich M, Greenleaf AL, Bautz EK (1977) Localization of RNA polymerase in polytene chromosomes of drosophila melanogaster. Proc Natl Acad Sci U S A 74:2079–2083
Jehle S, Vollmar BS, Bardiaux B et al (2011) N-terminal domain of alphaB-crystallin provides a conformational switch for multimerization and structural heterogeneity. Proc Natl Acad Sci U S A 108:6409–6414
Johnson FB, Sinclair DA, Guarente L (1999) Molecular biology of aging. Cell 96:291–302
Jovcevski B, Kelly MA, Rote AP et al (2015) Phosphomimics destabilize Hsp27 oligomeric assemblies and enhance chaperone activity. Chem Biol 22:186–195
Kallappagoudar S, Varma P, Pathak RU, Senthilkumar R, Mishra RK (2010) Nuclear matrix proteome analysis of drosophila melanogaster. Mol Cell Proteomics 9:2005–2018
Kampinga HH, Garrido C (2012) HSPBs: small proteins with big implications in human disease. Int J Biochem Cell Biol 44(10):1706
Kang SH, Kang KW, Kim KH et al (2008) Upregulated HSP27 in human breast cancer cells reduces Herceptin susceptibility by increasing Her2 protein stability. BMC Cancer 8:286
Kannan R, Sreekumar PG, Hinton DR (2012) Novel roles for alpha-crystallins in retinal function and disease. Prog Retin Eye Res 31:576–604
Kantorow M, Piatigorsky J (1998) Phosphorylations of alpha A- and alpha B-crystallin. Int J Biol Macromol 22:307–314
Kappe G, Verschuure P, Philipsen RL et al (2001) Characterization of two novel human small heat shock proteins: protein kinase-related HspB8 and testis-specific HspB9. Biochim Biophys Acta 1520:1–6
Kappe G, Leunissen JA, de Jong WW (2002) Evolution and diversity of prokaryotic small heat shock proteins. Prog Mol Subcell Biol 28:1–17
Kappe G, Franck E, Verschuure P, Boelens WC, Leunissen JA, de Jong WW (2003) The human genome encodes 10 alpha-crystallin-related small heat shock proteins: HspB1-10. Cell Stress Chaperones 8:53–61
Karlin S, Brocchieri L (1998) Heat shock protein 70 family: multiple sequence comparisons, function, and evolution. J Mol Evol 47:565–577
Kato K, Shinohara H, Kurobe N, Goto S, Inaguma Y, Ohshima K (1991) Immunoreactive alpha A crystallin in rat non-lenticular tissues detected with a sensitive immunoassay method. Biochim Biophys Acta 1080:173–180
Kayser J, Haslbeck M, Dempfle L et al (2013) The small heat shock protein Hsp27 affects assembly dynamics and structure of keratin intermediate filament networks. Biophys J 105:1778–1785
Kim KK, Kim R, Kim SH (1998) Crystal structure of a small heat-shock protein. Nature 394:595–599
King AM, MacRae TH (2012) The small heat shock protein p26 aids development of encysting Artemia embryos, prevents spontaneous diapause termination and protects against stress. PLoS One 7:e43723
King V, Tower J (1999) Aging-specific expression of Drosophila hsp22. Dev Biol 207:107–118
Ko E, Kim M, Park Y, Ahn YJ (2017) Heterologous expression of the carrot Hsp17.7 gene increased growth, cell viability, and protein solubility in transformed yeast (Saccharomyces cerevisiae) under heat, cold, acid, and osmotic stress conditions. Curr Microbiol 74:952–960
Kobayashi T, Kobayashi E, Sato S et al (1994) Characterization of cDNAs induced in meiotic prophase in lily microsporocytes. DNA Res 1:15–26
Kostenko S, Moens U (2009) Heat shock protein 27 phosphorylation: kinases, phosphatases, functions and pathology. Cell Mol Life Sci 66:3289–3307
Kriehuber T, Rattei T, Weinmaier T, Bepperling A, Haslbeck M, Buchner J (2010) Independent evolution of the core domain and its flanking sequences in small heat shock proteins. FASEB J 24:3633–3642
Labbadia J, Morimoto RI (2015) The biology of proteostasis in aging and disease. Annu Rev Biochem 84:435–464

Laganowsky A, Benesch JL, Landau M et al (2010) Crystal structures of truncated alphaA and alphaB crystallins reveal structural mechanisms of polydispersity important for eye lens function. Protein Sci 19:1031–1043
Lambert W, Rutsdottir G, Hussein R, Bernfur K, Kjellstrom S, Emanuelsson C (2013) Probing the transient interaction between the small heat-shock protein Hsp21 and a model substrate protein using crosslinking mass spectrometry. Cell Stress Chaperones 18:75–85
Landis GN, Abdueva D, Skvortsov D et al (2004) Similar gene expression patterns characterize aging and oxidative stress in Drosophila melanogaster. Proc Natl Acad Sci U S A 101:7663–7668
Landis G, Shen J, Tower J (2012) Gene expression changes in response to aging compared to heat stress, oxidative stress and ionizing radiation in Drosophila melanogaster. Aging (Albany NY) 4:768–789
Landry J, Lambert H, Zhou M et al (1992) Human HSP27 is phosphorylated at serines 78 and 82 by heat shock and mitogen-activated kinases that recognize the same amino acid motif as S6 kinase II. J Biol Chem 267:794–803
Lee GJ, Vierling E (2000) A small heat shock protein cooperates with heat shock protein 70 systems to reactivate a heat-denatured protein. Plant Physiol 122:189–198
Lee GJ, Roseman AM, Saibil HR, Vierling E (1997) A small heat shock protein stably binds heat-denatured model substrates and can maintain a substrate in a folding-competent state. EMBO J 16:659–671
Linder B, Jin Z, Freedman JH, Rubin CS (1996) Molecular characterization of a novel, developmentally regulated small embryonic chaperone from Caenorhabditis elegans. J Biol Chem 271:30158–30166
Lindner RA, Treweek TM, Carver JA (2001) The molecular chaperone alpha-crystallin is in kinetic competition with aggregation to stabilize a monomeric molten-globule form of alpha-lactalbumin. Biochem J 354:79–87
Lindquist S, Craig EA (1988) The heat-shock proteins. Annu Rev Genet 22:631–677
Litt M, Kramer P, LaMorticella DM, Murphey W, Lovrien EW, Weleber RG (1998) Autosomal dominant congenital cataract associated with a missense mutation in the human alpha crystallin gene CRYAA. Hum Mol Genet 7:471–474
Liu L, Chen J, Yang B, Wang Y (2015a) Oligomer-dependent and -independent chaperone activity of sHsps in different stressed conditions. FEBS Open Bio 5:155–162
Liu L, Chen JY, Yang B, Wang FH, Wang YH, Yun CH (2015b) Active-State Structures of a Small Heat-Shock Protein Revealed a Molecular Switch for Chaperone Function. Structure 23:2066–2075
Lu T, Pan Y, Kao SY et al (2004) Gene regulation and DNA damage in the ageing human brain. Nature 429:883–891
Lund AA, Rhoads DM, Lund AL, Cerny RL, Elthon TE (2001) In vivo modifications of the maize mitochondrial small heat stress protein, HSP22. J Biol Chem 276:29924–29929
Ma C, Haslbeck M, Babujee L, Jahn O, Reumann S (2006) Identification and characterization of a stress-inducible and a constitutive small heat-shock protein targeted to the matrix of plant peroxisomes. Plant Physiol 141:47–60
Maaroufi H, Tanguay RM (2013) Analysis and phylogeny of small heat shock proteins from marine viruses and their cyanobacteria host. PLoS One 8:e81207
MacRae TH (2016) Stress tolerance during diapause and quiescence of the brine shrimp, Artemia. Cell Stress Chaperones 21:9–18
Marin R, Tanguay RM (1996) Stage-specific localization of the small heat shock protein Hsp27 during oogenesis in Drosophila melanogaster. Chromosoma 105:142–149
Marin R, Valet JP, Tanguay RM (1993) hsp23 and hsp26 exhibit distinct spatial and temporal patterns of constitutive expression in Drosophila adults. Dev Genet 14:69–77
McCarroll SA, Murphy CT, Zou S et al (2004) Comparing genomic expression patterns across species identifies shared transcriptional profile in aging. Nat Genet 36:197–204
McClaren M, Isseroff RR (1994) Dynamic changes in intracellular localization and isoforms of the 27-kD stress protein in human keratinocytes. J Invest Dermatol 102:375–381

McHaourab HS, Godar JA, Stewart PL (2009) Structure and mechanism of protein stability sensors: chaperone activity of small heat shock proteins. Biochemistry 48:3828–3837

McVicar N, Li AX, Goncalves DF et al (2014) Quantitative tissue pH measurement during cerebral ischemia using amine and amide concentration-independent detection (AACID) with MRI. J Cereb Blood Flow Metab 34:690–698

Merck KB, De Haard-Hoekman WA, Oude Essink BB, Bloemendal H, De Jong WW (1992) Expression and aggregation of recombinant alpha A-crystallin and its two domains. Biochim Biophys Acta 1130:267–276

Michaud S, Lavoie S, Guimond MO, Tanguay RM (2008) The nuclear localization of Drosophila Hsp27 is dependent on a monopartite arginine-rich NLS and is uncoupled from its association to nuclear speckles. Biochim Biophys Acta 1783:1200–1210

Morrow GT, Robert M (2015) Drosophila small heat shock proteins: an update on their features and functions. The big book on small heat shock proteins

Morrow G, Tanguay RM (2003) Heat shock proteins and aging in Drosophila melanogaster. Semin Cell Dev Biol 14:291–299

Morrow G, Inaguma Y, Kato K, Tanguay RM (2000) The small heat shock protein Hsp22 of Drosophila melanogaster is a mitochondrial protein displaying oligomeric organization. J Biol Chem 275:31204–31210

Morrow G, Battistini S, Zhang P, Tanguay RM (2004a) Decreased lifespan in the absence of expression of the mitochondrial small heat shock protein Hsp22 in Drosophila. J Biol Chem 279:43382–43385

Morrow G, Samson M, Michaud S, Tanguay RM (2004b) Overexpression of the small mitochondrial Hsp22 extends Drosophila life span and increases resistance to oxidative stress. FASEB J 18:598–599

Mymrikov EV, Haslbeck M (2015) Medical implications of understanding the functions of human small heat shock proteins. Expert Rev Proteomics 12:295–308

Mymrikov EV, Seit-Nebi AS, Gusev NB (2012) Heterooligomeric complexes of human small heat shock proteins. Cell Stress Chaperones 17:157–169

Mymrikov EV, Daake M, Richter B, Haslbeck M, Buchner J (2017) The Chaperone Activity and Substrate Spectrum of Human Small Heat Shock Proteins. J Biol Chem 292:672–684

Narberhaus F (2002) Alpha-crystallin-type heat shock proteins: socializing minichaperones in the context of a multichaperone network. Microbiol Mol Biol Rev 66:64–93. table of contents

Ojha J, Karmegam RV, Masilamoni JG, Terry AV, Cashikar AG (2011) Behavioral defects in chaperone-deficient Alzheimer's disease model mice. PLoS One 6:e16550

Oya-Ito T, Liu BF, Nagaraj RH (2006) Effect of methylglyoxal modification and phosphorylation on the chaperone and anti-apoptotic properties of heat shock protein 27. J Cell Biochem 99:279–291

Pandey P, Farber R, Nakazawa A et al (2000) Hsp27 functions as a negative regulator of cytochrome c-dependent activation of procaspase-3. Oncogene 19:1975–1981

Pasta SY, Raman B, Ramakrishna T, Rao Ch M (2003) Role of the conserved SRLFDQFFG region of alpha-crystallin, a small heat shock protein. Effect on oligomeric size, subunit exchange, and chaperone-like activity. J Biol Chem 278:51159–51166

Pauli D, Arrigo AP, Vazquez J, Tonka CH, Tissieres A (1989) Expression of the small heat shock genes during Drosophila development: comparison of the accumulation of hsp23 and hsp27 mRNAs and polypeptides. Genome 31:671–676

Perng MD, Muchowski PJ, van Den IP et al (1999) The cardiomyopathy and lens cataract mutation in alphaB-crystallin alters its protein structure, chaperone activity, and interaction with intermediate filaments in vitro. J Biol Chem 274:33235–33243

Peschek J, Braun N, Franzmann TM et al (2009) The eye lens chaperone alpha-crystallin forms defined globular assemblies. Proc Natl Acad Sci U S A 106:13272–13277

Peschek J, Braun N, Rohrberg J et al (2013) Regulated structural transitions unleash the chaperone activity of alphaB-crystallin. Proc Natl Acad Sci U S A 110:E3780–E3789

Petko L, Lindquist S (1986) Hsp26 is not required for growth at high temperatures, nor for thermotolerance, spore development, or germination. Cell 45:885–894

Pletcher SD, Macdonald SJ, Marguerie R et al (2002) Genome-wide transcript profiles in aging and calorically restricted Drosophila melanogaster. Curr Biol 12:712–723
Powers ET, Morimoto RI, Dillin A, Kelly JW, Balch WE (2009) Biological and chemical approaches to diseases of proteostasis deficiency. Annu Rev Biochem 78:959–991
Preis W, Bestehorn A, Buchner J, Haslbeck M (2017) An alternative splice variant of human alphaA-crystallin modulates the oligomer ensemble and the chaperone activity of alpha-crystallins. Cell Stress Chaperones 22:541–552
Quinlan RA, Ellis RJ (2013) Chaperones: needed for both the good times and the bad times. Philos Trans R Soc Lond B Biol Sci 368:20130091
Rajagopal P, Tse E, Borst AJ et al (2015) A conserved histidine modulates HSPB5 structure to trigger chaperone activity in response to stress-related acidosis. Elife 4
Reis-Rodrigues P, Czerwieniec G, Peters TW et al (2012) Proteomic analysis of age-dependent changes in protein solubility identifies genes that modulate lifespan. Aging Cell 11:120–127
Richter K, Haslbeck M, Buchner J (2010) The heat shock response: life on the verge of death. Mol Cell 40:253–266
Ritossa P (1962) Problems of prophylactic vaccinations of infants. Riv Ist Sieroter Ital 37:79–108
Rodwell GE, Sonu R, Zahn JM et al (2004) A transcriptional profile of aging in the human kidney. PLoS Biol 2:e427
Rogalla T, Ehrnsperger M, Preville X et al (1999) Regulation of Hsp27 oligomerization, chaperone function, and protective activity against oxidative stress/tumor necrosis factor alpha by phosphorylation. J Biol Chem 274:18947–18956
Rossi JM, Lindquist S (1989) The intracellular location of yeast heat-shock protein 26 varies with metabolism. J Cell Biol 108:425–439
Sanmiya K, Suzuki K, Egawa Y, Shono M (2004) Mitochondrial small heat-shock protein enhances thermotolerance in tobacco plants. FEBS Lett 557:265–268
Satish Kumar M, Mrudula T, Mitra N, Bhanuprakash Reddy G (2004) Enhanced degradation and decreased stability of eye lens alpha-crystallin upon methylglyoxal modification. Exp Eye Res 79:577–583
Shashidharamurthy R, Koteiche HA, Dong J, McHaourab HS (2005) Mechanism of chaperone function in small heat shock proteins: dissociation of the HSP27 oligomer is required for recognition and binding of destabilized T4 lysozyme. J Biol Chem 280:5281–5289
Shi J, Koteiche HA, McDonald ET, Fox TL, Stewart PL, McHaourab HS (2013) Cryoelectron microscopy analysis of small heat shock protein 16.5 (Hsp16.5) complexes with T4 lysozyme reveals the structural basis of multimode binding. J Biol Chem 288:4819–4830
Siddique M, Gernhard S, von Koskull-Doring P, Vierling E, Scharf KD (2008) The plant sHSP superfamily: five new members in Arabidopsis thaliana with unexpected properties. Cell Stress Chaperones 13:183–197
Slingsby C, Wistow GJ, Clark AR (2013) Evolution of crystallins for a role in the vertebrate eye lens. Protein Sci 22:367–380
Sohal RD (1975) Mitochondrial changes in flight muscles of normal and flightless Drosophila melanogaster with age. J Morphol 145:337–353
Song NH, Ahn YJ (2011) DcHsp17.7, a small heat shock protein in carrot, is tissue-specifically expressed under salt stress and confers tolerance to salinity. N Biotechnol 28:698–704
Specht S, Miller SB, Mogk A, Bukau B (2011) Hsp42 is required for sequestration of protein aggregates into deposition sites in Saccharomyces cerevisiae. J Cell Biol 195:617–629
Stamler R, Kappe G, Boelens W, Slingsby C (2005) Wrapping the alpha-crystallin domain fold in a chaperone assembly. J Mol Biol 353:68–79
Stege GJ, Li GC, Li L, Kampinga HH, Konings AW (1994) On the role of hsp72 in heat-induced intranuclear protein aggregation. Int J Hyperthermia 10:659–674
Stromer T, Ehrnsperger M, Gaestel M, Buchner J (2003) Analysis of the interaction of small heat shock proteins with unfolding proteins. J Biol Chem 278:18015–18021
Studer S, Narberhaus F (2000) Chaperone activity and homo- and hetero-oligomer formation of bacterial small heat shock proteins. J Biol Chem 275:37212–37218

Sun Y, MacRae TH (2005) The small heat shock proteins and their role in human disease. FEBS J 272:2613–2627
Sun X, Fontaine JM, Rest JS, Shelden EA, Welsh MJ, Benndorf R (2004) Interaction of human HSP22 (HSPB8) with other small heat shock proteins. J Biol Chem 279:2394–2402
Taylor RC, Dillin A (2011) Aging as an event of proteostasis collapse. Cold Spring Harb Perspect Biol 3
Tissieres A, Mitchell HK, Tracy UM (1974) Protein synthesis in salivary glands of Drosophila melanogaster: relation to chromosome puffs. J Mol Biol 84:389–398
Tower J (2011) Heat shock proteins and Drosophila aging. Exp Gerontol 46:355–362
Treweek TM, Meehan S, Ecroyd H, Carver JA (2015) Small heat-shock proteins: important players in regulating cellular proteostasis. Cell Mol Life Sci 72:429–451
Uhlen M, Fagerberg L, Hallstrom BM et al (2015) Proteomics. Tissue-based map of the human proteome. Science 347:1260419
Van Montfort R, Slingsby C, Vierling E (2001a) Structure and function of the small heat shock protein/alpha-crystallin family of molecular chaperones. Adv Protein Chem 59:105–156
van Montfort RL, Basha E, Friedrich KL, Slingsby C, Vierling E (2001b) Crystal structure and assembly of a eukaryotic small heat shock protein. Nat Struct Biol 8:1025–1030
Verschuure P, Tatard C, Boelens WC, Grongnet JF, David JC (2003) Expression of small heat shock proteins HspB2, HspB8, Hsp20 and cvHsp in different tissues of the perinatal developing pig. Eur J Cell Biol 82:523–530
Volkov RA, Panchuk II, Schoffl F (2005) Small heat shock proteins are differentially regulated during pollen development and following heat stress in tobacco. Plant Mol Biol 57:487–502
Vos MJ, Zijlstra MP, Kanon B et al (2010) HSPB7 is the most potent polyQ aggregation suppressor within the HSPB family of molecular chaperones. Hum Mol Genet 19:4677–4693
Vos MJ, Carra S, Kanon B et al (2016) Specific protein homeostatic functions of small heat-shock proteins increase lifespan. Aging Cell 15:217–226
Voss OH, Batra S, Kolattukudy SJ, Gonzalez-Mejia ME, Smith JB, Doseff AI (2007) Binding of caspase-3 prodomain to heat shock protein 27 regulates monocyte apoptosis by inhibiting caspase-3 proteolytic activation. J Biol Chem 282:25088–25099
Wadsworth SC, Craig EA, McCarthy BJ (1980) Genes for three Drosophila heat-shock-induced proteins at a single locus. Proc Natl Acad Sci U S A 77:2134–2137
Walker DW, Benzer S (2004) Mitochondrial "swirls" induced by oxygen stress and in the Drosophila mutant hyperswirl. Proc Natl Acad Sci U S A 101:10290–10295
Walther DM, Kasturi P, Zheng M et al (2015) Widespread Proteome Remodeling and Aggregation in Aging C. elegans. Cell 161:919–932
Wang HD, Kazemi-Esfarjani P, Benzer S (2004a) Multiple-stress analysis for isolation of Drosophila longevity genes. Proc Natl Acad Sci U S A 101:12610–12615
Wang W, Vinocur B, Shoseyov O, Altman A (2004b) Role of plant heat-shock proteins and molecular chaperones in the abiotic stress response. Trends Plant Sci 9:244–252
Wang M, Zou Z, Li Q et al (2017) Heterologous expression of three Camellia sinensis small heat shock protein genes confers temperature stress tolerance in yeast and Arabidopsis thaliana. Plant Cell Rep 36:1125–1135
Waters ER (2013) The evolution, function, structure, and expression of the plant sHSPs. J Exp Bot 64:391–403
Waters ER, Vierling E (1999) Chloroplast small heat shock proteins: evidence for atypical evolution of an organelle-localized protein. Proc Natl Acad Sci U S A 96:14394–14399
Waters ER, Aevermann BD, Sanders-Reed Z (2008) Comparative analysis of the small heat shock proteins in three angiosperm genomes identifies new subfamilies and reveals diverse evolutionary patterns. Cell Stress Chaperones 13:127–142
Wehmeyer N, Vierling E (2000) The expression of small heat shock proteins in seeds responds to discrete developmental signals and suggests a general protective role in desiccation tolerance. Plant Physiol 122:1099–1108
White HE, Orlova EV, Chen S et al (2006) Multiple distinct assemblies reveal conformational flexibility in the small heat shock protein Hsp26. Structure 14:1197–1204

Wilhelm M, Schlegl J, Hahne H et al (2014) Mass-spectrometry-based draft of the human proteome. Nature 509:582–587
Wilhelmus MM, Boelens WC, Otte-Holler I, Kamps B, de Waal RM, Verbeek MM (2006) Small heat shock proteins inhibit amyloid-beta protein aggregation and cerebrovascular amyloid-beta protein toxicity. Brain Res 1089:67–78
Wotton D, Freeman K, Shore D (1996) Multimerization of Hsp42p, a novel heat shock protein of Saccharomyces cerevisiae, is dependent on a conserved carboxyl-terminal sequence. J Biol Chem 271:2717–2723
Xun W, Shi L, Cao T et al (2015) Dual functions in response to heat stress and spermatogenesis: characterization of expression profile of small heat shock proteins 9 and 10 in goat testis. Biomed Res Int 2015:686239
Yang H, Huang S, Dai H, Gong Y, Zheng C, Chang Z (1999) The Mycobacterium tuberculosis small heat shock protein Hsp16.3 exposes hydrophobic surfaces at mild conditions: conformational flexibility and molecular chaperone activity. Protein Sci 8:174–179
Young JC (2010) Mechanisms of the Hsp70 chaperone system. Biochem Cell Biol 88:291–300
Zahn JM, Sonu R, Vogel H et al (2006) Transcriptional profiling of aging in human muscle reveals a common aging signature. PLoS Genet 2: e115
Zhang J, Liu B, Li J et al (2015) Hsf and Hsp gene families in Populus: genome-wide identification, organization and correlated expression during development and in stress responses. BMC Genomics 16: 181
Zimmerman JL, Petri W, Meselson M (1983) Accumulation of a specific subset of D. melanogaster heat shock mRNAs in normal development without heat shock. Cell 32:1161–1170
Zou S, Meadows S, Sharp L, Jan LY, Jan YN (2000) Genome-wide study of aging and oxidative stress response in Drosophila melanogaster. Proc Natl Acad Sci U S A 97:13726–13731

MIX
Papier aus verantwortungsvollen Quellen
Paper from responsible sources
FSC® C105338
FSC
www.fsc.org

If you have any concerns about our products,
you can contact us on
ProductSafety@springernature.com

In case Publisher is established outside the EU,
the EU authorized representative is:
Springer Nature Customer Service Center GmbH
Europaplatz 3, 69115 Heidelberg, Germany

Printed by Libri Plureos GmbH
in Hamburg, Germany